Philosophical Studies in Contemporary Culture

Volume 27

Founding Editor
H. Tristram Engelhardt†, Jr., Department of Philosophy, Rice University and
Baylor College of Medicine, Houston, TX, USA

Associate Editor
Mark J. Cherry, Department of Philosophy, St. Edward's
University, Austin, TX, USA

Assistant Editor
James Stacey Taylor, College of New Jersey, Ewing, NJ, USA

Editorial Board Members
David Bradshaw, Department of Philosophy, University of Kentucky,
Lexington, KY, USA
Peter Jaworski, McDonough School of Business, Georgetown
University, Washington, DC, USA
Terry Pinkard, Department of Philosophy, Georgetown
University, Washington, DC, USA
C. Griffin Trotter, Center for Health Care Ethics & Emergency
Medicine, Saint Louis University, St. Louis, MO, USA
Kevin Wm. Wildes, S.J., President, Loyola University, New Orleans, LA, USA

This series explores the philosophical issues, concerns, and controversies framing our contemporary culture. The volumes address studies of the perennial philosophical questions and major thinkers whose works have shaped current intellectual debates. The series draws on the intellectual heritage of the past in order to illuminate current concerns. The goal is to offer volumes whose perspective and focus can contribute through philosophical studies to a better appreciation of the contemporary human condition.

More information about this series at http://www.springer.com/series/6446

Teri Merrick

Helmholtz, Cohen, and Frege on Progress and Fidelity

Sinning Against Science and Religion

 Springer

Teri Merrick
Department of Philosophy
Azusa Pacific University
Azusa, CA, USA

ISSN 0928-9518   ISSN 2215-1753   (electronic)
Philosophical Studies in Contemporary Culture
ISBN 978-3-030-57298-3   ISBN 978-3-030-57299-0   (eBook)
https://doi.org/10.1007/978-3-030-57299-0

This Springer imprint is published by the registered company Springer Nature Switzerland AG
The registered company address is: Gewerbestrasse 11, 6330 Cham, Switzerland

*To my husband Jeff Merrick and my parents
Ray and Jan Glaze*

Acknowledgements

This book interweaves answers to questions that I have been thinking and writing about for the past 20 or so years: Why was Frege so resistant to the development of non-Euclidean geometry? Why are my co-religionists so resistant to calls for progressive change within our tradition? What are some wise and workable solutions to the ever-present problem of epistemic injustice? Though these questions seem disparate and initially I treated them as such, I came to see that my answer to one informed my answers to the others. I also came to see Cohen as a mentor in reasonably responding to Kuhnian-scale changes in scientific and religious traditions. One hope then for the book is that my reader will come to see this as well.

Fortunately, my circle of mentors includes more than a neo-Kantian philosopher who died over a century ago. I am especially thankful for Penelope Maddy. She pushed and prodded me to make my thoughts more articulate and accessible. My career as a philosopher and the completion of this book is due in large measure to her patient and persistent support, as well as an oft-needed kick in the pants.

My interest in Frege began and was cultivated while a graduate student in the Department of Philosophy and Department of Logic and Philosophy of Science (LPS) at the University of California, Irvine. Upon entering the profession, I often heard colleagues, particularly my fellow female colleagues, describe their graduate school experience as one marked by alienation and marginalization if not downright harassment or abuse. In contrast, I cannot say enough about the steadfast encouragement and kind collegiality that I experienced in the Departments of Philosophy and LPS. Special thanks go out to Martin Schwab and Terry Parsons for paddling alongside me through my first deep dive into Kant and Frege scholarship, to David Malament for tutoring me in projective geometry, and to the members of my dissertation committee (Chair Penelope Maddy, Robert May, Alan Nelson, and P. Kyle Stanford) for their invaluable commentary on the raw material of much that appears in Chaps. 2–4 of the book. Thanks as well to my LPS traveling companions: Jason Alexander, Gary Bell, Sean Ebels-Duggan, Curtis Franks, Patricia Marino, Carol (Skrenes) Trabling, and Brian Woodcock for being exemplars of Aristotelian friendship.

   I am indebted to my faculty colleagues at Azusa Pacific University (APU) for inspiring my research on Cohen's philosophy of religion and its relevance to contemporary discussions about epistemic injustice. I want to thank Sarah Richart for introducing me to Katrina Karkazis' *Fixing Sex* and the controversy over the unjust treatment of intersex patients. Kay Higuera Smith introduced me to the postcolonial literature within biblical and religious studies and enabled me to see the relationship between Cohen's Marburg School of Neo-Kantianism and twenty-first century postcolonial and feminist epistemology. She proofread the penultimate draft of Chap. 5 and was my go-to consultant on questions concerning Cohen's philosophy of Judaism and his exegesis of the Hebrew Bible. Rico Vitz read the entire manuscript and urged me to submit it to Springer for their consideration. If not for the help and support of these and other faculty colleagues, the manuscript would have remained an incomplete and unpublished draft on my bookshelf.

   For both Kant and Cohen, beliefs in God and in genuine human progress are essentially related to hope. In Cohen's case, this is the hope that we can asymptotically arrive at the messianic age envisioned by the Hebrew prophets and the Kantian idea of a Kingdom of Ends. However, as Cohen realized, faithfully laboring at the tasks that hope inspires often requires repeated doses of joy. I do not know if the completion of this book or any of the other tasks I have labored at in the course of my 60 years have contributed to genuine human progress in Cohen's sense, but I do know that I have been blessed with many joy-bearers. So in addition to the folks already mentioned, I would like acknowledge and thank Christopher Wilby for guiding me through the submission process and the staff, faculty, and students at APU who often renew my hope. Finally, I want to thank my children and grandchildren—Brady, Austin and his wife Jen, Joel, Ian, Tilly, Nixon, and Ryder—for the continual joy they bring to my life.

Contents

1 Introduction 1
  References 12

2 An Ambiguous Reading of Kant's Epistemology 15
  2.1 Concepts, Intuitions, and a Cognitive Synthesis 17
  2.2 Deploying the Distinction Between Concepts and Intuitions 42
  2.3 Kant's Philosophy of Mathematics: The Good, the Bad, and the
       Notoriously Obscure 54
  References 61

3 Disambiguating Kant for the Sake of Science 63
  3.1 Helmholtz's Naturalized Kantianism 67
  3.2 Cohen's Contra-Helmholtz Kantianism 83
  References 102

4 Disambiguating Frege for the Sake of Charity 105
  4.1 What Frege Meant by 'Kant Is Right About Geometry' 109
  4.2 Frege's Kantianism vis-à-vis Helmholtz and Cohen 120
  4.3 Deploying the Distinction Between Concepts and Objects 131
  References 158

5 Some Sanctifying Precepts for Science and Religion 161
  5.1 Redeeming the Concept of Scientific Progress 167
  5.2 Abrahamic Traditions and Accumulating Knowledge 194
  5.3 Test Case – A Christian Community in Crisis 243
  5.4 Conclusion 261
  References 262

Conclusion 267

Bibliography 269

Chapter 1
Introduction

Abstract This chapter introduces the expected pay-offs for a close reading of the nineteenth century debate between Hermann Helmholtz, Hermann Cohen, and Gottlob Frege. For readers interested in the history of early analytic philosophy and Frege scholarship, the pay-off is providing a more charitable interpretation of Frege's notorious resistance to the development of non-Euclidean geometry and his insistence on the absolute distinction between concepts and objects. For readers interested in more contemporary questions about modern western science, religious traditions, and epistemic injustice, examining this debate and the epistemic crises surrounding it promises to yield precepts for wisely navigating revolutionary changes in scientific and religious praxis, changes called for in hope of making genuinely humane progress.

Keywords Hermann Helmholtz · Hermann Cohen · Gottlob Frege · Thomas Kuhn · Science and religion · Epistemic injustice

In the 1906 issue of *Jahresbericht der Deutschen Mathematike-Vereinigung*, Gottlob Frege wrote:

> Whoever willfully deviates from the traditional sense (Sinne) of a word and does not indicate in what sense he wants to use it, whoever suddenly begins to call red what otherwise is called green, should not be astonished if he causes confusion. And if this occurs deliberately in science, it is a sin against science (*FG*, 58)[1]

---

[1] The following abbreviations will be used for frequently cited Frege texts: *BGS – Begriffsschrift, a formula language, modeled upon that or arithmetic for pure thought* in *Frege and Gödel: Two Fundamental Texts in Mathematical Logic*; *BLA – Basic Laws of Arithmetic: Exposition of a System*; *CP - Collected Papers on Mathematics*; *FA – The Foundations of Arithmetic*; *FG – On the Foundations of Geometry*; *FR – The Frege Reader*; *PMC – Philosophical and Mathematical Correspondence*; *PW – Posthumous Writings*; and *TPW – Translations from the Philosophical Writings*

© Springer Nature Switzerland AG 2020

T. Merrick, *Helmholtz, Cohen, and Frege on Progress and Fidelity*,
Philosophical Studies in Contemporary Culture 27,
https://doi.org/10.1007/978-3-030-57299-0_1

Frege's accusation was explicitly targeted at the discursive practices of David Hilbert and Alwin Korselt.[2] This book aims to show, however, that Frege considered this to be the unifying sin of all mathematicians whom he pejoratively branded 'formalists' – all are guilty of failing to adequately retain the historically recognized cognitive content of a term.[3] We will see that he also traced the roots of this sin to Hermann Helmholtz's reading of Kant. Why should anyone care about a century-old intradisciplinary squabble over linguistic usage and Kantian exegesis? Because understanding the context and content of Frege's charge against his colleagues promises to deliver a deeper appreciation of what is at stake in current debates over the epistemic status of scientific propositions and their relationship to religious thought and practice.

Throughout the latter half of the nineteenth century, Helmholtz proposed and defended a naturalized reading of Kantian epistemology.[4] Given this reading, Kant's claims concerning the a priori conditions of scientific theorizing became susceptible to falsification by means of revolutionary advances in non-Euclidean geometry and discoveries in the emerging field of experimental psychology.[5] In 1871, Hermann Cohen, founder of the Marburg school of neo- Kantianism, countered that Helmholtz's naturalized epistemology rested on a mistaken interpretation of Kant's 'a priori.' In addition, Cohen maintained that Helmholtz's naturalized Kantianism threatened the very notion of scientific progress. I argue that Frege sided with Cohen in this debate. Like Cohen, Frege accused Helmholtz of misreading Kant's ground-laying work in the epistemology of science and thus illegitimately deviating from

---

[2] By 'discursive practices,' I mean the definitions, semantic content and licensed inferences of geometric terms and sentences that Hilbert and Korselt proposed or assumed.

[3] 'Cognitive content,' 'conceptual content' ['begrifflicher Inhalt'] and 'judgeable content' ['beurteilbarer Inhalt'] are highly charged terms of art for Kant, Frege and anyone they have influenced. Thus, correctly interpreting these terms is the subject of much debate. However, there is no debate that, for Frege, the cognitive, conceptual or judgeable content of a word or sentence includes what he refers to as its sense ['Sinn']. In the 1891 'On Sinn and Bedeutung,' Frege famously split the content of an expression into sense [Sinn] and reference [Bedeutung]. Frege's complaints against the discursive practices of the formalists begin at least as early as 1884 and last throughout his career. The extent to which the 1891 distinction between sense and reference must be factored into any account concerning his feud with the formalists is addressed in §IV.B "Frege's C&O Distinction Deployed."

[4] By claiming that Helmholtz offers a naturalized Kantian epistemology, I simply mean that Helmholtz applied the findings of natural science, specifically the findings of experimental psychophysiology, to assess the truth of Kant's thesis the Euclidean axioms were a priori propositions and to discover the 'transcendental' conditions of our knowledge. Hyder 2006 questions whether 'naturalized Kantianism' adequately captures the nuanced appropriation of Kantian epistemology contained in Helmholtz's work. But our goal is to see Helmholtz's Kantianism as Cohen and Frege did, and Cohen clearly saw Helmholtz mistakenly tethering Kant's conceptions of apriority and the transcendental conditions of knowledge to the experimental results of perceptual psychology (see Cassirer 2005, p. 96). Besides, there is ample textual evidence for reading Helmholtz has having anticipated Quine's proposal for naturalizing epistemology.

[5] Enriques (1929), Cohen and Elkana (1977), Poma (1997), Hatfield (1990), DiSalle (2006), Patton (2012).

the understood sense of primitive logico-epistemological and scientific terms.[6] For Frege, as for Cohen, Helmholtz's Kantianism ends up promulgating methods and discursive practices that erode the cognitive content of scientific propositions and hence the epistemic prestige accorded to the sciences.

The pay-off for adequately defending my thesis is twofold. First, it resolves long-standing puzzles in Frege scholarship. As late as 1925, Frege still insisted on the epistemic primacy and a priori status of the geometric axioms in their "old Euclidean sense" (*PW*, 273). He routinely claimed that the axioms, understood in this sense, expressed basic facts of our intuition and that intuition provided the justification from which the axioms' improvable truth was derived (*FA*, 20-1; *PMC*, 37). The problem is how to reconcile Frege's claims with the fact that there was increasing consensus among leading German mathematicians that the Euclidean axioms were neither necessities of thought nor intuition throughout the nineteenth century (Coffa (1991), 41–43). Matters are further complicated when one considers Frege's consistent condemnation of those who appeal to psychological sources of justification for the basic propositions of science or logic. The challenge is making sense of Frege's stance on Euclidean geometry without merely dismissing him as an aging mathematician unwilling or unable to make the requisite shift to a new paradigm.

A second puzzle has to do with Frege's insistence that the distinction between concepts and objects (hereafter 'the C&O distinction') must be recognized as exhaustive and absolute. Scholars have long acknowledged the importance of the C&O distinction relative to Frege's aim of reducing arithmetic to logic and the subsequent development of modern mathematical logic. Viewed from the perspective of Frege's logicism and the distinction's role in revolutionizing logical analysis, its value lies in Frege's insight that a conceptual or predicative expression should be conceived along the lines of the mathematical expression for a function. Combining this insight with his claim that the objective cognitive content of expressions must be determined within a propositional context, Frege was able to produce a finely grained analysis of propositional content revealing the higher order variables, multiplace predicates, and nested quantifiers required for formalizing mathematical reasoning.[7] The problem is that the fruit which the distinction bears for logicism and for modern logic can be gleaned without having to treat the distinction as absolute in the manner that Frege describes.

---

[6]While it is true that Frege is sometimes careful to distinguish a logical from an epistemological line of inquiry, this is not the case when it comes to his critique of the formalists. We will see Frege accuse the formalists of committing "logical" errors and trying to resolve "epistemologically" confused problems. I will also argue that Frege's tendency to interweave the logical with the epistemological when critiquing Helmholtz and other formalists illustrates his gravitation toward Cohen's Kantianism. Thus, I take 'logico-epistemological' to be the most appropriate modifier for Cohen's and Frege's inquiry into the foundations and methodology of the exact and natural sciences.

[7]See Burge (2005 [1986]) and Ricketts (1986, 2010). I am indebted to William Demopoulus for pointing out that a previous draft of this manuscript did not sufficiently stress the importance of the C&O distinction for Frege's logicism and the development of modern predicate calculus.

In *The Foundations of Arithmetic* [*Die Grundlagen der Arithmetik*], Frege sets out the syntactic and semantic rules governing his use of 'concept' and 'object,' rules subsequently defended in his 1892 response to logician Benno Kerry:

> The business of a general concept word is precisely to signify [bezeichnet] a concept. Only when conjoined with the definite article or a demonstrative pronoun can it be counted as the proper name of a thing, but in that case in ceases to count as a concept word. The name of a thing is a proper name (*FA*, 63).[8]

> A concept is for me that which can be predicate of a singular judgment-content, an object that which can subject of the same (*FA*, 77).[9]

Kerry challenged the absoluteness of the distinction and the correlated syntactic rules by means of a counter-example: 'The concept horse is a concept easily attained.' Frege notoriously responded by reaffirming his syntactic criteria and asserting "[T]he concept horse is not a concept" (*FR*, i185).

Michael Dummett (1981), Terence Parsons (1986), Crispin Wright (1998) and Tyler Burge (2005) mount persuasive arguments to the effect that Frege could have responded in either of two ways. First, he could have dismissed 'the concept horse' as an ill-formed expression, given that it cannot satisfy both Frege's syntactic specifications for a proper name and his mathematico-logical analysis of concept-words as inherently incomplete or unsaturated. Kerry's counter-example would then have been rendered a senseless and meaningless proposition.[10] Second, he could have affirmed the truth that 'the concept horse is a concept' simply by weakening the inferential links between an expression's superficial syntax (the conjoined 'the'), its logico-grammatical function (whether or not it functions predicatively in particular propositional context), and its denotation (whether or not it refers to a concept).[11]

In short, despite subtle differences in their critique of Frege's response to Kerry and their various proposals for resolving the paradox, Dummett and Wright would agree with Parson's that Frege should not have said, 'The concept horse is not a concept,' because "nothing in his theory forced him to say it, and because the reasons he had for saying it are bad ones" (Parsons 1986, 451). Burge thus nicely captures the state of the literature on Frege's C&O distinction when he says:

---

[8] In 1884, Frege did not distinguish between the sense and reference of expressions. By 1892, he takes care to identify a concept as the *Bedeutung* of a concept- word [*Begriffswort*] and an object as the *Bedeutung* of a proper name [*Eigenname*].

[9] Here Frege restricts the use of the term 'concept' to designate one-place, first-level concept. Elsewhere, 'concept' and 'function' are used to denote multi-place and higher order concepts. Throughout the book, I use 'concept' (or 'function') to refer to one-place, first-level concepts unless otherwise indicated.

[10] Dummett maintains that this was Frege's subsequent unpublished response to Kerry-like locutions, though he admits that he cannot marshal evidence to document this claim (Dummett (1981), 211–213).

[11] All four reject Frege's insistence that the presence of a definite article is a necessary and sufficient condition for treating an expression as a proper name, echoing Kerry's complaint that logical and semantic distinctions cannot merely be read off of ordinary linguistic usage.

> Let us lay aside Frege's view that no objects are functions and no functions are objects. I think that this view is extremely doubtful and that it probably does constitute an instance in which Frege allowed his sound conceptions of logical function to harden unnecessarily into a metaphysical doctrine (Burge 2005, 102).

However, the question remains: are there any additional fruits or benefits that Frege attributes to the distinction and its correlative syntax that could possibly outweigh the cost of imposing such linguistic infelicities as 'The concept F is not a concept' on his and his colleagues' discursive practices?

Chapter 4, 'Disambiguating Frege for the sake of charity,' shows how both of these interpretive puzzles can be resolved by reading Frege within the context of the debate between Helmholtz and Cohen. In Sect. 4.1, I argue that Frege's infamously long commitment to the thesis that Euclidean geometry is synthetic a priori is best explained by realizing Frege is operating with two notions of 'intuition.' Sometimes he uses the term as it is understood within the context of Helmholtz's naturalized, psychophysiological rendering of Kant. However, when using 'intuition' to denote the source of geometric knowledge, he is invoking Cohen's use of Kantian terminology. As explained in Chap. 3, Cohen reinterpreted Kantian notions so as to rid them of any psychophysiological connotation. His goal was recover the robust normativity of Kantian doctrines that he considered crucial methodological guarantees of the epistemic status enjoyed by the sciences. For Cohen, the Kantian modifier 'a priori' signals a normativity that cannot be grounded empirically, and it is this necessary normative element in properly scientific propositions that is lost under Helmholtz's naturalized Kantianism.[12] Reading Frege's remarks on the Euclidean axioms in the light of Cohen's sophisticated reading and defense of Kant's apriority thesis enables them to be viewed more favorably.

Next, I turn to the C&O distinction, focusing on its introduction in the *Foundations of Arithmetic*. Here we see that Frege assigns the distinction a very specific task: to establish that "widely-held formalist theory of fractional, negative, etc., numbers is untenable" (*FA*, x).[13] Section 4.2 shows that Frege ultimately traces this formalist theory to Helmholtz's naturalized Kantianism and its methods for underwriting the objectivity of mathematical concepts and propositions. I go on to argue that Frege

---

[12] Anderson (2005) proposes a useful test for picking out the "orthodox" Kantian complaints from among the multiple complaints against psychologism lodged in Germany in the latter half of the nineteenth century (291). For Anderson, the distinctive Kantian critique against conflating logic and psychology is that logical laws possess a normativity that cannot be grounded in empirical psychology. Orthodox neo-Kantian critiques of psychologism thus stress the normativity of logic, whereas non-orthodox critiques stress its objectivity. Applying this test, Anderson argues that Cohen's Marburg school of neo-Kantianism deserves not only its fame, but also its claim of being the most Kantian of the neo-Kantians (306). He also concludes that Frege's critique of psychologism shows him to be outside the pale of Kantian orthodoxy. An implication of the argument of this book is that applying Anderson's test to Frege, once his complaint against the formalists is fully considered, shows him to be much closer to Kantian orthodoxy than Anderson maintains.

[13] I am grateful to Penelope Maddy for first suggesting that my line of argumentation would be clearer and more compelling if its grounding in Frege's *Foundation of Arithmetic* was made more explicit.

accepts Cohen's criteria for fixing the meaning and justifying the introduction of primitive terms of a science. That is, for Frege, as for Cohen, terms may be introduced into scientific discourse or the meanings of previously used terms modified if and only if (1) the semantic continuity befitting a scientific discipline charged with accumulating a body of knowledge over a specified domain can be sustained; and (2) the fruitfulness of these new or modified terms is demonstrated relative to the epistemic goals of the practice itself or those of another discipline dependent on the practice as a source of justification. Jamie Tappenden (1995) has already explained how Frege employs the fruitfulness criterion to introduce and fix terminology for use in mathematical practice. Lydia Patton (2005) and Alan Richardson (2006) tell us that Cohen's goal of "extricating Kant from Helmholtz's interpretation" was motivated by his interest in establishing epistemology as a science in its own right (Patton, 111). Thus, my aim in this section is to show that Cohen and Frege accept the same criterion of fruitfulness for fixing the terminology to be used in the emerging field of scientific epistemology, the field that Cohen designates 'Erkenntniskritik.'

Section 4.3 is devoted to establishing that Frege's C&O distinction should be interpreted as Cohen-like reworking of Kant's distinction between concepts and intuitions (hereafter 'the C&I distinction'). Frege's use of the C&O distinction to expose and forestall the purported fallacies of the formalists parallels Kant's use of the C&I distinction to expose and forestall the exact same fallacies. Having aligned Frege with Cohen's Kantianism, I conclude that Frege's use of the C&O distinction against formalists should be read as part and parcel of his justification for the distinction. In sum, Frege's use of the C&O distinction parallels Kant's use of the C&I distinction, and, given Cohen's precedent, this suffices for introducing a modified Kantian doctrine into the lexicon that circumscribes the domain of faithful scientific speech and practice. Thus, pace Burge, we need not interpret Frege's hardline on the absoluteness of the C&O distinction as an unnecessary lapse into metaphysical dogmatism, but rather an attempt to reinstate a methodological precept aimed at blocking the discursive moves licensed in Helmholtz's naturalized epistemology. It follows that deciding whether to accept this hardline hinges on deciding whether Helmholtz's naturalized Kantianism posed the threat to mathematical practice that Frege thought it did and whether countering this threat warranted adopting the linguistic rules aimed at enforcing the distinction.

Chapter 5 presents the second pay-off for examining the Helmholtz-Cohen-Frege debate, namely, showing what it can contribute to current discussions about whether the notion of progress is a viable one for scientific and religious traditions.[14] Readers more interested in interdisciplinary questions within philosophy of

---

[14] For introductory purposes, it suffices to understand 'tradition' in the broad sense defined in the Oxford English Dictionary: "A long established and generally accepted custom or method of procedure, having almost the force of a law; an immemorial usage; the body (or any one) of the experiences and usages of any branch or school of having almost the force of a law; an immemorial usage; the body (or any one) of the experiences and usages of any branch or school of art or literature, handed down by predecessors and generally followed." Given this definition, the two religions that I will be focusing on, Judaism and Christianity, qualify as distinct traditions, as would

science and religion than in puzzles pertaining to Fregean exegesis may want to start with Chap. 5 and then skim Chaps. 2, 3 and 4 for the relevant background material.

As generally recognized, Thomas Kuhn's *The Structure of Scientific Revolutions* is credited with first challenging the idea that the modern sciences have progressed in the sense of having accumulated knowledge over a fixed domain of entities. On Kuhn's telling, the history of science is marked by revolutionary paradigm shifts. These shifts—the shift say from Newtonian to Einsteinian physics—involve such radical changes in the meaning of the terms employed by each theory that their respective ontologies have effectively shifted as well. As a result, the successor theory is better conceived as having changed the subject than as having accumulated knowledge about the same subject matter as its predecessor (Kuhn 1996 [1962], 92–110; Friedman 2008, 239–240). For P.K. Feyerabend, this semantic variance is an inescapable relation between any two sufficiently comprehensive scientific theories, and it is this relation that he refers to as their "incommensurability" (Feyerabend 1962, 74–82; Oberheim and Hoyningen-Huene 2013).

Since Kuhn and Feyerabend's publications, historians and philosophers of science have offered various responses to the so-called incommensurability thesis and its two most serious implications: (1) that scientific progress is neither a historically nor a theoretically viable concept; and (2) that the choice to adopt a revolutionary theory cannot be rationally determined. More recently, Bas van Fraassen and Michael Friedman have conceded that successive revolutionary theories are not entirely incomparable insofar as a successor theory is often retrospectively recognized as having recovered and expanded on the predictive and explanatory successes of its predecessor. Even so, they claim, there is the problem of explaining how the choice to accept a new paradigm or conceptual framework can viewed as a rational decision to take a step forward in better understanding the topic at hand at the very time the decision is being called for. Van Fraassen puts the point as follows:

> [Scientific] practice lives by the criterion that it must retrospectively account for past success and remain undogmatic about future vindication. This phenomenon of our cognitive life poses a major problem for philosophy. In retrospect, we can see the changes we went through—from suffering through reversal to recognition, to use the Aristotelian phrase—as fortunate. But was the trauma overcome by reason or by lucky epistemic mutation? (van Fraassen 2002, 66)

For van Fraassen, Kuhnian paradigm shifts are real, not merely apparent, phenomena in our cognitive lives and constitute an epistemic crisis or trauma. Therefore,

---

several sub-branches, e.g. the Methodist or Wesleyan branch of Christianity. A scientific practice also qualifies as a tradition if it meets Michael Friedman and Thomas Kuhn's description of paradigm-guided normal science: an inquiry governed by a "received" model providing a "basis for a 'firm research consensus,'" that is, a consensus about a "single set of rules" setting parameters "for all practitioners of the discipline from that point on (at least for a time)" (Friedman 2001, 19). As we will see, Frege considers Euclidean geometry as a tradition in precisely this sense, which is why he views the discursive practices of some of his colleagues as a sin against the scientific tradition of geometry.

any adequate epistemology must take into account such crises and explain how they can be rationally navigated (van Fraassen 2002, 67–74).

As Chaps. 3 and 4 will make clear, Helmholtz, Cohen, and Frege are confronting the very type of epistemic crisis that van Fraassen describes. Developments in non-Euclidean geometries not only challenged the presumed a priori status of the Euclidean axioms, but also Kant's claims about the necessary conditions of proper scientific reasoning. Helmholtz, Hilbert and others who treated research in non-Euclidean geometry as perfectly proper scientific research were essentially calling for a revolutionary shift in the mathematical sciences and in Kant's transcendental philosophy and epistemology of science. Both Cohen and Frege worried that the discursive practices increasingly employed by nineteenth century mathematicians and logicians and countenanced by a naturalized Kantianism were introducing semantic discontinuities threatening the status of mathematics, logic and epistemology as knowledge-gathering enterprises. In other words, Cohen and Frege anticipated the concerns expressed in Kuhn's and Feyerabend's incommensurability thesis. Moreover, Helmholtz, Cohen, and Frege all maintained that their means of navigating this crisis were the most rational means of moving forward. Critically examining their debate can thus help us identify the oars one needs to best navigate the rough passage from one conceptual paradigm to another.

Given where Cohen's Marburg School of Neo-Kantianism sits in the history of science and history of the philosophy of science and given its sophisticated rereading of Kant in response to revolutionary developments in math, logic and physics, it is not surprising that philosophers are reexamining the work of Cohen and his students, Paul Natorp and Ernst Cassirer for help in crafting an adequate post-Kuhnian epistemology of science.[15] Indeed, what is more surprising is that this reexamination has only begun to pick up steam within English-speaking philosophical circles in the past decade or so.[16] Michael Friedman was among the first to initiate a 'Back to the Neo-Kantians' movement in the history and philosophy of science.[17] He looks to the work of Cassirer for clues on how to modify Kant's account of the synthetic a priori such that the transitions between paradigms can be seen, both prospectively and retrospectively, as rationally-guided and progressive.

According to Friedman, Cassirer attempts to ground scientific progress on the historical fact that scientific revolutions involve successive transitions to increasingly rich and refined mathematical reconstructions of nature. Progressive scientific

---

[15] See Richardson (1996, 1997, 2003, 2006), Kim (2004), Patton (2004, 2005, 2009, 2012), Carus (2007) Edgar (2008), Heis (2010, 2011).

[16] In the introduction to a 2008 volume of *The Philosophical Forum* dedicated to "Neo-Kantianism and Its Relevance Today," Andrew Chignell writes: "Despite their massive importance in the first part of the 20th century, most Neo-Kantian texts have fallen out of favor in the last 50 years, and the movement itself is barely mentioned in contemporary surveys of the era. It is quite common for Anglo-American students to encounter a history of philosophy that goes through Hegel, Kierkegaard, Mill and Nietzsche and then on to Frege, Moore, Russell, and Wittgenstein without any mention of the once-mighty Neo-Kantian schools" (Chignell 2008, 121).

[17] Friedman (1996, 1997, 1999, 2000a, b, 2001, 2006, 2008)

rationality is thus "made possible" and "secured" in reference to a "continuously evolving sequence of mathematical structures" (Friedman 2008, 246). For Friedman, Cassirer's emphasis on the a priori, constitutive role of mathematical and logical principles in our empirical theorizing is a necessary part of any adequate response to Kuhn's incommensurability thesis. However, it is not enough: "I agree with Kuhn that purely mathematical continuity and convergence is not sufficient" (Friedman 2008, 249). One also needs to account for continuity at the level of the interconnection between "mathematical concepts and sensible experience." Or, to invoke Kant's own terminological distinctions, Cassirer secures continuity and rational progress in the domain of *"logically"* or *"mathematically"* possible theories, but what is needed is to secure these within the domain of *"empirically,"* *"physically"* or *"really"* possible theories (Ibid.). To say that a relationship of rationally-guided progress holds in this latter domain, says Friedman, we must *"relativize* the Kantian *a priori* to a given scientific theory in a given historical context" and thus *"historicize* the notion of transcendental philosophy itself" (Friedman 2008, 251 italics in the original).

In response to Friedman's proposal, one might ask: how is this relativizing and historicizing supposed to account for the continuity of the semantic content of a priori concepts and principles across scientific revolutions and the progressive march of rational human thought across differing frameworks of normative scientific reasoning? In a recent reconsideration and defense of his proposal, Friedman suggests that satisfactorily addressing this question will involve constructing a historical narrative that locates the development of science within the broader context of pursuing extra-scientific cognitive interests. Such interests include trying to further grasp, to the extent that one can, conceptions of the infinite as initially characterized in pre-Kantian metaphysics and theology, as well as better understanding religio-ethical conceptions of the good (Friedman 2012, 52). Friedman describes our post-Kantian, post-Kuhnian predicament as follows:

> [B]oth our science and our technology have now rapidly evolved and proliferated to the point of being completely unsurveyable by any human being. And as the result of this evolution, in particular, modern science and its associated technologies have become inextricably entangled with the very fabric of everyday life—directly impacting its moral, political and spiritual dimensions in a way that could not have been imagined by even the most sophisticated thinkers of the late eighteenth century Enlightenment (Ibid.)

In hopes of making rational sense of how we arrived here and how to move forward, he urges his colleagues to craft a historicized "Kantian philosophy" that is "directly relevant" to the multifaceted "predicament of our own time" (Ibid.). For van Fraassen too, the philosophical and epistemological problem posed by Kuhn and Feyerabend is not just a matter of the epistemic crises experienced in our scientific inquiries:

> These conceptual overturnings are not peculiar to the empirical sciences alone. There have been great changes in our conception of what it means to be a person, or conscious, or moral. Some of these changes are at least loosely associated with those in science. That is not nearly all there is to them.... Other illuminating examples of other-than-scientific conceptual revolutions occur in religion (van Fraassen 2002, 70).

So, for Friedman and van Fraassen, responding to the incommensurability thesis and its attendant concerns necessitates telling a historically informed story about our cognitive lives, a story that interweaves scientific, political, moral and religious cognitive interests and the various epistemic norms governing our pursuit of those interests. Chapter 5 is aimed at contributing one chapter to this story.

Section 5.1 presents a detailed account of van Fraassen's and Friedman's diagnosis of the problem that Kuhn poses for the concept of scientific progress. I show that their description of the problem finds substantial historical support in my exposition of Helmholtz's, Cohen's and Frege's reactions to the epistemic crises of their day. This exposition also provides the basis for evaluating van Fraassen's and Friedman's differing proposals for retaining a viable notion of scientific progress in a post-Kuhnian age. In the course of this evaluation, we learn that there is no uniquely rational response to a successful resolution to an epistemic crisis of Kuhnian proportions, but that we can identify certain tactical rules aiding us in this effort. Moreover, there is significant overlap in the rules for successfully resolving crises in scientific, ethical and religious traditions. The section ends with a summary of those rules.

Section 5.2 offers a brief look at Cohen's efforts to extend his work on the conditions for accumulating and transmitting scientific knowledge to the case of religio-ethical knowledge. As we will see, Cohen maintains that a religious tradition like Judaism must retain its status as a communal epistemic endeavor. For Cohen, ethics is and must remain an ongoing task or "problem of cognitive pursuit," a pursuit that cannot be divorced from Jewish moral teaching on "religious service" and the "awe and love of God" (*EM*, 3 and 27).[18] It follows that the philosophical ethics emerging from the German Enlightenment must be made "compatible with ethical norms that seem to be inherently religious" and an "integral part of religious doctrines" (Ibid., 26). Moreover, ethical and religio-ethical norms retain their epistemic significance only by maintaining their connection to "any other conscientious intellectual endeavor" (Ibid., 3). This then leads to an apparent dilemma, since Cohen ascribes to religio-ethical concepts and propositions a normative necessity stricter than that accorded to a priori scientific concepts and propositions:

> Only as an idea will the Good constitute an object of cognition—but is the Good admissible as an idea? …It is this very question that serves as the testing ground for scientific idealism. Here we arrive at a crossroad which we feel prompted to compare to the prophetic metaphor: Heaven and Earth, nature and science, may pass away, if only God's word, if only ethics shall remain. On the other hand, if ethics itself must become a science, and nothing but a science—will it not thus become subject to the destiny of all natural sciences, by being submitted to scientific methodology? (Ibid. 9)

---

[18] The following abbreviations will be used for frequently cited Cohen texts: *KTE*[1] - *Kant's Theorie der Erfahrung* (1871) 1st edition; *KTE*[2] – *Kant's Theorie der Erfahrung* (1885) 2nd edition; *PIM* – *Das Princip Infinitesimal-Method* (1883); *EM* – *Ethics of Maimonides* (1908); *RoR* – *Religion of Reason: Out of the Sources of Judaism* (1919); *RH* – *Reason and Hope: Selections from the Jewish Writings of Hermann Cohen* (1993). *KTE*[1], *KTE*[2] and *PIM* have not been translated into English and so are mine unless otherwise indicated. I will also indicate when I am substituting my own translation for the English translations referenced above.

His *Ethics of Maimonides* aims at resolving this dilemma. Here Cohen demonstrates and articulates the methodological and hermeneutical precepts that can steer Abrahamic traditions and communities through the "Charybdis" of an isolationist dogmatism that ends up depriving religio-ethical concepts and propositions of any cognitive content and the "Scylla" of a scientism similarly divesting them of epistemic significance or making them entirely derivative on extant scientific theories (Ibid.). I will argue that Cohen's attempt to resolve this dilemma is an attempt to wisely navigate an epistemic crisis confronting his Jewish community. My primary goal in this section is unearthing additional tactical rules for Kuhnian crises in religious traditions, specifically Abrahamic traditions or communities with a high view of biblical authority.

Section 5.3 applies the rules articulated in Sects. 5.1 and 5.2 to a particular case: the debate currently raging within the United Methodist Church (UMC) over the definition of marriage. The Book of Discipline of the United Methodist Church states: "We support laws in civil society that define marriage as the union of one man and one woman" (United Methodist Church 2016, §161.C). At the 2016 General Conference, the UMC decided not to adopt language that would require modifying this definition. However, findings in biomedical research and changes to healthcare protocols—not to mention feminist philosophy, theology and biblical studies—are increasingly calling for radical changes in our inherited conceptions of sex and gender. This section intends to show how the concept of Methodist marriage might be modified so as to apply to unions not fitting the description of one man and one woman and still be recognized as a rationally endorsable conceptual successor of its traditional predecessor. My hope is to go some way in convincing my reader that Cohen was right in thinking that religious traditions are capable of incorporating knowledge from the natural sciences and other conscientious intellectual endeavors without losing their distinctive religio-ethical integrity and vision. Assuming that the rapidity of scientific and technological revolutions will only increase in the foreseeable future, I maintain that would-be faithful practitioners of both scientific and religious traditions must become better prepared to confront the trans-paradigm situation that van Fraassen describes: "On the one hand, we are forced to acknowledge a chasm between the old and new, and on the other, we must be able to see our present as a rationally endorsable continuation of the past" (van Fraassen 2002, 112). Learning to identify and deploy the tactical rules for resolving epistemic crises is part of this preparation.

As stated above, this book aims at filling in one small chapter in the story of how those of us living in an entangled state of rapidly evolving scientific, technological, socio-political and religious trajectories arrived in this position with an eye towards learning what it has to teach us about becoming more responsible epistemic agents. This particular chapter begins with the richness and vagaries of Kant's epistemology, and so I begin with an intentionally ambiguous reading of the *Critique of Pure Reason*.

References

Anderson, L. (2005). Neo-Kantianism and the roots of anti-psychologism. *British Journal for the History of Philosophy, 13*(2), 287–323.

Burge, T. (1986). 'Frege on truth' in Haarparanta and Hintikka, 97–154.

Burge, T. (2005). *Truth, thought, reason: Essays on Frege*. New York: Oxford University Press.

Carus, A. W. (2007). *Carnap and twentieth-century thought*. Cambridge: Cambridge University Press.

Cassirer, E. (2005 [1912]). Hermann Cohen and the renewal of Kantian philosophy (L. Patton, Trans.). *Angelaki* (Abingdon: Routledge), *10*(1), 95–108, repr. in Luft (2015), 221–235.

Chignell A. (2008). Introduction: On going back to Kant. *The Philosophical Forum, xxxix*(2) (Summer 2008), 109–124.

Coffa, J. A. (1991). In L. Wessels (Ed.), *The semantic tradition from Kant to Carnap*. Cambridge: Cambridge University Press.

Cohen, H. (1871). *Kant's Theorie der Erfahrung* (1st ed) (selected portions trans: Prasse, J., Gallagher, K., Merrick, T.). Berlin: F. Dummler.

Cohen, H. (1876). Introduction to 1902 edition Friedrich Lange's *Geschichte des Materialismus* (Leipzig: J. Baedeker).

Cohen, H. (1881). Preface to the 1887 edition of Friedrich Lange's *Geschichte des Materialismus* (Iserlohn: J. Baedeker).

Cohen, H. (1883). *Das Princip der Infinitesimal-Methode und seine Geschichte* (selected portions trans. Hildebrand, K., Merrick, T.). Berlin: F. Dummler.

Cohen, H. (1885). *Kant's Theorie der Erfahrung* (2nd ed.) (selected portions trans. Prasse, J., Gallagher, K., Merrick, T.). Berlin: F. Dummler.

Cohen, Hermann (1907a), *Religiöse Postulate*, repr. in B. Strauss (Ed.), *Jüdische Schriften* Vol. 1, 1–14.

Cohen, H. (1907b). *Religion und Sittlichkeit: Eine Betrachtung zur Grundlegung der Religionphilosophie*. Berlin: verlag von M. Poppelauer.

Cohen, H. (1993 [1890–1917]). *Reason and hope: Selections from the Jewish writings of Hermann Cohen* (Ed., and Trans. by Eva Jospe). Cincinnati: Hebrew Union College Press.

Cohen, H. (1995 [1919]). *Religion of reason: Out of the sources of Judaism* (S. Kaplan, Trans.). The American Academy of Religion.

Cohen, H. (2004 [1908]). *Ethics of Maimonides* translated with commentary by Almut Sh. Bruckstein. Madison: University of Wisconsin Press.

Cohen, R. S., & Elkana, Y. (1977). *Hermann von Helmholtz: Epistemological writings*. Dordrecht: D. Reidel Publishing Co..

Disalle, R. (2006). 'Kant, Helmholtz and the Meaning of Empiricism' in Friedman and Nordmann (Eds.), (2006), 123–139.

Dummett, M. (1981). *Frege philosophy of language* (2nd ed.). Cambridge: Harvard University Press.

Edgar, S. (2008). Paul Natorp and the emergence of anti-psychologism in the nineteenth century. *Studies in the History of the Philosophy of Science, 39*, 54–65.

Enriques, F. (1929). *The historic development of logic* (J. Rosenthal, Trans.). New York: Henry Holt and Company.

Feyerabend, P. K. (1962). Explanation, reduction, and empiricism. *Minnesota studies in philosophy of science, III*, 28–97. Retrieved from http://www.mcps.umn.edu/philosophy/complete-Vol3.html

Frege G. (1984). *Collected papers on mathematics, Logic and philosophy (CP)* (B. McGuinness, Ed., and M. Black, Trans.). New York: Basil Blackwell.

Frege, G. (1970). *Begriffsschrift, a formula language, modeled upon that of arithemetic, for pure thought (BGS)* (J. van Heijenoort, Ed., and Trans.). In *Frege and Godel: Two fundamental texts in mathematical logic*. Cambridge: Harvard University Press.

Frege, G. (1952). *Translations from the philosophical writings of Gottlob Frege (TPW)* (P. Geach, and M. Black, Eds.). Oxford: Basil Blackwell.

Frege, G. (1997). *The Frege Reader (FR)* (Ed., and introduction by M. Beaney). Oxford: Blackwell.
Frege, G. (1979). *Posthumous writings (PW)* (H. Hermes, Kambartel, F., & Kaulbach, F. Eds., P. Long, & White, R. Trans.). Oxford: Basil Blackwell.
Frege, G. (1980). *Philosophical and Mathematical Coorespondence (PMC)* (B. McGuinness, Ed., and H. Kaal, Trans.). Chicago: University Chicago Press.
Frege, G. (1953). *Die Grundlagen der Arithmetic* (J. L. Austin, Trans.), as *The Foundations of Arithmetic (FA)*. Evanston: Northwestern University Press.
Frege, G. (1885–1908). *On the foundations of geometry (FG)* (Trans. with an introduction by Eike-Henner W. Kluge. New Haven: Yale University Press, 1971.
Frege, G. (1964). *Basic laws of arithmetic: Exposition of a the system (BLA)* (Ed., and Trans. with introduction by Montgomery Furth). Berkeley: University of California Press.
Frege, G. (1977). *Logical investigations* (P. Geach, Ed., and P. Geach, & Stoothoff, Trans.). Oxford: Basil Blackwell.
Friedman, M. (1996). Overcoming metaphysics: Carnap and Heidegger. In Giere & Richardson (Eds.), 45–79.
Friedman, M. (1997). Helmholtz's *Zeichentheorie* and Schlick's *Allgemeine Erkenntnislehre*: Early logical empiricism and its nineteenth-century background. *Philosophical Topics, 25* (2): 19–50.
Friedman, M. (1999). *Reconsidering logical positivism*. Cambridge: Cambridge University Press.
Friedman, M. (2000a). *A parting of the ways*. Chicago/La Salle: Open Court.
Friedman, M. (2000b). Geometry, construction, and intuition in Kant and his successors. In G. Sher & R. Tieszen (Eds.), *Between logic and intuition*. Cambridge: Cambridge University Press.
Friedman, M. (2001). *Dynamics of reason*. Stanford: CSLI Publications.
Friedman, M., & Nordmann, A. (Eds.). (2006). *The Kantian legacy in nineteenth-century science*. Cambridge, MA: The MIT Press.
Friedman, M. (2008). Ernst Cassirer and Thomas Kuhn: The neo-Kantian tradition in history and philosophy of science. *The Philosophical Forum, xxxix*(2) (Summer 2008), 239–252.
Friedman, M. (2012). Reconsidering the dynamics of reason. *Studies in History and Philosophy of Science, 43*, 47–53.
Hatfield, G. (1990). *The natural and the normative*. Cambridge, MA: MIT Press.
Heis, J. (2010). Critical philosophy begins at the very point where logistic leaves off: Cassirer's response to Frege and Russell. *Perspectives on Science, 18*(4), 382–408.
Heis, J. (2011). Ernst Cassirer's neo-Kantian philosophy of geometry. *British Journal for the History of Philosophy, 19*(4), 759–794.
Hyder, D. (2006). Kant, Helmholtz and the determinacy of physical notes. In V.F. Hendricks, K. Jørgensen, J. Lützen, S. Pedersen (Eds.), *Interactions: Mathematics, physics and philosophy, 1860–1930* (pp. 1–44), Boston Studies in the Philosophy of Science, Vol. 251, Berlin: Springer Verlag.
Kim, A. (2004). Paul Natorp. In E. N. Zalta (Ed.), *The Stanford Encyclopedia of Philosophy* (Summer 2004), http://plato.stanford.edu/archives/sum2004/ entries/natorp/
Kuhn, T. (1996 [1962]). *The structure of scientific revolution*, 3rd edition (Chicago: University of Chicago Press).
Oberheim, E., & Hoyningen-Huene, P. (2013). The incommensurability of scientific theories. In E. N. Zalta (Ed.), *The Stanford Encyclopedia of Philosophy* (Spring 2013 Edition), http://plato.stanford.edu/archives/spr2013/entries/incommensurability/
Parsons, T. (1986). Why Frege should not have said "The Concept *Horse* is not a Concept", *History of Philosophy Quarterly*, 3, 449-465.
Patton, L. (2004). *Hermann Cohen's history and philosophy of science*. Ph.D dissertation, McGill University, Montreal.
Patton, L. (2005). The critical philosophy renewed: the bridge between Herman Cohen's early work on Kant and later philosophy of science. *Angelaki, 10*(1), 109–118.
Patton, L. (2009). Signs, toy models, and the a priori: From Helmholtz to Wittgenstein. *Studies in the History and Philosophy of Science, 40*(3), 281–289.

Patton, L. (2012). Hermann von Helmholtz. In E. N. Zalta (Ed.), *The Stanford Encyclopedia of Philosophy* (Winter 2012 Edition), http://plato.stanford.edu/archives/win2012/entries/hermann-helmholtz/

Poma, A. (1997). *The critical philosophy of Hermann Cohen*, trans John Denton, (Albany: State University of New York Press, 1997).

Richardson, A. W. (1996). From epistemology to the logic of science: Carnap's philosophy of empirical knowledge in the 1930s, in Giere & Richardson (Eds.), 309–332.

Richardson, A. W. (1997). *Carnap's construction of the world*. Cambridge: Cambridge University Press.

Richardson, A. W. (2003). Conceiving, experiencing, and conceiving experiencing: Neo-Kantianism and the history of the concept of experience. *Topoi, 22*, 55–67.

Richardson, A. W. (2006). "The Fact of Science" and critique of knowledge: Exact science as problem and resource in Marburg neo-Kantianism, in Friedmann & Nordmann (Eds.), 211–226.

Ricketts, T. G. (1986). Objectivity and objecthood: Frege's metaphysics of judgment. In Haapparanta & Hintakka (Eds.) (1986), 65–98.

Ricketts, T. G. (2010). Concepts, objects, and the context principle. In Ricketts & Potter (Eds.), *The Cambridge Companion to Frege* (pp. 149–219). Cambridge: Cambridge University Press.

Tappenden, J. (1995). Extending knowledge and "Fruitful Concepts": Fregean themes in the foundations of mathematics. *Nous, 29*(4), 427–467.

United Methodist Church. (2016). *The book of discipline of the United Methodist Church 2016*. Nashville: United Methodist Publishing House.

Van Fraassen, B. C. (2002). *The empirical stance*. New Haven: Yale University Press.

Wright, C. (1998). Why Frege did not deserve his 'Granum Salis': A note on the paradox of 'The Concept Horse' and the ascription of Bedeutungen to predicates. *Grazer Philosophische Studien: Internationale Zeitschrift für Analytische Philosophie, 55*, 239–263.

Chapter 2
An Ambiguous Reading of Kant's Epistemology

Abstract This chapter lays the foundation for the thesis, more fully supported in Chaps. 3 and 4, that the debate between Helmholtz, Cohen, and Frege is best understood as a debate between Kantian exegetes. The first section (Sect. 2.1) presents a textually grounded yet ambiguous rendering of crucial Kantian terminology, including 'cognitive synthesis,' 'concept,' and 'intuition.' The second section (Sect. 2.2) shows how Kant deploys his distinction between concepts and intuitions to highlight the fallacies that he identifies with an insufficiently critical rationalism and overweening empiricism. The importance of illustrating Kant's use of the distinction becomes evident in Chap. 4, where I argue that Frege's insistence on the absolute distinction between concepts and objects is an attempt to recover the significance of Kant's distinction between concepts and intuitions as a norm for rigorous scientific thought. The final section (Sect. 2.3) ends with a summary of Kant's philosophy of mathematics. This summary helps explain why Helmholtz, Cohen, Frege and other nineteenth century practitioners and philosophers of science were intent upon reinterpreting and repurposing Kant-inspired terms and dicta, as opposed to ditching them altogether.

Keywords Immanuel Kant · Cognitive synthesis · Intuitive representations · Conceptual representations · Philosophy of science · Philosophy of math

The main point of this chapter is to say enough about Kant's account of concepts, intuitions and a cognitive synthesis for my reader to see that Helmholtz's and Cohen's differing interpretations of these Kantian notions—interpretations fully explained in Chap. 3—can each lay claim to a textual basis. And since I argue in Chap. 4 that Frege's C&O distinction is a Cohen-like reworking of Kant's C&I distinction and that this entails deriving the meaning of Kantian distinctions from their use as norms of scientific reasoning, I need to show how Kant deploys his C&I distinction against the fallacies that he takes to typify a pseudo-science, namely, pre-critical epistemology and metaphysics. In other words, I intend to present a reading of Kant's C&I distinction and its use in combating the mistakes that he ascribed to those whom he identified as rationalists and empiricists, a reading that comes as close as can be textually supported to resembling Frege's C&O distinction and its use against the formalists.

Let me also clarify what I am not trying to accomplish. I do not intend to present the most charitable, most accurate, or most novel reading of Kant's C&I distinction and critique of the empiricists and rationalists. Quite the opposite and this manifests itself in several ways. First, I do not carefully distinguish Kant's analysis of concept formation as it occurs at transcendental level from that which occurs at the level of empirical psychology. Many German interpreters prior to Cohen were not careful in drawing this distinction,[1] and our aim is to try to see in Kant what they saw. Second, my analysis of Kant's critique of his opponents focuses exclusively on the role that the C&I distinction plays in Kant's accusation that they commit two mistakes: an illicit appeal to abstraction and the fallacy of inferring truth and existence from mere logical consistency. Again, my narrow focus is defensible in this case, given my interest in showing that if someone were inclined to view the C&I distinction as highlighting and blocking these errors, there is some textual support for doing so. My thesis, of course, is that Frege was that someone.

The fact that the book is designed in part to contribute to Frege scholarship and not Kant scholarship also explains my use of the secondary literature throughout in this chapter. Most often, I cite authoritative Kant scholars, e.g. Henry Allison, simply to support and flesh out my account of what Kant meant by 'concept' and 'intuition,' thus providing additional evidence that this account does indeed have some basis in the original texts. Since I am not claiming that this reading remains closest to the texts, I devote little space defending it against contrary views.

Having clarified the aims of this chapter, let me present its structure. Sect. 2.1, "Concepts, Intuitions, and a Cognitive Synthesis," addresses the question: what are Kantian concepts and intuitions? Here I present the criteria that must be satisfied for representations to qualify as full-fledged concepts or intuitions. Terminology is also introduced that will be referred to throughout the remainder of the text: 'subjective perception', 'objective perception', 'logical possibility', 'objectively valid', 'objectively real', 'nominal definition', and 'real definition.' Some of the interpretive points teased out in this section are the following. Concepts and intuitions are objective perceptions; as such they are cognitively significant representations comprising the content of a judgment whose truth or falsity is both inter-subjectively ascertainable and binding. The paradigmatic intuitive representation is an ordinary visual perception of an empirical object, visual perception itself being a type of judgment. Kant denies that we are capable of intellectual intuition, and abstract mathematical objects find no place in his epistemological or ontological framework. This last point is especially important because anyone wanting to claim that Frege is committed to any kind of Kantian epistemological framework owes it to her reader to explain how Frege could presume that explicitly non-spatial, atemporal, sensibly unintuitable numbers might be housed there. Here again, my reading of Frege as

---

[1] See Hatfield (1990), 111, and Poma (1997), 3.

endorsing Cohen's Kantianism intends to provide that explanation. Lastly, the section draws attention to Kant's concern with articulating the conditions for arriving at inter-subjectively binding judgments and perceptions, as opposed to articulating the conditions under which a particular subject's representations can be said to mirror external reality. This provides the basis for showing, in the next section, that Kant's conception of the epistemological task is quite different from his opponents'.

Section 2.2, "Deploying the Distinction between Concepts and Intuitions," examines the role the C&I distinction plays in Kant's critique of the empiricists and rationalists. In his critique, Kant charges them with failing to recognize the distinct and necessary roles that concepts and intuitions play in grounding objectively valid judgments (A271/B327).[2] He also holds them accountable for perpetuating illusory knowledge claims and for encouraging a pervasive skepticism, bordering on solipsism. Here we will see that, according to Kant, both the cognitive illusions and skepticism are the inevitable result of neglecting his C&I distinction. I present Kant's specific allegations against his opponents and show how each allegation is related to his broad complaint that the C&I distinction is being ignored. The importance of this section becomes apparent in Chap. 4, where we will witness Frege lodging these same allegations against the formalists, only this time the erroneous practices are treated as a failure to appreciate the distinction between concepts and objects.

Section 2.3, "Kant's Philosophy of Mathematics: the Good, the Bad, and Notoriously Obscure," summarizes the basic tenets of Kant's philosophy of mathematics and describes to what extent his C&I distinction derives its content and justification from Euclidean geometry. This section is intended to present the reader with enough background information that she can appreciate the pressure that nineteenth century scientific developments exerted on Kant's philosophy, his C&I distinction in particular. Recounting the strengths of Kant's philosophy of mathematics also explains why nineteenth century debates concerning the appropriate foundations and methods of the mathematical sciences were conducted within a broadly Kantian framework.

2.1 Concepts, Intuitions, and a Cognitive Synthesis

In the *Critique of Pure Reason,* Kant presents his taxonomy of the various types of representations. Concepts and intuitions are classified as two distinct types of objective perceptions or cognitions (A320/B376–77). An objective perception is a representation that the subject is conscious of having and that refers to an inter-subjectively accessible object, rather than "to the subject as the modification of its state" (Ibid.).

---

[2] Unless noted otherwise, all citations to Kant's *Critique of Pure Reason* refer to 1998 Cambridge Edition, translated and edited by P. Guyer and A. Wood. Because different nineteenth century interpreters focused on either the A or B edition relative to their aims in mining Kant's texts, my citations will indicate what appears in both editions and what appears in only one.

For further clarification on the notion of an objective perception, we turn to the *Prolegomena,* where Kant contrasts two types of judgments. Subjectively-valid judgments are those that "express only a relation of two sensations in the same subject, namely myself, and this only in my present state of perception" (§19, 4:299). These judgments are valid only for a particular subject concerning her state at a particular time (Ibid.). By contrast, objectively valid judgments concern the properties or relations adhering in an object(s). These judgments concern something that is inter-subjectively accessible, and we expect their validity to hold at "all times for us and for everyone else" (§18, 4:298). It is only through objectively-valid judgments that we obtain knowledge about objective experience (§19, 4:299). Henry Allison (1983) characterizes Kant's contrast between subjectively-valid and objectively-valid judgments as the contrast "between judgments of the form 'It seems to me that *p*', and those of the form 'It is the case that *p*'" (Ibid., 150). We will soon see that Kantian concepts and intuitions are identified as representations comprising the content of objectively-valid judgments.[3]

Let me thus introduce the following distinction between objective and subjective perceptions. Subjective perceptions are representations that must be related to the modification of the mental or physical state of the subject. These include imaginings or psychologically associated representations that comprise the content of a particular subject's mental history.[4] Such representations can only ground subjectively-valid judgments, e.g., 'It seems to me that Fido is a dog'. They do not qualify as cognitively significant representations. Objective perceptions represent an inter-subjectively accessible content and ground objectively-valid judgments, e.g., 'It is the case that Fido is a dog.' Describing concepts and intuitions as objective perceptions is thus tantamount to describing them as cognitive representations that fix the references of a sentence such that its truth is considered inter-subjectively verifiable and binding.[5]

As previously stated, Kant classifies intuitions and concepts as the two *distinct* types of objective perceptions: an intuition is "immediately related to the object and

---

[3] See Sect. 2.1.2.2, "The Characteristics of a Proper Kantian Concept" and Sect. 2.1.2.4, "Proper Kantian Intuitions".

[4] See also Allison (1983), 72–3. This is essentially how Allison characterizes those representations that, according to Kant, are incapable of being the constitutive representations in an objectively-valid judgment. I want to stress, however, that Allison does not use the expressions 'subjective perception' or 'objective perception' and that my use of this contrast throughout the book does not necessarily track with Allison's views on related topics.

[5] Those familiar with Frege's philosophy should notice the similarity here between an objective perception and what Frege describes as the sense associated with the terms in a proposition (*FR*, 148–154). Clinton Tolley (2014) argues that what Kant means by the content ['Inhalt'] of a cognition ['Erkenntnis'] comes much closer to what Frege means sense ['Sinne'] than has been commonly recognized. Though I agree with Tolley's thesis, it is important to note that his thesis is stronger than the one that I am defending. Tolley claims that Kant's view *rightly* interpreted is significantly similar to Frege's, whereas I am arguing the Kant's view *can be* interpreted as significantly similar to Frege's and that Frege could expect his audience to recognize this similarity based on the prompts that he provides in the *Foundations of Arithmetic* and sprinkled throughout his other works.

is singular;" a concept is related mediately "by means of a mark, which can be common to several things" (A320/B377). Intuitions represent a particular object as immediately present or given, e.g. the experience of Fido sitting before me. By contrast, concepts represent an object in general and mediately, by means of the properties commonly associated with that kind of object, e.g., the representation of a four-footed, barking, mammal (A141/B181).

Kant accounts for the difference in the representational character of our intuitions and concepts by maintaining that they are derived from two faculties, the sensibility and the understanding, with differing capabilities. The capacity for "bringing forth representations itself" belongs to the understanding, and concepts are included in that class of spontaneously produced representations (A51/B75). Thus, concepts not only represent objects by means of conceptual marks, but also represent them as "**thought** through the understanding" (A19/B33). Sensibility is a passive receiver of various representations. Unlike a divine intellect, beings with our cognitive capacities cannot spontaneously produce an immediate representation of a particular, given object. We obtain intuitive representations only if something "affects the mind is in a certain way," and this capacity to be so affected "is called **sensibility**" (Ibid.).[6] For us, objects can only be represented as "**given** to us by means of sensibility, and it alone affords us **intuitions**" (Ibid.). Therefore, intuitions not only represent objects as singular and immediately given, but also as, and only as, "things that we take as **objects of our senses**" (A34–5/B51).

When highlighting the difference between conceptual and intuitive representations, Kant sometimes makes it sound as if they are generated or received by each faculty, operating independently. If this were the case, the understanding and sensibility could be recognized as two self-contained sources of knowledge. Up until fairly recently, most contemporary Kant scholars held that a main tenet of *Critique* is that both faculties, each with their distinctive mode of representing an object, must be involved in any one act of cognition.[7] This tenet is most clearly expressed

---

[6] Bold in the original.

[7] Thus, Allison (1983) states that a "the central tenet of [Kant's] epistemology" is that sensibility alone cannot deliver full-fledged, cognitively significant intuitive representations (67–8). Friedman (2000a) makes same point concerning the understanding and concepts: "Pure forms of thought" do not possess cognitive significance, unless they are "given a determinate spatio-temporal content in relation to the pure forms of sensible intuition" (27). And it is precisely here that he sees Cassirer and the entire Marburg School making a "profound interpretive mistake" (Friedman (2008), 247). They ignore the role of the faculty of sensibility, maintaining that the a priori structures of space-time presupposed in Newtonian and then Einsteinian physics can be derived almost exclusively from pure conceptual rules of the understanding. Thus, they fail to sufficiently take into account the conditions necessary for ascribing to concepts real possibility, their "real empirical meaning," in contrast to ascribing mere logico-mathematical possibility (Ibid., 249). As Chaps. 3 and 4 will make clear, I take issue with Friedman on this last point, at least insofar as Cohen is concerned. While I agree that Cohen's de-psychologized reading of the *Critique* downplays any reference to faculties almost to the point of extinction, I maintain that both Cohen and Frege were well aware of the need to build into the definition of logical and mathematical concepts their necessary relevance to empirical concepts, inferences and observations. That is, all parties to the Helmholtz-Cohen-Frege exchange understood the crucial distinction between the logically possible and the

in Kant's: "Intuition and concepts therefore constitute the elements of all our cognition, so that neither concepts without intuition corresponding to them in some way, nor intuition without concepts can yield a cognition" (A50/B74). Because this tenet plays such a crucial role in Kant's use of his C&I distinction to show that the representations so prized by rationalists and empiricists are lacking in cognitive content, we need to focus our attention on those passages where Kant develops and defends it.[8] That is, we need to examine his description of the act giving rise to objective perceptions, the act that I refer to as a 'cognitive synthesis.'

2.1.1 The Three Necessary Features of a Cognitive Synthesis

My exposition of a cognitive synthesis draws extensively from its description in the A edition of Kant's *Critique* and here is why. First, the passage from A96 to A114 contains a clear account of the subtasks involved in arriving at an objectively-valid judgment and the necessary contributions made by intuitions and concepts. Here, for example, Kant explicitly tells us what is indicated by the "the word 'concept'" (A103). Secondly, Kant introduces both the A and B editions of the Transcendental Deduction by citing the need to show "how **subjective conditions of thinking** should have **objective validity**, i.e., yield conditions of the possibility of all cognition of object" (A89/B122). He is especially interested in showing that the pure concepts of the understanding are necessarily employed in any intuitive representation of spatio-temporal objects. Therefore, the cognitive synthesis described in the A edition of the Deduction is of the most general and fundamental kind. Kant expects, however, that the subtasks described will carry over in more specific settings. Analyzing cognitive synthesis in its most general form facilitates understanding Kant's account of cognizing mathematical and empirical entities as well. Thirdly, unlike the B edition, the A edition does not stress the difference between the cognitive activity of the empirical knowing subject and the cognitive activity of the transcendental knowing subject, thus putting us in a better position to appreciate Helmholtz' subsequent naturalized reading of Kant's C&I distinction.

Kant breaks down the performance of a cognitive synthesis into three necessary and interrelated acts: (1) the synthesis of apprehension in intuition; (2) the synthesis of reproduction in imagination; and (3) the synthesis of recognition in a concept.

---

empirically possible; the debate was over how best to establish the latter for presumptive a priori concepts and propositions.

[8] In Anglophone Kant scholarship, this tenet is referred to as the 'togetherness' principle. For a summary of the 21 century debate on how to best interpret this principle and its implications for understanding Kant's epistemological commitments, see Hanna 2018. This debate hinges, in part, on what Kantian texts are given pride of place. Given my purposes, this choice is determined by which texts demonstrate most clearly how Kant employs the togetherness principle, along with the C&I distinction, to highlight the purported fallacies of his opponents. I leave it to others to decide how Kant's use of the principle in within this specific context factors into the debate over his conceptualism.

Each of these three must be performed successfully for the unified representation to count as an objective perception, i.e. a representation of an intuited conceptualized object.

The synthesis of apprehension is defined as the act whereby the various representations which make up the manifold of the intuition are run through and held together (A99). Even though an intuition represents its object as singular, the object is still complex, capable of being analyzed into diverse parts. Thus Kant claims, "Every intuition contains a manifold in itself...." (A99). Recall that, for Kant, sensibility is the passive receptor of various representations. His proposed Copernican revolution in metaphysics and epistemology begins with the identification of space and time as our form of sensibility. Allison (1983) thus describes Kantian space and time as "as subjective conditions in terms of which alone the human mind is capable of receiving the data for thought or experience" (7). Space orders received representations insofar as they are represented as outside and alongside one another, "thus not merely different but as in different places" (A23/B38). Time orders received representations insofar as they are represented as successive, coexisting, or enduring (B67). The synthesis of apprehension is completed when it yields the representation of a complex object having a determinate spatial and temporal form within the spatio-temporal manifold.

Kant identifies temporality as the most universal and necessary sensible condition associated with the diversity and unity embodied in intuitive representations:

> Every intuition contains a manifold in itself, which however would not be represented as such if the mind did not distinguish the time in the succession of impressions one on another; for **as contained in one moment** no representation can ever be anything other than absolute unity. Now in order for **unity** of intuition to come from this manifold (as, say, in the representation of space), it is necessary to run through and then take together this manifoldness, which action I call the **synthesis of apprehension**...(A99).

The mind distinguishes the diversity of representations via the temporal relation of succession, one representation following another. To represent something as both single and complex, it must be possible to run through the representations and recognize them as holding together in a complete temporal sequence. Consider the apprehension of a dog standing in front of me. According to Kant, the apprehension of Fido as the union of differentiated sensible impressions immediately presented to my senses requires, in addition to determining a relation of temporal succession, the determination of a temporal whole. I must not only successively run through Fido's furriness, brownness, four legs, etc., but also "take together this manifold" in a spatio-temporal here and now (A99). The furriness, brownness, etc. must be apprehended as amenable to two temporal relations, succession and coexistence. The successive relation grants representations their diversity, while coexistence enables us to recognize them as cohering in a unified whole. Therefore, Kant can maintain that "appearances... in the end come down to determinations of inner sense [time]" (A101).

He goes on to argue that the possibility of representing something amenable to succession and coexistence, the possibility of representing something as enduring, presupposes another synthesis, the synthesis of reproduction in the imagination.

Even apprehending pure spatial and temporal forms as singular, complex manifolds requires a synthetic act by the imagination:

> Now it is obvious that if I draw a line in thought, or think of the time from one to noon to the next, or even want to represent a certain number to myself, I must necessarily first grasp one of these manifold representations after another in my thoughts. But if I were to lose the preceding representations (the first parts of the line, the preceding parts of time, or the successively ordered units) from my thoughts and not reproduce them when I proceed to the following ones, then no whole representation and none of the previously mentioned thoughts, not even the purest and most fundamental representations of space and time, could ever arise. (A102).

Obtaining a complete representation requires representing prior representations as coexistent with those following after them. Even the representation of a determinate period of time requires the reproduction of past moments; I must be able to recognize each moment as contained in a single, determinate representation of time. Or, consider a triangular image: it is made of up parts—three lines, three vertices, etc.— Kant maintains our recognition of these as parts depends on our ability to attend to each one successively. However, to see these as parts of a single, complete representation, we must also be able to reproduce them while advancing to those that follow.

It seems as if Kant is relying on memory to play the key role here: we must keep in mind, i.e. not forget, the successively intuited parts such that that they are represented as coexisting in a determinate spatial form and a unified temporal sequence. He argues, however, that our psychological capacities to recollect and associate diverse images cannot account for the necessary relation holding between the constituent parts of an objective perception. It is not good enough to say 'whenever a triangle is represented, three lines are always recalled as parts of the representation'; rather, we must be able to say 'whenever a triangle is represented, three lines *must* be recalled as parts of that representation. The coexistence and coherence required for the apprehension of an intuitive representation of an object thus display a certain necessity, and this necessity can only derive from rules of a particular kind:

> It is, to be sure, a merely empirical law in accordance with which representations that have often followed or accompanied one another are finally associated....This law of reproduction, however, presupposes that appearances themselves are actually are subject to such a rule, and that in the manifold of their representations an accompaniment or succession takes place in according to certain rules. (A100).

Kant's point here is that appealing to spatio-temporal proximity, sensory input, and empirical laws of psychological association cannot account for the necessary unity manifest in objective perceptions. Moreover, if objective perceptions did not exhibit this necessary unity, then the fact of our psychologically associating or recollecting certain representations would be inexplicable: "If cinnabar were now red, now black, now light, now heavy...then no empirical synthesis of reproduction could take place" (A101).

Kant concludes that rules accounting for the necessary unity characteristic of objective perception must be ones viewed as deriving from the nature of the object itself: "Thus we think of a triangle as an object by being conscious of the composition of three straight lines in accordance with a rule according to which such an intuition can always be exhibited" (A105). Recall the distinction between the subjectively-valid "It seems me like an *F*' and the objectively-valid 'This is an *F.*' This distinction is now explained and justified relative to the distinction between a subjective and objective perception. When apprehending an intuitive representation, the unity exhibits an objective, not simply subjective validity, a unity "combined in the object, i.e., regardless of any difference in the condition of the subject" (B142). An objective perception is thus a judgment that the unity of representations is determined by a given object:

> [T]he relation of all cognition to its object carries something of necessity with it, since namely the latter is regarded as that which is opposed to our cognitions between determined at pleasure or arbitrarily rather than being determined *a priori* (A104).

The justification for my assertion that 'This is a triangle' is the recognition that the intuited unity is a collection of sensibly given representations produced and held together according to the rule for constructing or representing triangle-type entities, a rule that cannot be derived entirely a posteriori.

Insofar as we are conscious of these rules and the role they play in imparting the requisite necessity and generality to their respective intuited representations, we are justified in identifying the rules as concepts:

> The word 'concept' itself could already lead us to this remark. For it is this **one** consciousness that unifies the manifold that has been successively intuited, and then also reproduced, into one representation. This consciousness may often be only weak, so that we connect it with the generation of the representation only in the effect, but not in the act itself, i.e., immediately; but regardless of these differences one consciousness must always be found, even if it lacks conspicuous clarity, and with that concepts, and with them cognition of objects would be entirely impossible (A103–4).[9]

Concepts express a necessary unity; they represent objects as they must be thought in accordance with the understanding. They are thus perfectly suited to serve as rules determining the synthesis of reproduction. Even empirical concepts express a necessity and generality that cannot be wholly derived from the strictly receptive faculty of sensibility:

> The concept of a dog signifies a rule in accordance with which my imagination can specify a shape of a four-footed animal in general, without being restricted to any particular shape that experience offers me or any possible image that I can exhibit *in concreto* (A141/B181).

The concept 'dog' expresses what characteristics a representation must have, if I am to recognize it as a dog. Certain representations may come to me via sensibility, but I am entitled to the claim 'This is a dog' only if I discover in the unified form of these representations the very same form I associate with a 'dog-like' figure in

---

[9] This passage is especially ripe with the ambiguities that will later give rise to Helmholtz's naturalized reading of Kant, whereby the necessary consciousness of the act giving rise to concepts is interpreted as a cognitive psychological claim and to Cohen's methodological reading of Kant, whereby the necessary consciousness of the act is interpreted as a normative epistemological claim asserting that a priori logical and mathematical concepts should be recognized as bequeathing the objectivity accorded to the facts of the empirical sciences.

general. Thus, for Kant, concepts are rules that both explain and justify the objective validity of judgments concerning an immediately-given presentation of a singular, but complex object.

Notice too that we must be conscious that the intuitive representation was generated in accordance with the concept-rule. Recognizing the intuition as falling under a certain concept presupposes that I am, to some degree, conscious of the concept-guided act of the imagination that produced the intuition (A103–4). We become conscious of this activity insofar as it is a spontaneous, "productive" act of the imagination, which "is an effect of the understanding on sensibility" (B151–2). This activity, coupled with our awareness of it, enables us to apprehend the resulting figure as an intuitive representation of an object falling under the appropriate concept(s). We know the triangular figure is a single, complex representation consisting of successively, given parts—lines, vertices, etc.—which are reproduced and combined in accordance with a rule for generating triangles because we somehow detect the activity that constructed it. Recognizing the concept is being conscious, at some level, of the act whereby the manifold was combined and reproduced into a single representation. This is the third thing, *the synthesis of recognition in a concept*, which must be accomplished in any one cognitive synthesis.

2.1.2   So What Are Kantian Concepts and Intuitions?

2.1.2.1   Concepts Qua Pure Spontaneous Productions Are Empty Conceptual Forms

We are now in a better position to appreciate the magnitude of Kant's claim that concepts considered solely as spontaneous productions of the understanding and devoid of any relationship to sensibility are "empty" (A51/B75). They do not qualify as mediate representations of objects, not even a priori representations of mathematical entities: "We cannot think a line without **drawing** it in thought, we cannot think of a circle without **describing** it" (B154). Concepts must be related to sensibility not only to show that they are actually instantiated, but also to show that they possess any cognitive content or significance whatsoever. J. Michael Young (1992a) confirms this as a moral Kant draws from his account of a cognitive synthesis:

> Kant is not merely stipulating that concepts will be said to have content only if things of the kind conceived are given in intuition. His claim is rather that things intuited somehow figure in an essential way in the concept itself. Apart from this relation to things intuited, the concept would be merely an empty shell, which could not serve as the basis of knowledge. In thinking it we would merely have 'played with representations' (A155/B195)....The notion of synthesis is supposed to make clear how it is that intuition enters into concepts and provides them with content that they would otherwise lack. (Young (1992a), 113).

A conceptual unity that is nothing more than a spontaneous production of the understanding "signifies nothing at all" (B306).[10]

So if the pre-synthesized a priori concepts and pre-schematized pure categories are not concepts in the sense of mediate representations of a possibly-given object, why refer to them as concepts at all? To answer this question, we must consider Kant's distinction between a concept's form and matter, which leads in turn to his distinction between a concept's logical and real possibility. In the *Jäsche Logic*, we read: "With every concept we are to distinguish *matter* and *form*. The matter of concepts is their capacity to represent an object, their form is *universality*" (§1, 9:91).[11] Even if something fails to qualify as a concept in virtue of its lack of matter or content, it may nevertheless possess the form of a concept:

> We demand in every concept, first, the logical form (of thought) in general, and secondly, the possibility of giving it an object to which it may be applied. In the absence of such object, it has no meaning and is completely lacking in content, though it may still contain the logical function which is required for making a concept out of any data that may be presented (A239/B299).

This holds for the categories as well:

> Now when this condition [the general condition of sensibility] has been omitted from the pure category, it can contain nothing but the logical function for bringing the manifold under a concept (A245).

Pre-synthesized spontaneous productions of thought are concepts only in the sense that they may manifest the appropriate form or function.

What is the form or function associated with a concept? Kant answers, "…a concept is always, as regards its form, something universal, which serves as a rule" (A106). We have seen how concepts function as rules in performing a cognitive synthesis. The concept-rule guides the productive activity of the imagination, which affects the manifold of sensibility and thereby enables us to apprehend an intuitive representation corresponding to the concept. In this context, the form of a concept serves as a rule determining those properties that a sensibly-given, spatio-temporal object must manifest to warrant being subsumed under a certain concept.

Not every conceptual form, however, qualifies as a rule in the sense of governing a cognitive synthesis. Rather, Kant uses the expression 'logical form of concept' more broadly to indicate any spontaneous act of the understanding whereby distinct representations are united under a common notion, irrespective of the source of those representations (A68/B93 and *Jäsche Logic*, §5, 9:94). When considering the activity of the understanding from this broader perspective, one should not assume

---

[10] See too Kant's conclusion on the need to schematize the pure concepts of the understanding: "Without schemata, therefore, the categories are only functions of the understanding for concepts, but do not represent any object. This significance comes to them from sensibility, which realizes the understanding at the same time as it restricts it" (A147/B186–7).

[11] Contemporary Kant scholars balk at taking the *Jäsche Logic* as indicative of Kant's views. However, as Michael Kremer notes, "throughout the nineteenth century it was simply known as 'Kant's Logic,' and was taken to represent Kant's views by Frege and his contemporaries" (Kremer 2000, 558 note 27).

that it is constrained by the manifold of sensibility. In fact, as we will see, given the relative ease with which the understanding generates conceptual forms, Kant warns that simply because an idea has the form of a concept is no guarantee that it is a vehicle of thought about a possible *thing*.

Kant recognizes only two constraints on the generation of conceptual forms. First, conceptual forms must be generated by the understanding:

> To make concepts out of representations one must thus be able *to compare, to reflect* and *to abstract*, for these three logical operations of the understanding are the essential and universal conditions for generation of every concept whatsoever (*Jäsche Logic* §6, 9:94).

There is, therefore, no such thing as an entirely empirically given concept; its content may be given, but not its form:

> An empirical concept arises from the senses through comparison of objects of experience and attains through the understanding merely the form of universality....The form of a concept, as that of a discursive representation, is always made (*Jäsche Logic* §3, 9:92 and §4, 9:93).

By stressing the necessary role of the understanding in generating conceptual forms, one can see Kant driving home the point that the mere accumulation of passively received sensible impressions never yields a concept.[12]

Whatever comes to us by means of the senses is simply too particular and contingent to perform the job that Kant assigns to concepts. Recall his description of the concept 'dog' as rule determining the "shape of four-footed animal in general" (A141/B180). The generality attributed to concepts enables them to function as rules. They will be applicable to a variety of possible experiences and thus not "restricted to any single particular shape that experience offers me or any possible image that I can exhibit *in concreto*" (Ibid.). Similarly, the necessity accruing to conceptual form tells against its being given by experience, which is why the form of a concept qua rule is always made. The spontaneous activity of the understanding is required for taking up particular sensible impressions and conferring upon them the generality and necessity characteristic of a rule. To recognize something as a product of the understanding's spontaneous activity is tantamount to recognizing it as a rule. Kant, therefore, refines the initial description of the understanding as "a spontaneity of cognition" and designates it a "faculty of rules," maintaining that this latter "designation is more fruitful and comes closer to its essence" (A126).

The second constraint on the generation of conceptual forms is that their constituent marks must be mutually consistent:

> To **cognize** and object, it is required that I be able to prove its possibility (whether by the testimony of experience from its actuality or *a priori* through reason. But I can **think** whatever I like, as long as I do not contradict myself, i.e., as long as my concept is a possible thought, even if I cannot give assurance whether or not there is a corresponding object somewhere in the sum total of all possibilities (Bxxvi).

---

[12] In Chaps. 3 and 4, we will see that this is a Kantian claim about cognitively significant concepts that Cohen and Frege retain and press against their colleagues.

If the constituent marks are consistent, we have established the logical possibility of a conceptual form: "The concept is always possible if it does not contradict itself. That is the logical mark of possibility…" (A596/B624). The principle of contradiction is thus a legitimate means of establishing the logical possibility of conceptual forms (*Jäsche Logic*, 9:51). Kant thus grants that the "the concept of a highest being" is a logically possible idea, given that its constituent marks "do not generate a contradiction" (A602/B30).

In recounting Kant's description of a threefold cognitive synthesis, I have intended to describe, albeit generally and vaguely, the conditions that must obtain to grant that a concept has cognitive content, i.e., that the possibility of relating it to a given object has been guaranteed. I will conclude by showing how this relates to Kant's discussion, in the A edition, of the contrast between a real and merely nominal definition of concepts. But first, it will be helpful to fix some of relevant terminology discussed thus far. We will say that a concept is *logically possible*, if it does not contain or give rise to a logical contradiction. As we have seen, for Kant, 'God' qualifies as a logically possible concept. A concept is *really possible* or *objectively valid*, if it has the capacity to represent a *possible* object, namely an object that can be given to knowers with cognitive capacities like ours. To say that a concept is objectively valid is equivalent to saying that it has been characterized sufficiently such that one can grasp "what sort of thing is really intended by concepts of that sort" (A241/B299). This, in turn, is equivalent to saying that a concept is potentially applicable and thus has cognitive content. Finally, we will say that a concept is *objectively real*, if it is objectively valid and the sort of thing represented by the concept actually exists.[13]

According to Kant, to guarantee that concepts possess objective validity it is not enough to show that they consist of mutually consistent marks; "something more is required" (Bxxvi). We must be able to conceive of those marks as potential properties or relations, and Kant likens this to providing a real, as opposed to a nominal, definition:

> I here mean a real definition, which does not merely supply other and more intelligible words for the name of a thing, but rather contains in itself a clear **mark** by means of which the **object** (*definitum*) can always be securely cognized and that makes the concept that is to be explained usable in application (A241–2)

Since, for us, whatever qualifies as a possible object must also be capable of coming to us via sensible intuition (the spatio-temporal manifold) conceiving of constituent marks as the composite parts or properties of a possible object presupposes that they are invested with some sensible, spatio-temporal meaning: "Hence it is also requisite for one **to make** an abstract concept **sensible**, i.e., display the object that corresponds to it in intuition, since without this the concept would remain (as one says)

---

[13] Kant does not always distinguish the terms 'objectively valid' and 'objectively real', which is why I wanted to fix the usage for purposes of clarity. For proof that the distinctions I have drawn here have a textual basis, see (A218/B225–B266), (A224/B271–A226/B273), (A234/B286) and (A290/B346–A292/B349). For further evidence that such distinctions are textually warranted, see Allison (1983), 145–6.

without **sense,** i.e., without significance" (A240/B299). This holds for the categories as well: "[W]e cannot even define a single one of them without immediately descending to the conditions of sensibility…(A241/B300).

The objective validity of a concept is thus demonstrated by showing that it not only functions as a rule determining a logically possible idea, but also as a rule for uniting sensibly-given representations resulting in the apprehension of a corresponding intuitive representation.[14] Since 'God' cannot name a rule stipulating that a necessary relation holds among sensibly-given representations, it cannot represent a possible thing and thus fails to achieve objective validity.[15] Since 'pegasus' can name such a rule, Kant would seem to have no grounds for denying its objective validity. Unlike 'God', the sensible content of 'pegasus' does make it applicable to experience. Based on the properties expressed by this concept, I know, for instance, that I have never seen one.[16] The take-away point is that all pure spontaneous productions are not full-fledged Kantian concepts until their objective validity, their capacity to represent objects, is guaranteed: "An *a priori* concept that was not related to the latter [the manifold of sensible intuition] would be only the logical form of a concept, but not the concept itself through which something is thought" (A95).

2.1.2.2   The Characteristics of a Proper Kantian Concept

So full-fledged, objectively-valid concepts are rules of the understanding that bear a cognitive relation to possibly-given objects, which is equivalent to saying that concepts are predicates of possible judgments:

> We can…trace all actions of the understanding back to judgments, so that the **understanding** in general can be represented as a **faculty of judging**. For according to what has been said it is a faculty of thinking. Thinking is cognition through concepts. Concepts, however, as predicates of possible judgments, are related to some representation of a still undetermined object (A69/B94).

For Hans Sluga (1980), this passage expresses Kant's doctrine of "the priority of judgments over concepts" and anticipates "the Fregean doctrine of concepts" as

---

[14] This is why, in the previous account of a cognitive synthesis, the recognition of a concept is simultaneous with and does not precede the apprehension of the intuition.

[15] This is at least true within the domain of theoretical reason. 'God' is not a rule determining the spatio-temporal manifold. For this reason it bears no cognitive relation to any possible objects and does not achieve objective validity. In the first *Critique,* Kant leaves open the possibility that such a concept obtains objective validity in the practical realm (Bxxvi). It is worth noting, however, that we are barred from ascribing the same meaning to the terms 'God' and 'soul' when used in this practical/moral discursive context (Bxxvii).

[16] Allison (1983) grants that fictional, empirical concepts achieve objective validity for Kant: "To claim that a concept has objective reality is to claim that it refers or is applicable to an actual object. Thus, a fictional concept, such as 'unicorn,' would not have objective reality, although it could very well function as a predicate in an objectively valid judgment, such as 'unicorns do not exist' (135).

essentially predicative.[17] For Allison (1983), the formation of a judgment and the conceiving of a possible object are, on Kant's view, identical activities:

> Kant's claim that 'all judgments are functions of unity among our representations' is intended to underscore the point that every judgment involves a unification or 'collection' of representations under a concept, that is, an act of conceptualization....Kant is thus saying that the essential task of every judgment is to produce a unity of representations under a concept. (Ibid., 70)

Given Allison's description of judgment-formation and my previous description of a cognitive synthesis, performing a cognitive synthesis just is the forming of a judgment.

As we have seen, however, not just any collection of representations will represent a possible object. Thus, a judgment relates to a possible object only when one of its constituent concepts unites representations supplied via sensibility. So while Sluga is certainly right in saying that much of what Kant writes about the nature of a proper concept anticipates what Frege will say, there are some significant differences as well. Still, our present concern is to present the textual support for the various and conflicting Neo-Kantian doctrines espoused by Helmholtz, Cohen and Frege. With that in mind, let us press on in seeing how close one can get in presenting a textually grounded, scholarly supported exegesis of Kant's account of proper concepts that anticipates the Fregean account presented in Chap. 4.

To explain how a concept determines the sensible content for a judgment, Allison (1983) unpacks Kant's brief description of 'All bodies are divisible.' Kant maintains that the concept 'divisible' is applied to subject concept 'body,' and the subject concept determines the intuitive representations, which stand in an immediate relation to the object (A69/B93). The concept 'body' thus functions as a "real predicate," meaning that it "provides the initial description under which the subject x is to be taken in the judgment" (Ibid., 71). The subject x is the real subject, i.e. the sensible content or object(s), of the judgment. The judgment 'All bodies are divisible' then asserts that every object "thought through the predicate 'body' is also thought through the predicate 'divisibility'" (Ibid.). Given Allison's explanation, Kantian concepts—understood as mediate representations and constituent elements in an objectively valid judgment—are necessary functions enabling us to take up objects as judgeable subjects.

Allison also notes that Kant usually defines judgment in terms of objective validity; a judgment possesses objective-validity, as opposed to mere subjective-validity. He thus interprets Kant's claim that every judgment is objectively valid as equivalent to the claim that every judgment has a truth-value (Allison (1983), 73). Furthermore, based on my exposition of Kant's cognitive synthesis, a judgment is objectively valid if and only if its constituent concepts are objectively-valid. Therefore, one might say that establishing the objective validity of a concept is

---

[17] Sluga (1980), 91. To my knowledge, Sluga never draws out the full-force of the similarities between Kant's and Frege's views on the nature of a concept. That is, he never argues that Frege's C&O distinction is an orthodox Neo-Kantian descendant of Kant's C&I distinction.

tantamount to showing that it can be used in connecting the real subjects of a judgment, i.e., objects, with a truth-value. If so, it is not too much of a stretch to think of proper Kantian concepts as functions taking objects to truth-values. And this is about as close as one can get to bridging the gap between Kant's *Critique* and the *Jäsche Logic* and the Fregean insight that concepts are necessarily predicative, objective constituents of a "judgeable content" and ought to be conceived along the line of a mathematical function.[18]

2.1.2.3 Blind Intuitions

Just as concepts bearing no relation to the manifold of sensible intuition are empty, so too, Kant claims, "intuitions without concepts are blind" (A51/B75). Remember, concepts function as the rules uniting sensory representations where the result is the apprehension of an object and not a collection of haphazard sense data. By labeling unconceptualized intuitions 'blind,' Kant's intends to stress that sensible *impressions* are not sensible *representations* of an object. Because his predecessors and contemporaries thought otherwise, they missed the spontaneous activities of the imagination and the understanding necessarily involved in the empirical perception of objects:

> No psychologist has yet thought the imagination is a necessary ingredient in perception itself. This is so partly because this faculty has been limited to reproduction, and partly because it has been believed that the senses do not merely afford us impressions but also put them together, and produce images of objects, for which without doubt something more than the receptivity of impressions is required, namely a function of the synthesis of them (A120).

The imagination, employing the appropriate concept-rules, combines impressions and only then do we obtain an image of an object. Moreover, as this quote suggests, pre-synthesized sense impressions cannot represent an object because our faculty for receiving these impressions is also blind.

According to Kant, our intuitive faculty, the faculty responsible for delivering an objective perception of immediately-given particular, does not accomplish this task alone. It cannot yield full-fledged intuitions because it is entirely passive or receptive. This is why he identifies our faculty of intuition with sensibility: "It comes along with our nature that **intuition** can never be anything other than **sensible**, i.e., that it contains only the way in which we are affected by objects" (A51/B75). So, when he writes that "Objects are...**given** us by means of sensibility, and it alone yields **intuitions**," this is somewhat misleading. The mere reception of sensible impressions does not give us an object. The sensibility alone may deliver the immediate affectations of some undetermined source, but not an intuitive representation of an object.

---

[18] See Frege's 1882 'Letter to Marty' (*FR*, 79–33), as well as *FA*, §46–§48, §51, §62 note 2.

To put ourselves in a better position to evaluate Helmholtz's subsequent critique of Kant's claims about our forms of sensibility, it will help to address the question whether Kant thought that pre-synthesized, pre-conceptualized sense impressions possessed any spatial or temporal characteristics. That is, do the pure forms of sensibility possess any structure of their own? Friedman (2008) argues that they do and faults the Marburg School's interpretation for failing to note this:

> Kant himself, on the contrary, takes the faculty of pure sensibility to have an independent *a priori* structure of its own—given by the Euclidean structure of space and the Newtonian structure of time (more precisely, space-time—and this is the reason, for Kant, that all our sensible or perceptual experience must necessarily be in accordance with these forms…(247).

As we will see, Helmholtz's interpretation matches Friedman's. This is not the place to fully present Helmholz's and Cohen's differing views on the matter.[19] Besides, my stated aims in examining their debate over how to interpret Kant are not so much concerned with whose exegesis remained most faithful to the texts, but rather who provides us with the best tools for navigating the shifting semantics within scientific and religious traditions. Therefore, I will bracket the issue as to whether Helmholtz and Friedman are right that, for Kant, the pure forms of sensibility contain a structure as well-articulated as the space-time manifold presupposed in Newtonian physics and simply agree that they contain some structure. In other words, it is consistent with what Kant says here and elsewhere to assume that pre-conceptualized sense impressions possess some spatial or temporal characteristics. So, for example, a modification of my inner sense, say my anxiety, can be temporally indexed as in 'anxiety occurring now.' It would be inconsistent, however, to say that for Kant pre-conceptualized sense impressions possessing spatio-temporal characteristics, e.g. 'red here now,' qualify as intuitions in the proper sense: "[T]he sensations of colors, sounds, and warmth…since they are merely sensations and not intuitions, do not in themselves allow any object to be cognized, least of all *a priori*" (B44).

2.1.2.4   Proper Kantian Intuitions

The best way to get at the notion of a proper Kantian intuition is to begin by way of contrast with a proper Kantian concept. Intuitions are singular, immediate representations. They represent an individual object as given or presented: "[T]he representation which can be given only through a single object is intuition" (A32/B47). Intuitions relate "immediately to the object". Concepts are mediate representations of objects, representing an object in general by means of conceptual marks that can be conceived as properties (A320/B377).

Some scholars have suggested that Kant's notion of an intuitive representation and its contrast with conceptual representations can be adequately explicated

---

[19] This will be done in Part II.

without having to ground it in any correlative distinction in faculties.[20] Moreover, as I hope to show in Chap. 4, Frege's elucidation of what he means by the term 'object' is precisely an attempt to retain Kant's notion of an intuition sans any talk of faculties. Still, I take this to be a significant modification of what Kant says in the *Critique,* and it is worth noting why here. Recall that Kant's Copernican revolution is intimately connected with his refusal to grant us intellectual intuitive capabilities: "An understanding, in which through self-consciousness all of the manifold would at the same time be given, would **intuit;** ours can only think and must seek the intuition in the senses (B135).[21] For cognizers like us, intuitive representations are possible only if something "affects the mind" (A19/B34). Since sensibility is our capacity to be affected by objects, it follows that the intuition of an object represents that object as given via sensibility.

Kant maintains that an intuitive representation presupposes sensible affectation even when intuiting our own mental states: "[W]e intuit ourselves only as we are internally **affected** ..." (B153). He admits that this may seem paradoxical, given that one must now posit two ways of viewing the self, namely as a spontaneously acting self and the self as the "**passive** subject" whose sensibility is being acted upon (B153). But as readers of the *Critique* are well aware, Kant embraces the implication that the self can and must be viewed under two aspects or given a "**two-fold meaning**" (Bxxviii). For this is equivalent to the claim: "I therefore have **no cognition** of myself as I am, but only as I **appear** to myself" (B158), which allows the possibility that who I really am is free in sense required for autonomous rational and practical agency. I trust this suffices to show that any strict, textually faithful exegesis of Kantian intuitions must make mention of our sensible faculty.

Although sensible affectation is a necessary condition for obtaining an intuitive representation, it is not sufficient. An intuitive representation is an objective perception, implying that the source of the sensible affectations must be referenced to an

---

[20] See J. Hintikka (1967) and Michael Friedman (1992). Friedman (2000b) describes the Hintikka interpretation of Kantian intuitions that he used to defend as follows: "On this approach the primary role of Kantian intuition is formal or inferential: it serves to generate singular terms in the context of mathematical reasoning in inferences such as we would represent today by existential instantiation" (186). It does not follow from this approach that intuitions are non-sensible. It does, however, "downplay" the immediacy of intuitions that Kant himself contrasts with the mediacy of concepts. And since the immediacy of intuitions is explained by relating intuitions to the faculty of sensibility, this approach similarly downplays any reference to space or time as forms of sensibility (Ibid., 187). Friedman has since abandoned this interpretation in favor of one recognizing that, for Kant, intuitions exhibit an immediacy that derives from their relationship to the faculty of sensory perception. Further, as evidenced in his critique of the Marburg interpretation of Kant, Friedman now maintains that any attempt to downplay Kant's reference to sensibility significantly modifies his transcendental idealism.

[21] Allison (1983) argues that Kant's Copernican revolution should be seen as a rejection of a theocentric model of cognition, which, on the rationalists' version of the model, assumes we are capable of intellectual intuition. Allison's point can be further supported if one takes into consideration that Kant was raised within the context of Lutheran pietism. Insisting on the difference between divine and human cognition is commonly recognized within this and other religious traditions as the properly pious expression of intellectual humility.

inter-subjectively ascertainable, particular something. As we have seen, this is accomplished by recognizing that the given particular falls under a concept.[22] So, intuitions are singular, immediate representations of objects only if they arise within the context of a judgment:

> [A]lthough intuitions do not in fact represent or refer to objects apart from being 'brought under concepts' in a judgment, they *can* be brought under concepts and when they are they *do* represent particular objects (Allison (1983), 67–8).

For Kant, the visual perception of empirical objects also involves the synthetizing activities of the imagination and the understanding (A120–122; B129–130; Kitcher (1990) 60).[23] Ordinary visual perception is thus a kind of objectively-valid judgment. The paradigmatic intuitive representation is, therefore, the empirical perception of a particular, spatio-temporal object, e.g. my perception of the keyboard in front of me. This perception qualifies as a singular, immediate, objective representation insofar as I am thereby presented with an individual thing, which I also apprehend as an object given to my senses and as the inter-subjectively accessible particular referenced in the objectively-valid claim 'That is a keyboard.'

Given all the theoretical machinery built into the notion of an intuition, Kant's tendency to conflate the intuition *of* an object with the intuition *as* an object is at least understandable.[24] Objecthood, for Kant, is determined by the nature of the representation. This is first announced as a hypothesis in his proposed Copernican revolution in metaphysics:

> If intuition has to conform to the constitution of objects, then I do not see how we can know anything of them *a priori*; but if the object (as an object of the senses) conforms to the constitution of our faculty of intuition, then I can very well represent this possibility to myself (Bxvii).

Kant's revolution effectively made epistemology explanatorily and justificatorily prior to metaphysics. Furthermore, given Kant's analysis of the synthetic, i.e., judgmental, character of visual perception, his hypothesis can be read as a research agenda for a scientific investigation into perception of empirical objects, and he suggests as much in *Prolegomena* §14 and §16. This is why, as we will see, Helmholtz can legitimately claim his work in perceptual psychology is furthering Kant's epistemological project. For now, we simply need note that Kant's Copernican revolution results in "a radically new conception of an object":

---

[22] See Sect. 2.1.1.

[23] I take this to be an implication of Kant's analysis of the necessary cognitive synthesis giving rise to a proper intuition.

[24] There are thus competing interpretations extant in the literature regarding whether Kantian intuitions are best viewed as representational mental states, the object related or referred to by that state, or something more akin to a Fregean sense, i.e. a "distinctive and objective way in which a cognition representationally *relates* the subject to a particular object" (Tolley (2014), 201). Fortunately, for our purposes, we do not have to decide which of these competitors gets Kant right. If my thesis holds, however, we will see that Helmholtz opted for the first interpretation and Frege, inspired by Cohen's reading of the *Critique*, proposed and developed the third.

An object is now understood as…whatever conforms to the mind's conditions (both sensible and intellectual) for the representation of it as an object. Consequently, an object is by its very nature something represented; and in that sense a reference to the mind and its cognitive apparatus is built into the definition of the term (Allison (1983), 30).

In sum, whenever we are in possession of a proper Kantian intuition qua representation, the subject of that intuition (the intuited) is, by definition, an object.

Since obtaining an intuitive representation presupposes a determination of space and time involving pure concepts, the intuited will conform to certain a priori specifications dictated by the structure of our cognitive capacities. Hence, another moral of Kant's cognitive synthesis is that objects of intuition should not be identified with objects considered independently of our mode of cognition[25] or the intuited objects of someone with entirely different cognitive capacities. When I point to the keyboard, I can be assured that what I am pointing at is similarly perceived by all those sharing my spatio-temporal intuitive faculty and employing my logical categories. I cannot be certain whether the objectivity of my intuitive representations—or my conceptual ones for that matter—extends beyond that: "For we cannot judge at all whether the intuitions of other thinking beings are bound to the same conditions that limit our intuition and that are universally valid for us (A27/B43). Even so, Kant maintains that the scope of validity accorded to intuitive representations is large enough to warrant defining them as one of the two types of objective perceptions, i.e. cognitions (A320/B377).

Having said what proper intuitive representations are, let me explain the role they play in grounding the objective reality of concepts. Remember, a concept is considered objectively valid if it represents a possible object, one that *could* be sensibly given. Kant leaves open the possibility that objective validity of certain concepts can be established completely independently of perceiving an empirical object corresponding to the concept.[26] The objective reality of concepts, however, has to do with whether the conceived object actually exists. And, for Kant, the existence of a thing cannot be established independently of empirical perception.

Kant clarifies what he means by real possibility in contrast to actuality in "The postulates of empirical thinking in general":

1. Whatever agrees with the formal conditions of experience (in accordance with intuition and concepts) is **possible.**

---

[25] Kant's notorious term for an object considered independently of our mode for cognition is a 'thing in itself.' What he means by this expression and whether what he says about it is consistent with his transcendental idealism has long been the subject of debate. In Chap. 3, we will see that Helmholtz tended to interpret Kant's thing-in-itself as the unknowable causes of what appears to us. Hermann Cohen then accused Helmholtz of misreading Kant on this point and failing to appreciate Kant's Copernican turn. Based on the Cohen's reading of Kant, we can infer that for the Marburgians the notion of a thing-in-itself equates to 'object apart from any scientific understanding of it.'

[26] For example, Pegasus is an empirical concept that is potentially applicable to experience even though a winged horse has not been perceived, so far as we know.

2. That which is connected with the material conditions of experience (of sensation) is **actual** (A218/B265)
3. That whose connection with the actual in accordance with general conditions of experience is (exists) **necessarily** (A218/B265–6)

He stresses that these "principles of modality" only to pertain to the "empirical use" of a concept and hence its relationship to possible, actual or necessary "**things**" in the domain of "possible experience," (A219/B266–7).[27] Thus, to say that a concept is objectively valid is to say that it is either an "**empirical concept**" or a "**pure concept**," the latter being an a priori concept containing a synthesis enabling it to represent a possible object of cognition (A220/B267). Kant acknowledges that it may look as if the existence of some things might be established without recourse to empirical perception, mathematical things for instance, given that they can be constructed "entirely *a priori*" (A224/B271). He denies this, however, and explains why:

> In the **mere concept** of a thing no characteristic of its existence can be encountered at all. For even if this concept is so complete that it lacks nothing required for thinking of a thing with all of its inner determinations, still existence has nothing in the least to do with all of this, but only with the question of whether such a thing is given to us in such a way that the perception of it could in any case precede the concept. For that the concept precedes the perception signifies its mere possibility; but perception, which yields material for the concept, is the sole characteristic of actuality (A225/B272)

Here is what I take to be his point. In the process of synthesizing or schematizing an a priori concept, mathematical concepts in particular, we may generate a perception or image of an object corresponding to the concept. For example, when constructing the concept 'triangle' by thinking of the spatial figure being generated via the connection of three straight lines, we may produce a singular immediate representation of a triangular shape.[28] However, insofar as we are conscious of this shape, and the perception of it, as mere by-products of an act of mathematical construction, we do not recognize it as a thing, the perception of which "can, if need be, precede the concept." By contrast, empirical perceptions of things requiring an actual, sensory

---

[27] Chignell (2008) points out that Kant's use of modal notions throughout the *Critique* is not as clear as one would like. This is why I decided to fix this terminology from the outset (see Sect. 2.1.2.1). In fact, throughout Kant's discussion of these postulates, he uses the term 'objective reality' to refer to a concept's real possibility, whereas I stipulated that we would reserve the term 'objective validity' for this reference. However, even this fixing of Kant's terminology does not remove all of the ambiguities in Kant's discussion of the postulates. For, according to Chignell, there are "actually *two* different" kinds of empirically real possibility at work in Kant's discussion of the postulates. Given the second and third postulate and Kant's claim in the Analogies that all empirical actualities are causally connected, it follows that "the empirically actual and the empirically necessary (and *ergo* the empirically possible) are coextensive" (Ibid., 259–60). So, the notion of empirically real possibility that I mean to hone in on here is the one that Chignell associates with the first postulate, where to say that a concept is really possible is to say that it agrees "with the formal conditions" of the "objective form of experience," implying that is empirically applicable, but allowing that it may not be actually, empirically instantiated (A220/B267).

[28] See Kant's own discussion of this example in A220–1/B267.

encounter with the intuited do strike us as perceptions that could have preceded the concept.[29]

In any case, Kant insists that the actual existence of things is not something that can be settled a priori:

> The postulate for cognizing the **actuality** of things requires **perception**, thus sensation of which one is conscious—not immediate perception of the object itself the existence of which is to be cognized, but still its connection with some actual perception in accordance with the analogies of experience, which exhibit all real connection in an experience in general (A225/B272)

I have said that the paradigmatic intuitive representation is empirical perception. Here we see that the paradigmatic Kantian object, an object to which he grants existence, is either the subject of an empirical perception or related to such a perception by means of an empirical law. He thus limits the domain of full-fledged, knowable, actually existing objects to those inhabiting the spatio-temporal, causal nexus of empirical experience. This leaves us with the question: what is the ontological status of mathematical entities within Kant's system?

2.1.2.5 Pure (Non-empirical) Intuitions

What Kant means by 'pure intuition' and the role it is supposed to play in his philosophy of mathematics has been the subject of much controversy. By the nineteenth century, 'pure intuition' had no settled meaning, and depending on how it was interpreted, it either was or was not considered relevant to mathematical practice. In Chaps. 3 and 4, I explain the different meanings that Helmholtz, Cohen and Frege attached to the term 'pure intuition.' My aim here is to answer the following: (1) Do Kant's own remarks concerning pure intuition warrant the assumption that he allows for the possibility that we have cognitive access to abstract mathematical objects? (2) If not, what is the mathematician supposed to be intuiting by means of a pure, non-empirical, intuitive representation?

Based on what I have said concerning Kant's views on existence, it should obvious that I believe the answer to (1) is no. However, this answer is difficult to reconcile with other passages where Kant explicitly identifies pure intuitions as the

---

[29] Based on Kant's notion of a cognitive synthesis, empirical perceptions are generated as well and arise simultaneously with the recognition of some concept under which the perceived object falls. Therefore, it does not seem as if Kant can allow that empirical perceptions ever *precede* a concept. Still, he maintains that there are greater and lesser degrees of awareness concerning the intellectual activity involved in generating intuitive representations: "This consciousness may often only be weak, so that we connect it with the generation of the representation only in the effect, but not in the act itself…." (A103). It is certainly plausible to think that this awareness is highest when the mathematician is constructing a concept and lowest when she perceives the keyboard in front of her. So, Kant can consistently maintain that human cognizers can distinguish between perceptions of mathematically constructed particulars and perceptions of particular entities which are not consciously recognized as constructions. The latter type of perception is thus given to them as if it could, in principle, precede a concept, whereas the former is not.

objects of mathematical inquiry. Therefore, I will present the relevant passages and show how these can be reconciled with Kant's refusal to treat mathematical entities as full-fledged objects. The key here will be to distinguish the *procedure* or schema employed in generating an image corresponding to mathematical concepts from the *product* or intuited image generated thereby. I will argue that Kant thinks of a mathematical entity qua pure intuition as the schema used to generate the images or forms corresponding to the concept. Thus, my answer to (2) is that the primary object of the mathematician's intuition is the rule-guided activity necessary for yielding a determinate spatial or temporal figure, rather than figure itself or an abstract object corresponding to figure.

I will start with Kant's contrast of the synthetic activity yielding an empirical intuition with one yielding a pure intuition. With empirical entities, we begin with sensible impressions ('red here now') coming to us from some unknown source, something out there so to speak.[30] The imagination, working with rules supplied by the understanding, say the concept-rule 'apple,' synthesizes these impressions so as to generate an image of an object (A121). In this case, the object or immediately-given particular content of the intuition is the image or intuited apple. When apprehending empirical particulars, the primary intentional object is not the imaginative activity, but only "the effect" (A103).[31] When constructing the pure intuitions associated with mathematical entities, e.g. a triangle, we begin with a spontaneously generated idea and a pure sensible manifold, a manifold devoid of anything "that belongs to sensation" (A20/B34). In this case, the primary intentional object is the rule-guided activity resulting in the figure or image: "Thus we think of a triangle as an object by being conscious of the composition of three straight lines in accordance with a rule according to which such an intuition can always be exhibited" (A105).

There is substantial textual evidence that, for Kant, the desired intuited content when synthesizing the a priori mathematical concepts is a schema or rule-governed procedure for constructing figures related to the concept and not the figures themselves. Consider this long passage on the "mathematical cognition" that arises from the "**construction** of concepts":

> [T]o construct a concept means to exhibit *a priori* the intuition corresponding to it. For the construction of a concept, therefore, a **non-empirical** intuition is required, which consequently, as intuition, is an **individual** object, but that must nevertheless, as the construction of a concept (a general representation), express in the representation universal validity for all possible intuitions that belong under that same concept. Thus I construct a triangle by exhibiting an object corresponding to this concept, either through mere imagination, in pure

---

[30] For Kant, it is the presence of sensations or sensible impressions (color, heat, etc) that serves to contrast an empirical intuition from a pure intuition (A20–1/B34–5).

[31] Throughout my discussion of the synthesis yielding an empirical intuition, I will use the term 'intentional object' for the presented content of the intuition with the aim of remaining neutral as to whether this content ought to be identified with a mental image, an non-mental object represented by the image or something in between. Once I turn to the synthesis yielding a pure intuition, I will be more specific. This specificity pushes me towards Tolley's position that the content of the intuition is akin to a Fregean sense, a distinctive and objective mode or method by which a particular object is given. .

> intuition, or on paper, in empirical intuition, but in both cases completely *a priori*, without having had to borrow the pattern for it from any experience. The individual drawn figure is empirical, and nevertheless serves to express the concept, to which many determinations, e.g. those of the magnitude of the sides and angles are entirely indifferent, and thus we have abstracted from these differences, which do not alter the concept of the triangle (A713–4/B741–2)

Notice that the figure represented in our imagination or on paper cannot serve as the pure, non-empirical intuition, since it contains too many particularities that would preclude it from adequately expressing the generality contained in the concept 'triangle.' Rather, to represent the triangle qua pure intuition, we must disregard these particularities and focus on "the representation of a general procedure" or "schema" whereby the imagination can construct images or instances of this concept:

> No image of a triangle would ever be adequate to the concept of it. For it would not attain the generality of the concept, which makes this valid for all triangles, right or acute, etc., but would always be limited to one part of this sphere. The schema of the triangle can never exist anywhere but in thought, and signifies a rule of the synthesis of the imagination with regard to pure shapes in space (A140–1/B179–80)

Whenever I point to a particular image and say 'That is a triangle', the intentional object of the requisite non-empirical intuition underwriting that assertion is the universal procedure or schema producing triangular shapes. Again, "we think of a triangle as an object by being conscious of the composition of three straight lines in accordance with a rule…(A104).

So too, a line, a circle, spatial dimension, time itself, and numbers all qualify as pure intuitions only insofar as we can identify them with an a priori procedure generated by the imagination:

> We cannot think of a line without **drawing** it in thought, we cannot think of a circle without **describing** it, we cannot represent the three dimensions of space at all without **placing** three lines perpendicular to each other at the same point, and we cannot event represent time without, in **drawing** a straight line…, attending to merely to the action of the synthesis of the manifold through which we successively determine the inner sense, and thereby attending to the succession of this determination in inner sense (B154)

> The pure **schema of a magnitude** (*quantitatis*)…as a concept of the understanding, is **number**, which is the representation that summarizes the successive addition of one (homogeneous) unit to another. (A142/B182)

> The schema is in itself as always only a product of the imagination….Now this representation of a general procedure of the imagination for providing a concept with its image is what I call the schema for this concept (A140/B179)

Kant is especially careful to distinguish the schema underlying particular number concepts with an image corresponding to that concept: "[I]f I place five points in a row,…, this is an image of the number five" (A140/B178). We should not identify this image as the mathematical entity corresponding to the concept 'five', which becomes evident once we consider larger numbers: "For with such a number as a thousand the image can hardly be surveyed and compared with the concept" (A140/B179). Rather, representing five or a thousand as the mathematical magnitudes or quantities corresponding to their respective concepts requires that we represent them as the sums of a particular successive addition of homogenous units. That is, we must intuit them as the completion of a procedure whereby the imagination yields a determinate, pure temporal magnitude (A241/B300).[32]

I have argued that when Kant identifies mathematical entities as pure intuitions, he is identifying them with a schema. I have also maintained that a schema is the blue-print for a rule-guided activity performed by the imagination and that this activity can serve as the content of an intuitive representation. I now maintain that recognizing this helps account for Kant's tendency to talk about schemas or pure intuitions as objects. As previously noted, objecthood, for Kant, is determined relative to a mode of representation. If something can serve as the content of an intuitive representation, it qualifies as an object. The question then is whether Kant thinks that acts of the imagination can qualify as the content of an intuitive representation, that is, can they serve as particular, inter-subjectively accessible entities given via sensibility. The answer is yes and no.

Kant clearly states that the productive acts of the imagination yielding the schemata underlying pure geometric concepts are affectations of sensibility:

> [T]he transcendental synthesis of the **imagination** is an effect of the understanding on sensibility…Now insofar as the imagination is spontaneity, I also occasionally call it the **productive** imagination, and thereby distinguish it from the **reproductive** imagination….
> [M]otion as **description** of a space is a pure act of the successive synthesis of the manifold in our intuition in general through productive imagination (B152–55).

He also indicates that these acts are particular in the sense that they present themselves to us as unified wholes that are susceptible to identity claims. Recall that the unity manifested in content of intuitions general is accounted for by the cognitive synthetic acts, which necessarily involve or presuppose the imagination's productive activity:

> [I]n order for the **unity** of intuition to come from this manifold (as, say, in the representation of space), it is necessary first to run through and then take together this manifoldness, which action I call the **synthesis of apprehension** (A99).

We cognize all objects as a singular, complex entities because we recognize them as the productions or reproductions of a complete, unique, rule-governed synthetic acts (A105).

Not only is the singularity of objects explained via the particularity of the acts responsible for their intuitive representation, but also the unity of the self and its

---

[32] Kant's account of what is involved in the apprehension of numbers is notoriously obscure and will be discussed in more detail in Sect. 2.3. My interest here is just to show that intuiting a number, for Kant, does not presuppose obtaining an image. Instead, he maintains that we simply need to articulate a procedure whereby the number could be intuited as the sum of successively-given units. I am claiming that, for Kant, articulating this procedure is tantamount to intuiting a rule-guided act of the imagination that yields a particular determination of time. For more on why Kant's account of number suggests that numbers themselves must have temporal content, see Friedman (1992), 115–122.

field of conscious representations is also explained by way of identifiable acts of the imagination:

> [F]or the mind could not possibly think of the identity of itself in the manifoldness of its representations, and indeed think this *a priori*, if it did not have before its eyes the identity of its actions, which subjects all synthesis of apprehension (which is empirical) to a transcendental unity, and first makes possible their connection with *a priori* rules (A108).

If we do not attribute unity to the a priori, productive act(s) of the imagination whereby the sensible manifold is affected and determined relative to pure concepts, we could not account for the unity of intuitive representations or a unified self capable of having such representations.

Finally, it is the re-identification of synthetic acts of the imagination within the context of empirical experience that explains and justifies subsuming empirical objects under pure mathematical concepts.

> [T]hat this very same formative synthesis by which we construct a figure in imagination is entirely identical with that which we exercise in the apprehension of an appearance...[I]t is this alone that connects with this concept the representation of the possibility of such a thing (A224/B271).

So, Kant equates mathematical entities with schemata, affirms that these schemata and their corresponding spatial or temporal forms can be obtained completely a priori and even suggests that we can intuit both the schemata and the corresponding forms independently of empirical experience

Yet, he still refuses to treat constructed mathematical entities as objects in any full-bodied sense:

> It may look, to be sure, as if the possibility of a triangle could be cognized from its concept in itself (it is certainly independent of experience), for in fact we can give it an object entirely *a priori*, i.e., construct it. But since this is only the form of an object, it would still always remain only a product of the imagination, the possibility of its object would be doubtful... (A224/B271).

What does Kant mean by referring to constructed mathematical entities as "only the form of an object" and a dubitable "product of the imagination"? We must return to his distinction between real possibility and actuality. An object is really possible, if it agrees with the formal conditions of experience, that is, with the structure of our intuitive framework and our stock of a priori concepts. Kant never doubts that the structure of outer sense, space, is Euclidean. Therefore, to say that something conforms to outer sense means that it is constructible within Euclidean space.

When the geometer obtains a procedure for constructing an entity within Euclidean space, she has established the real possibility of this object and the objective validity of its concept. At this point, the geometer possesses a mathematical entity about which she can form objectively valid judgments, for example, that the sum of the angles in a triangular figure is equal to the sum of two right angles. Because this construction can be accomplished completely a priori, whatever "follows from the general conditions of the construction must hold generally of the object of the constructed concept" (A716/B744). However, this procedure and its corresponding schematic form were spontaneously generated; they were not borrowed from experience, but are the result of the pure imaginative activity of the geometer. Therefore, unless it can be show that these pure mathematical forms are instantiated in empirical objects, then the geometer is merely rendering judgments about spontaneously generated forms or structures, i.e., products of the imagination.

Kant insists that mathematical concepts and judgments produce knowledge only if there are empirical objects manifesting the same form as the pure, constructed entities:

> **Things in space** and **time**...are given insofar as they are perceptions (representations accompanied with sensation), hence through empirical representations. The pure concepts of the understanding, consequently, even if they are applied to *a priori* intuitions (as in mathematics) provide cognition only insofar as these *a priori* intuitions, and by means of them also the concepts of the understanding, can be applied to empirical intuitions (B147).

> The [mathematical] concept is always generated *a priori*, together with the synthetic principles or formulas from such concepts; but their use and relation to supposed objects can in the end be sought nowhere but in experience...(A240/B299).

> Pure mathematics, and especially pure geometry, can have objective reality only under the single condition that it refers merely to objects of the senses...(*Prolegomena*, 4:287).

How can we be assured that the concepts and judgments made relative to mathematicians' freely-generated structures will hold with respect to empirical objects? Because the space that comprises the subject matter of geometry and the space by which we are given empirical entities are identical, namely, the form of outer sense:

> [S]pace is nothing other than the form of outer appearances, under which along objects of the senses can be given to us. Sensibility, whose form lies at the foundation of geometry, is that upon which the possibility of outer appearances rests: these, therefore, can never contain anything other than what geometry prescribes to them (*Prolegomena*, 4:287)

Because the procedure that the mathematician employs in constructing for herself a pure spatial or temporal form is "entirely identical with that which we exercise in the apprehension of appearance" manifesting that form, the objectively valid judgments made within the context of pure mathematics and pure geometry automatically translate into knowledge claims (A224). The real objects referenced in mathematical judgments qua knowledge claims will be empirical entities, but the source of the judgment is rooted in our pure concepts and our pure forms of sensibility. Therefore, this knowledge is necessarily and universally valid for all empirical entities falling under the relevant mathematical concepts.

Summarizing Kant's C&I Distinction

Concepts represent objects generally and immediately, representing them by means of conceptual marks that more than one object may share. Their cognitive significance or objective validity depends upon them possessing a content enabling them to represent possible object(s), which cognizers like us can experience. Pure concepts obtain this content by means of a process of construction or schematization.

Such constructions are productive acts of the imagination upon the pure forms of sensibility, space and time, and governed by the a priori concept-rules spontaneously generated by the understanding. Thus, pure concepts are rules determining what properties and relations an intuited object must manifest to qualify as a sensibly-given instantiation of the concepts. These concept-rules, therefore, provide the normative necessity that make it possible to assert inter-subjectively true or false claims of 'This is an F.' However, just because a concept is capable of being applied to an object of experience and functioning as a predicate in an objectively-valid judgment does not entitle us to infer that there exists an actual instantiation of the concept; examining conceptual content alone cannot tell us whether the represented object(s) exist.

By contrast, intuitions represent objects as particular and immediately-given. The paradigmatic intuitive representation is the ordinary sense perception of an actual, spatio-temporal particular. Since perception itself requires an act of the imagination whereby passively received sensible information is taken up and united via conceptual rules, this perceived particular will also be recognized as following under certain concepts. In this way, the intuited object is also presented to us as the real subject of a singular judgment. Figuring out precisely what Kant means by the 'pure' intuitions associated with mathematical is a matter of long-standing debate and various interpretations. I have argued that identifying pure intuitions with the constructive procedures or schemata of pure concepts finds substantial textual support within the *Critique*. Regardless of the precise meaning of 'pure intuition,' Kant insists that obtaining such intuitions is the necessary and sufficient condition for ensuring the objective validity and potential applicability of a priori mathematical concepts. Pure intuitions should not be thought of as the typical mathematical abstracta, namely atemporal, nonspatial, acausal objects, nor viewed as insuring the truth of mathematical propositions in a robust sense. Mathematical propositions are truths only when applied to existing objects and, for Kant, existing objects are those inhabiting the spatio-temporal causal nexus of our shared empirical experience.

2.2 Deploying the Distinction Between Concepts and Intuitions

Now that we have a handle on the distinction between concepts and intuitions, it is time to see how it functions in Kant's critique of his opponents. My goal is to show that the C&I distinction can be read as a methodological principle designed to highlight certain errors that Kant associated with empiricism and rationalism. Therefore, we need to view the empiricists and rationalists through Kantian eyes, rather than as they might be perceived by a more sympathetic commentator. For Kant, empiricism is epitomized in the work of Locke and Hume, whereas rationalism is best represented by Leibniz and Descartes. Concerning Locke and Hume, Kant writes, "The first of these two famous men opened the gates wide to **enthusiasm**, since reason, once it has authority on its side, will not be kept within the limits by indeterminate recommendations of moderation; the second gave way entirely to **skepticism**...(B128). Leibniz and Descartes are similarly chastised for offering fallacious proofs for the existence of abstract objects given to thought alone, e.g. God and lesser immaterial souls.[33] So, according to Kant, despite their different epistemological strategies, empiricists and rationalists both produce unsatisfactory results.

In this section, I will argue: (1) Kant sees Lockean enthusiasm rooted in a mistaken belief in the power of abstraction; and (2) he identifies the fallacy giving rise to the rationalists' pseudo-proofs as an illegitimate inference from logical possibility to objective reality. I then show that these two alleged logico-epistemological errors can traced to an initial error of neglecting the C&I distinction.

2.2.1 Lockean Enthusiasm and Illicit Appeals to Abstraction

There is no succinct, clear-cut passage where Kant directly links Locke's theory of abstraction to the accusation that Locke opened the door to enthusiasm. In fact, as we will see, it is the Leibnizian-Wolffian philosopher J.A. Eberhard that Kant specifically accuses of mistakenly appealing to Lockean abstraction as a means of gaining epistemic access to abstract entities having no sensible or spatio-temporal properties. So, it is fair to say that while Kant sees Locke's account of abstraction opening the door to enthusiasm, it is Eberhard that he sees walking through it.

I begin with a brief description of enthusiasm and Lockean abstraction. Although a more charitable reading of Locke might portray him as a nominalist with respect conceptual representations, I will show that his immediate successors read him as saying that abstraction resulted in the image of an abstract entity manifesting only those properties associated with the marks of a concept. I then turn to the Eberhard-Kant exchange on the power of abstraction. Finally, I explain how the C&I distinction reinforces Kant's precept that Lockean abstraction does not entitle us to claim that there are abstract objects manifesting only the right sort of properties or that we have cognitive access to such objects.[34]

The first thing to notice is the irony in Kant's accusing Locke of paving the way for enthusiasm. Locke himself devoted an entire chapter to the danger of enthusiasm, a danger which was presumably avoided by showing that our entire stock of well-founded ideas can be derived a posteriori. He describes enthusiasm as trusting in "the conceits of a warmed or overweening brain" whereby one is carried away in

---

[33] Cf. Kant on Leibniz (A270/B326) and on Descartes (B274–5). See also Kant's assessment of the ontological proof of the existence of God which he attributes to both Leibniz and Descartes (A602/B630).

[34] In Chap. 4, I argue that Frege similarly accuses his opponents, the formalists, of illicitly appealing to a Lockean account of abstraction as the means for underwriting existence claims about abstract mathematical entities. I go on to show that Frege's use of the C&O distinction in his exchange with the formalists mirrors Kant's use of the C&I distinction in his exchange with Eberhard.

their opinions, as if by a "natural motion", beyond "all restraint of reason and check of reflection" (Locke (1690) *Essay* IV.xix.7). Thus, Kant sees something in Locke's own method of accounting for the objectivity of our representations that encourages the conceits of an overweening brain and misleads us into thinking we can transcend the legitimate constraints and checks imposed by our cognitive capacities.

So how does Kant characterize Locke's method? Instead of recognizing that sensibility and the understanding each make a positive contribution in our acquisition of cognitively significant ideas, Locke privileged sensibility and thus "**sensitivized** the concepts of understanding" (A271/B327). On Kant's reading, Locke viewed sensibility as "immediately related to things in themselves" and assumed the understanding did "nothing but confuse or order the representations" that sensibility delivers (A271/B327). The full irony of Kant's criticism is that while Locke intended to downplay the role of the understanding, treating it as "merely passive" and limited to combining and abstracting from the features contained in sensibly-given particulars (*Essay,* II.ii.22–25). Kant and others perceived Locke's account of abstraction as ascribing to the understanding a power capable of rising above the domain of sensible appearances and affording us cognitive access to objects without any sensibly detectable properties.

Let me present a bit of the textual basis for Kant's negative assessment of Lockean abstraction and its propensity to induce metaphysical flights of fancy. On Locke's model of perception and cognition, both sensibility and the understanding are treated as passive in the sense that we are capable of obtaining mirror images of outer objects, without having to construct appearances conformable to our forms of intuition or a priori concepts:

> For objects of the senses do, many of them, obtrude their particular ideas upon our minds…These simple ideas, when offered to the mind, the understanding can no more refuse to have, nor alter when they are imprinted, nor blot them out and make new ones itself, than a mirror can refuse, alter, or obliterate the images or ideas which the objects set before it do therein produce. As bodies that surround us do diversely affect our organs, the mind is forced to receive impressions; and cannot avoid the perception of those ideas that are annexed to them. (*Essay,* II.i.25)

Passages such as this, filtered through the prism of Kant's judgmental model of perception, motivate his claim that Locke viewed sensibility as in an immediate relation to things-in-themselves. However, though Locke and Kant disagree on the conditions giving rise to the sensible representations of external objects, they agree that the content of sensible representations is irredeemably particular. Locke's self-appointed task then is to explain how we arrive at the general ideas signified by terms like 'man' or 'triangle,' without contradicting his empiricist stance that the understanding, when functioning properly, is passive and adds nothing to represented content designated by general terms (see *Essay,* II.xii.1 and III.iii.6).[35]

---

[35] In the first passage, Locke refers to the general ideas obtained via abstraction as "made" ideas. However, he does not intend for this to mean that the understanding spontaneously contributes elements that add to and modify the represented content of general terms.

According to Locke, general ideas are produced by means of abstraction, our presumed capacity to erase certain properties or relations of sensible particulars and retain only those features specified by the general term:

> Words become general by being made the sign of general *ideas*; and *ideas* become general by separating them from the circumstances of time and place and any other ideas that may determine them to this or that particular existence. By this way of abstraction they are made capable of representing more individuals than one; each of which, having in it a conformity to that abstract *idea*, is (as we call it) of that sort. (*Essay*, III.iii.6)

To show how this works, he walks us through an example, explaining how we derive ideas of the most abstract kind from very the humble beginnings. Children initially receive particular ideas mirroring the people they encounter: "The *ideas* of the nurse and mother are well-framed in their minds and, like pictures of them there, represent only those individuals" (Ibid., III.iii.7). As their social circle enlarges, they observe certain similarities and form the idea *man*. In forming this idea, they "make nothing new, but only leave out" of the particular ideas they had of "*Peter* and *James*, *Mary* and *Jane* that which is peculiar to each, and retain only what is common to them all" (Ibid.). By means of this same mental operation, "leaving out the shape and some other properties signified by the name *man*, and retaining only a body, with life, sense, and spontaneous motion," we form the idea *animal*. Ultimately, "by the same way the mind proceeds to *body*, *substance*, and at last to *being* and *thing*" (III. iii.8–9).

Regardless of whether Locke himself actually appealed to abstraction to underwrite the existence of abstract entities referred to by the terms 'man', 'substance' or 'being', Berkeley and Hume certainly thought he did and argued that this was inconsistent with the empiricist precept that our knowledge of existent entities does not extend beyond sensed particulars.[36] As they read it, Locke's account of abstraction suggests that we have the intellectual might and, more importantly, the intellectual right to erase away certain sensible properties of an object(s) in order to present ourselves with an image of something having only those properties specified in a particular concept.[37] Consider Berkeley's (1710/1988) critique of the abstract idea denoted by 'man':

> And after this manner it is said we come by the abstract idea of *man* or, if you please, humanity or human nature; wherein it is true there is included color, but then it can be neither white, nor black, nor any particular color; because there is no particular color wherein all men partake. So likewise there is stature, but then it is neither tall stature nor low stature, nor yet middle stature, but something abstracted from all of these....Whether others have this wonderful faculty of *abstracting their idea*, they best can tell....[T]he idea of man that I frame to myself, must be either of white, or a black, or a tawny, straight, or a crooked, a tall, or a low, or a middle-sized man. (*Principles of Human Knowledge*, 40)

---

[36]Thus, for example, it is Hylas' (Locke's) presumed belief in an existent material substratum, abstracted from all sensible features that is under attack in Berkeley's *Three Dialogues*. See also Hume's *Enquiry concerning the Human Understanding* (XII.ii.125 note).

[37]As to its being our intellectual privilege to form abstract ideas, Locke claims that it is this "excellency" that distinguishes "man and brute" (II.xi.10).

The specific complaint lodged here is the impossibility of imagining anything like a colorless color or something with incompatible properties, but it is also clear that Berkeley assumes the aim of Lockean abstraction is to eradicate features so as to form an image of some abstract entity.

So, despite the fact that a careful reading of Locke's *Essay* might show that Locke himself did not say that abstraction alone entitled us to lay claim to knowledge of suprasensible entities, this is what Berkeley, Hume and presumably Kant interpreted him as saying. Notice, for example, in the previously quoted passages that Locke places no restriction on what sensible features, pure or empirical, may be eradicated in the attainment of an abstract idea. By "leaving out" enough properties, the mind can present itself with the image of an abstract entity referred to by the term 'being as such' or 'immaterial substance'. Now compare the bare abstracted ideas or images associated with Locke's notions of a being as such or immaterial substance with Kant's description of the noumenal notion of substance:

> We have shown above: that the purity of the categories from all admixture with sensible determinations can mislead reason into extending their use entirely beyond all experience to things in themselves....Now hyperbolical objects are what are called **noumena** or pure beings of the understanding (or better yet, beings of thought)—such as, e.g., *substance*, but thought *without permanence* in time, or a *cause*, which would however *not* act in *time*, and so on.... (*Prolegomena* 4:332)

Here we see that, by Kant's lights, Lockean abstraction does impute to the understanding a "natural motion" whereby it propels itself "beyond all restraint of reason and check of reflection." Erase away enough sensible characteristics, including a determinate spatial or temporal form, and one is liable to be carried away into thinking he can present himself with the image of an object that would instantiate the concept of an atemporal substance with causal powers.

Fortunately, my argument that Kant identified a mistaken belief in the power of abstraction as the root cause of Lockean enthusiasm does not rest entirely on a highly selective exegesis, uncharitable commentaries presented by Berkeley and Hume, and attempting to get inside Kant's head while reading the *Essay*. In 1788, J.A. Eberhard founded a journal devoted to showing that "whatever is true in Kant is already found in Leibniz, and that wherever Kant differs from Leibniz he is wrong" (Allison (1973) 9). Specifically, Eberhard argued that, pace Kant, we do have cognitive access to non-sensible "universal things (*allgemeine Dinge*)" instantiating purely intelligible concepts, e.g. Leibniz's concept of a simple being (Ibid, 24). Allison describes Eberhard's surprisingly empiricist strategy for securing the objective content and validity of these Leibnizian concepts:

> [H]ow does the mind arrive at these general concepts through which the non-sensible is cognized? Here, Eberhard remarks in a note of triumph, the answer is so simple that he wonders how Kant could have failed to notice it. It is nothing more than abstraction, understood essentially in the way in which Locke described the process in the *Essay* (Ibid., 25).

Starting with sensed particulars and removing their distinctively sensible properties provides sufficient warrant for asserting the existence of the things falling under supra-sensible, purely intelligible concepts:

'In this way', Eberhard writes, 'the understanding rises from lower to higher concepts, and thus from the sensible to the intelligible.' Furthermore, this sensible or intuitive origin serves to guarantee the objective validity of these concepts. In proceeding by way of abstraction, the understanding keeps in contact with the real. It only loses the features pertaining to individuals in their particularity, what Eberhard calls the imageable or intuitive features, but it keeps the unimageable features yet still conceivable features pertaining to all reality (Ibid.)

Thus, in Eberhard, we have a concrete example of a Lockean enthusiast who appeals to the power of abstraction to ground our knowledge of supra-sensible things.

Furthermore, Kant specifically connects Eberhard's view of abstraction to his enthusiastic impulses. As Kant reads him, Eberhard seeks to ensure the objective reality of the supra-sensible concept 'simple being' using the following method. One begins with our sensory experience of time, i.e., a continuous flowing succession of temporal representations. One then abstracts away the successive property of time, leaving discrete, simple temporal moments. Similarly, one removes the continuity of the spatial manifold in order to abstract its simple components. One then infers that since these simple spatial and temporal components are not perceptible, they are non-sensible. Finally, one concludes that these simple elements are nothing other than Leibnizian monads, presumably by washing away their spatial or temporal properties altogether (Kant (1973), 117).

First, Kant counters that the characterization of pure time and space as composed of discrete elements is contradicted both by geometric proof and Newtonian physics (Ibid., 119). Second, he insists that the fact that these simples cannot be perceived is no justification for equating them with supra-sensible things, i.e., monads (Ibid., 127–128). Kant agrees that we may have sufficient warrant for asserting the existence of indivisible components of matter, despite our inability to perceive them. As component parts of the sensibly intuited, however, they remain sensible in nature.

Finally, and most importantly for our purposes, Kant contrasts his own method for establishing the objective validity of a priori concepts with Eberhard's illicit appeal to Lockean abstraction:

Mr. Eberhard…bases his system on the claim: 'We cannot have any general concepts which we have not *abstracted* from the things which we perceive through the senses or from those of which we are conscious in our own soul.'… The understanding therefore proceeds by means of *abstraction* (from sensible representations) to the categories, and now it advances from these to the essential characteristics of things to their properties (Ibid., 130).

We have seen that, for Kant, an essential function of concepts and categories is to order the sensible manifold making it possible for us to obtain intuitive representations of objective-valid, judgeable states of affairs. So if Eberhard is right, retorts Kant, and abstraction affords us immediate cognitive access to entities sans any sensible characteristics, why speak of concepts or categories at all:

For a truly *genuine* ascent, namely, to another species of being than can in general be given to the senses, even the most perfect, another mode of intuition, which we have named intellectual …would be demanded. With such an understanding not only would the categories no longer be required, but they would have absolutely no use. But who can provide us with such an intuitive understanding, or, if it lies concealed within us, who can acquaint us with it? (Ibid., 131)

Lockean abstraction, insofar as it supposedly allows us to cognize objects unmediated by any forms of sensibility and a priori logical categories, is tantamount to intellectual intuition. An intuitive intellect is one whose thought is sufficient for the appearance of an object. Kant maintains that a divine or intuitive intellect has no need of categories or concepts. For God, to think is to know because divine thought creates its objects (B145). The upshot of Kant's swipe at Eberhard is clear: the belief that we are capable of abstraction in Locke or Eberhard's sense is equivalent to the belief that we possess divine-like, creative cognitive capacities.

Having shown that Kant locates the error of a Lockean strategy for ensuring the objective reality of a priori concepts in a mistaken understanding of abstraction, I will now show how his C&I distinction can be interpreted as a means of preventing this kind of mistake. Kant issues at least two separate warnings regarding the legitimate input and output of abstraction. One appears appears in the *Jäsche Logic*:

> The expression *abstraction* is not always used correctly in logic. We must not speak of abstracting something (*abstrahere aliquid*), but rather of abstracting *from something* (*abstrahere ab aliquo*). With a scarlet cloth, for example, if I think only of the red color, then I abstract from the cloth; if I abstract from this too and think the scarlet as material stuff in general, then I abstract from still more determinations, and my concept has in this way become still more abstract (§6, 9:95).

Apparently a mistaken view of abstraction was epidemic among philosophers and logicians, and the mistake was thinking that abstraction operated on sensibly-intuited things and yielded the representation of a thing having only a specific set of sought after properties.

§6–§16 of the *Jäsche Logic* are devoted to delineating the precise role that abstraction plays in grounding various cognitions and inferences. By comparing, reflecting and abstracting sensibly-given singular representations, e.g. a spruce, a willow, and a linden, one may acquire an empirical concept, e.g. a concept of a tree (§6, 9:94). By abstracting from the marks contained in a concept, one may acquire a less determinate concept under which the prior one falls. Abstraction can thus precede a judgment about logical subordination (§6, 9:94–5; §9–§11). However, abstraction never licenses the inference that one has acquired a pure concept: "A *pure* concept is one that is not abstracted from experience but arises rather from the understanding even *as to content*." (§5, 9:92). Abstraction operating in isolation from comparison and reflection does not even entitle us to claim that we have obtained an empirical concept:

> Abstraction is only the *negative* condition under which universal representations can be generated, the *positive* condition is comparison and reflection. For no concept *comes to be* through abstraction; abstraction only perfects it and encloses (§6, 9:95).[38]

---

[38] See 'Of Concepts' in Kant's *Vienna Logic* for similar warnings to logicians and philosophers on the types of cognitions and inferences licensed by appeals to abstraction (Kant (1992), Ak 904–13, 348–56).

Finally, no amount of "logical abstraction" or "logical determination" ever yields "completed" and "thoroughly determinate" cognitive representations, a.k.a., "intuitions," of "individual things" ['einzelne Dinge].

These warnings about illicit appeals to abstraction are reiterated in Kant's critique of Eberhard. He reminds Eberhard and his readers that abstracting "something" in the sense of *abstrahere aliquid*—"removing a liquid from other matter in order to isolate it"—is a procedure that chemists alone are entitled to perform (Kant (1973), 117). Abstraction understood as a logical operation is *abstrahere ab aliquo*, abstracting from something (Ibid.). So understood, it is not even accurate to say that logical abstraction yields "*a concept* as a common mark, rather one abstracts *in the use* of a concept" (Ibid.)[39] For example, one may use either "the concept of a child *in abstracto*" or the "the concept of a child in civil society (*in concreto*)" to "formulate rules for education," but this contrast in the use of concepts neither requires nor warrants introducing a distinction between "abstract and concrete children" (Ibid.). Here as in the *Jäsche Logic*, Kant insists that the distinction between abstract and concrete understood in the proper logical sense is a relative distinction between concepts, a distinction relevant to judgments concerning logical subordination. It does not refer to an essential difference in the nature of concepts themselves,[40] and it is clearly irrelevant to the distinction that Eberhard intended to draw between concrete sensibly distinguishable particulars and abstracted universal things lacking undesirable sensible features.

The exhaustive and absolute nature of the C&I distinction serves to explain, justify and reinforce these methodological precepts. The *Jäsche Logic* emphatically states that, at best, abstraction yields empirical concepts, never pure mathematical concepts with objective, scientifically fruitful content and it never entitles us to posit the existence of cognitively accessible things with increasingly fewer distinguishing sensible characteristics. These same claims are implicitly, if not explicitly, repeated in Kant's 1790 response to Eberhard and the *Vienna Logic*, another well-circulated transcript of Kant's logic lectures. Kant's C&I distinction serves to highlight his opponents' illicit appeals to abstraction by reinforcing the precept that abstraction operates on concepts and their constituent marks, not on intuited things and their properties.

To lay the foundation of my argument that Frege's C&O distinction is a Cohen-like reworking of Kant's C&I distinction aimed at preventing what Frege takes to be mistaken appeals to abstraction to ground the objective validity of mathematical concepts, I needed to address two questions. First, is there is any textual evidence of Kant invoking the C&I distinction to underscore the precept that abstraction operating on sensibly-given, actual individuals can at best yield an empirical concept, but never individual things possessing fewer to no sensible features? Second, were Frege and his intended readers in a position to believe that Kant had? Given that the

---

[39] I take it Kant is repeating the same point elaborated in §6 of *JL* that comparison and reflection are the essential, positive conditions for concept formation.

[40] Kant (1973), 117 and *Jäsche Logic* §16).

Jäsche Logic was widely recognized as 'Kant's Logic' by Frege and his contemporaries, given that Frege cites the *Jäsche Logic* in the *Foundations of Arithmetic* and given that Kant-inspired critiques of abstraction as means of concept-formation were commonly used to challenge empiricist-leaning epistemologies like Helmholtz's, I hope to have gone some way in convincing my reader that the answer to these two questions is yes.

2.2.2   The Rationalist's Fallacy: Logical Consistency Implies Truth and Existence

Just at Kant's C&I distinction serves to highlight illicit appeals to abstraction, so too it serves to block any inference from the logical consistency of a concept to the existence of an object falling under it. While Kant rejects an empiricist strategy that seeks to account for the objective validity and reality of a priori concepts by appealing to Lockean abstraction, this strategy seems like it tries to comply with his maxims that "in the **mere** concept of a thing no mark of its existence can be encountered at all found" (A225/B272) and that "research into the existence of anything" must "begin with experience" (A226/B274). Leibniz and Descartes, on the other hand, are accused of disregarding these maxims entirely, and the evidence is their joint commitment to the ontological argument.

Kant maintains that proponents of the ontological argument "abstract from all experience and infer the existence of a highest cause entirely from *a priori* concepts" and, in doing so, allow reason to spread "its wings in vain" as if it could "rise above the world of sense through the mere might of speculation" (A591/B619). The philosophical enthusiasm ascribed to the rationalists, in contrast to that ascribed to Locke or Eberhard, reportedly stems from assuming that the objective reality of certain concepts can be established without having to exhibit, construct or abstract from sensibly-given particulars at all. Kant thus negatively characterizes the ontological argument as an attempt to "take an idea contrived quite arbitrarily and extract from it the existence of the object corresponding to itself" (A603/B631). In short, the error that Kant sees driving the rationalist's strategy for showing that we can gain cognitive access to supra-sensible things, e.g. God, is the false assumption that establishing the logical possibility of a concept suffices for showing that it is objectively valid and, what is worse, that we have sufficient grounds for asserting the existence of an object instantiating the concept.

Most often, when discussing what Kant thinks is wrong with the ontological argument, the focus is on his charge that it trades on a misunderstanding of the nature of existence. For Kant, a concept can function as a "real," not merely "logical," predicate in a judgment only if its constituent marks contain more information about the properties of an object than are expressed in the marks of the subject concept (A598/B262). The constituent marks of a real predicate must thus contain sufficient information about the properties of an object(s) that one is able to determine which objects do and do not fall under it. As Kant reads them, advocates of the ontological argument assume that existence is a constituent mark in our conception of an absolutely perfect being, existence being more perfect than nonexistence. One is then forced, on pain of contradiction, to conclude that such a being exists. Kant counters that 'existence' does not provide any further information or determination about an object's properties. To say that something exists does not add or subtract from its conceived properties: "the actual contains nothing more than the merely possible" (A599/B627). Hence, existence cannot serve as a constituent mark of any concept serving as a real predicate, and the argument's conclusion is unwarranted.

It is important to note, however, that Kant's objections to the ontological argument go beyond the specific charge that existence is not a real predicate. He begins his critique by stating that providing a "nominal definition of this concept [a something whose non-being is impossible] is quite easy" (A593/B621). He thus begins granting that existence or, to be precise, necessary existence may function perfectly well as a constituent mark of a concept. This is because the mistake that he is initially concerned with is thinking that providing a strictly nominal definition is enough to guarantee the concept's objective validity, that "through this concept we are thinking anything at all" (A594/B621). This mistake is not about the nature of existence per se, but about the practice of defining a concept with logically consistent marks and inferring that the concept represents a possibly given something bearing those marks as properties.

Throughout his critique, Kant points out that proponents of the argument fail to notice the difference between defining a logically consistent concept and establishing its objective validity. As Kant describes it, the argument sets out by declaring that there is one concept *ens realissimum* (a.k.a. *total reality*, a.k.a. *absolutely necessary being*) that contains existence as one of its marks and that one is "justified in assuming such a being is possible" (A596/B624). He then inserts the following parenthetical remark: "[E]ven though a non-contradictory concept falls far short of proving the possibility of its object" (Ibid.). To which he appends the footnote:

> A concept is always possible if does not contradict itself. That is the logical mark of possibility....Yet, it can nevertheless be an empty concept, if the objective reality of the synthesis through which the concept is generated has not been established in particular, but as was shown above, this always rests on principles of possible experience and not on the principles of analysis (on the principle of contradiction). This is a warning not to infer immediately from the possibility of the concept (logical possibility) to the possibility of the thing (real possibility) (Ibid).

A logically consistent concept, Kant reminds his readers, may nevertheless be void of cognitive content. To show that our concepts possess the latter, we need to prove that they are applicable to our shared experience of the spatio-temporal causal nexus of nature. This is accomplished, relative to a priori concepts, by demonstrating "the synthesis through which the concept is generated." Such demonstrations, as previously discussed, are exemplified in Kant's schematization of pure logical categories and the geometric constructions associated with pure mathematical concepts.

Kant's particular point about existence comes up only after he has granted his opponents their illicit inference from logical possibility to objective validity. He also acknowledges that the temptation to treat existence as a real predicate does not arise when dealing strictly with empirical concepts and their corresponding objects:

> If the issue were an object of sense, then I could not confuse the existence of the thing with the mere concept of the thing. For through its concept, the object would be thought only as in agreement with universal conditions of a possible empirical cognition in general, but through its existence it would be thought as contained in the context of the entirety of experience....(A600–1/B628–9).

Here see Kant relate his opponent's confusion regarding the nature of existence to an illicit inference from the objective validity of a concept to its objective reality, the illicit inference from a concept representing an object that conforms to the "universal conditions of a possible empirical cognition" to a concept with an object falling under it that exists within the context of experience. The fallacy identified here is assuming that obtaining an objectively valid concept and then narrowing its scope by introducing additional constituent marks entitles us to infer that it must be instantiated. Kant maintains to the contrary: "[H]owever much our concept of an object may contain, we have to go out beyond it in order to provide it with existence" (A601/B629).

Let me explain more fully why Kant thinks the slide from logical possibility to objective validity to objective reality is a more frequent and grievous occurrence in cases where a priori concepts are concerned. First of all, with concepts such as 'pegasus' and 'golden mountain' conflating logical possibility with real possibility does not really matter, since defining a concept by listing mutually consistent, empirically verifiable constituent marks is all that is required to ensure that a concept is potentially applicable, i.e., objectively valid. That is, when it comes to defining empirical concepts, logical inconsistency is all that stands in the way of showing that they meet the universal conditions of a possible empirical cognition. Secondly, the corresponding objects of empirical concepts are already conceived of as entities whose existence can be verified directly or indirectly via sense perception. In these cases, he claims, it is readily apparent that all that has happened is that the very thing conceived of has, in fact, been given to us. So, while we now see ourselves in a position to assert the objective reality of the concept, we are not prone to think that the content of the concept has in any way been altered (A601/B629). It is only when one is dealing with "objects of pure thinking" and sensible perceptibility seems completely out of the question that confusion about what grounds an existential claim arises and the correlative fallacy that logical consistency implies objective reality becomes a serious methodological concern (A601–2/B629–30).

Kant then offers a third reason as to why, in the case of a priori concepts, logical consistency was assumed to imply objective validity: it was thought to be the method employed by mathematicians in generating knowledge claims about their respective subject matter. In diagnosing why anyone would think that simply applying the law of non-contradiction to an arbitrarily stipulated, logically consistent concept suffices for positing the existence of an object bearing the relevant properties, Kant blames "the illusion" of logical necessity arising from the misunderstanding of mathematical proofs (A594/B623). For Kant, the philosophical enthusiasm typical of the rational tradition founded by Plato allows reason to soar like a dove in the "empty space of pure understanding" because it is entranced by a faulty conception of mathematics:

> Mathematics gives us a splendid example of how far we can go with *a priori* cognition independently of experience. Now it is occupied, to be sure, with objects and cognitions only so far as these can be exhibited in intuition. This circumstance, however, is easily overlooked, since the intuition in question can itself be given *a priori*, and thus can hardly be distinguished from a mere pure concept. Captivated by such a proof of the power of reason, the drive for expansion sees no bounds (A4–5/B8)

He maintains that rationalists and empiricists alike believed that the truth of mathematical concepts and propositions was grounded on logical possibility alone:

> *Hume*, when he felt the call…to cast his gaze over the entire field of *a priori* cognition…lopped off a whole (and indeed the most considerable) province of the same, namely pure mathematics, by imagining that the nature and so to speak legal constitution of this province rested…solely on the principle of contradiction….[I]t was as if he had said: Pure mathematics contains only *analytic* propositions, but metaphysics contains synthetic propositions *a priori* (*Prolegomena*, 4:272, 165–6)[41]

Since the method of defining logically consistent concepts and then subjecting them to the law of non-contradiction was believed to give rise to mathematical knowledge and sufficed for positing the existence of mathematical entities, it is not so surprising that a similar approach was deemed acceptable by metaphysicians wanting to ground our knowledge of the supra-sensible. Or, so says Kant.

While acknowledging the foregoing excuses, Kant still insists that anyone trying to "extract the existence of an object" from an "idea contrived quite arbitrarily" is guilty of engaging in a "mere novelty of scholastic wit" (A603/B631). And with his C&I distinction, Kant brings to fore the disingenuous maneuverings of the accused. By drawing such a sharp contrast between a logically possible conceptual form, an objectively valid concept with genuine cognitive content and the paradigmatic intuitive representation of a single, actually existing object instantiating a concept, Kant exposes the sleight of hand at work within the ontological argument or any attempt to move directly from logical consistency to potential applicability to existence. In sum, the Kantian notion of a properly valid concept serves as a warning against arguing directly from logical possibility of concepts to the real possibility of things, whereas the distinction between a proper Kantian concept and a proper Kantian intuition serves as a warning against arguing directly from the mediate representation of a conceived possible something with certain properties to the actual givenness of the particular entity having those properties.

Let us recap. Kant identifies two methodological errors prevalent in his day: (1) an illegitimate appeal to abstraction to assert the existence of an abstract object manifesting all and only the right sort of properties; and (2) the fallacious inference

---

[41] See Hume's *Enquiries* §IV, Part I, 20–21 and Part II, 30 for his expressed views on mathematics.

from logical consistency to objective validity to existence. Drawing on what Kant says about the distinction between concepts and intuitions in his *Critique* and the *Jäsche Logic,* it is not difficult to see how attending to this distinction would serve to underscore these mistakes and prevent their future occurrence. As we will see in Chap. 3, Cohen set a precedent for interpreting Kant's *Critique* as concerned with establishing precepts that ought to be followed by any practice laying claim to scientific status, i.e., a practice laying claim to having made continual epistemic progress.[42] When Frege introduces his distinction between objects and concepts in the *Foundations*, he introduces it as an exhaustive distinction among the class of objective ideas [*objective Vorstellungen*], maintaining that he is thus recovering the "true view" of a Kantian doctrine that needed to be extracted from the psychological connotations that Kant himself is responsible for. In Chap. 4, we will see Frege deploy the C&O distinction in accusing the formalists of committing these same mistakes. Therefore, I hope to have shown that if someone were inclined to look to Kant for a fundamental principle [*Grundsätze*] aimed at preventing such errors, recovering a modified version of the C&I distinction is what you would expect him to do.[43]

2.3   Kant's Philosophy of Mathematics: The Good, the Bad, and the Notoriously Obscure

In this section I tie up some loose ends concerning Kant's philosophy of mathematics, so my reader will be in a position to understand why nineteenth century developments in non-Euclidean geometry and perceptual psychology initiated a raging debate among Frege's contemporaries over the parts of Kant's system that needed to be abandoned, retained or modified. The very fact that such a debate was initiated at all introduces the first question I want to address: What is so attractive about Kant's theory of mathematical cognition that someone would try to defend or reinterpret it in the face of mounting anomalies and counter-examples?

Kant offers an account that allows mathematical inquiry to be a spontaneous and creative intellectual activity on the part of the mathematician, while simultaneously ensuring its applicability. By identifying space and time as our forms of intuition, the medium through which objects must be represented insofar as they are given to us, Kant can argue that the mathematician's spontaneously generated constructions have the potential for applicability necessarily built into them. On Kant's account, mathematics is composed of necessary truths about the world we experience. This is no small feat. Generally speaking, there are two standard ways of accounting for

---

[42] See *Prolegomena* 4:255–6 for Kant's criteria of a properly functioning scientific inquiry and his claim that metaphysics as practiced by his predecessors and contemporaries fails to meet this criteria.

[43] The third fundamental principle that Frege cites as guiding his inquiry in the *Foundations of Arithmetic* is "never to lose sight of the distinction between concept and object" (*FA*, x).

the apparent necessity of mathematical propositions: (1) Formalist accounts[44]—mathematical propositions are necessary because they refer to conventionally stipulated definitions or laws. Hume, modern-day fictionalists, and "true relative to a set of stipulated axioms" advocates offer such accounts; and (2) Platonic accounts—mathematical propositions are necessarily true because they refer to necessarily existing, immutable entities and relations. Plato, Descartes, and modern-day mathematical Platonists give this explanation. Both accounts have difficulty explaining why propositions about the totally invented or the entirely abstract should be so successful in describing the entities and relations of this world.[45]

In contrast, Kant tells a story that weaves together his notion of a threefold cognitive synthesis, the constructive method employed in traditional geometric proofs, a judgmental theory of sense perception, and a compromise between Newton and Leibniz on the status of space. The moral of this story is that faithful adherence to the method and fundamental assumptions of Euclidean geometry will yield universally valid concepts and propositions representing objects or states of affairs that are empirically realizable. All mathematical concepts and propositions are thus potentially applicable.

A second positive feature of Kant's philosophy of mathematics, and of his epistemology as a whole, is its sensitivity to actual scientific practice. His notion of a proper intuition, for example, is self-consciously developed in reference to the current debate regarding the psychophysiology of sense perception. This sensitivity also manifests itself in his critique of the rationalists. Recall Kant's claim that the objective validity of a priori concepts depends on showing that they can be schematized and that this more closely tracks the practice of mathematicians than does the proof technique employed by proponents of the ontological argument. In arguing for the contrast between logical possibility and objective validity, Kant points out that "in the concept of a figure that is enclosed by two straight lines there is no contradiction" (A221/B268). Yet, no self-respecting eighteenth century mathematician would deny that a trilateral figure is the rectilinear figure containing the least number of sides. The notion of a two-sided rectilinear figure may be logically consistent, but this would not have sufficed for a geometer to treat it as legitimate mathematical concept. Historian of mathematics Morris Kline (1972) confirms the significance of constructability for the geometers of Kant's time:

> Up to the middle of the nineteenth century no mathematician would have considered such lines [lines trisecting an angle] because only those elements and figures that are constructible

---

[44] Not to be confused with Frege's much more specific use of the term 'formalism.' For Frege, 'formalist' is a pejorative term and, broadly construed, refers to anyone accused of engaging in practices that deprive mathematical terms and sentences of their appropriate sense (*Sinn*) or reference (*Bedeutung*).

[45] This tension between mathematical certainty and applicability is so strong that some mathematicians, Einstein for instance, conclude that the two are mutually exclusive: "As far as the laws of mathematics refer to reality, they are not certain; and as far as they are certain, they do not refer to reality" (cited in Friedman (1992), 56).

> were regarded as having legitimacy in Euclidean geometry. Constructibility guaranteed existence (Ibid, 839–40)

As we have seen, Kant would not say that constructability in Euclidean space guaranteed existence in a robust sense, but it does guarantee the real possibility of mathematical concepts, that these concepts "pertain to possible things" and can serve as real predicates in an objectively-valid judgment (A221/B268).

Kant also suggests that his description of the threefold cognitive synthesis is modeled on the mathematician's method for establishing the objective validity of their concepts.[46] He insists that the mathematics and physics of his day "maintain their old reputations for well-groundedness" (Axi). To motivate his Copernican approach to metaphysics and epistemology, Kant likens it to the methods employed in math and physics and credits an unknown Greek geometer for discovering the method making synthetic a priori knowledge a viable possibility:

> For he found that what he had to do was not to trace what he saw in this figure, or even trace its mere concept, and read off, as it were, from the properties of this figure; but rather that he had to produce the latter from what he himself thought into the object and presented (through construction) according to *a priori* concepts, and that in order to know something securely *a priori* he had to ascribe to the thing nothing other except what followed necessarily from what he himself put into it in accordance with its concept (Bxii).

Just as mathematicians obtain well-grounded synthetic a priori knowledge by bringing their objects into conformity with a priori concepts, so too Kant will endeavor to obtain synthetic a priori knowledge by bringing objects into conformity with our pure forms of sensible intuition and the pure concept-rules supplied by our faculty of judgment.

Consider the symmetry in Kant's description of what grounds mathematical knowledge and what grounds synthetic a priori knowledge in general. Mathematical knowledge depends on constructing concepts, obtaining a universally valid procedure, i.e. a pure intuition, for subsuming objects under the concepts. Non-mathematical a priori knowledge depends on schematizing categories, obtaining a universally valid procedure, i.e., a pure intuition, for subsuming objects under them. Mathematical concepts and intuitions are the result of a transcendental synthesis of the imagination. Non-mathematical pure concepts and intuitions are the result of a transcendental synthesis of the imagination.

Finally, Kant looks to actual mathematical practice to support his notion of a pure intuition. Eberhard assumed that what Kant meant by 'pure intuition' was an actual sketch of the sought after mathematical entity. He then used this interpretation to argue, pace Kant, that mathematicians complete "the entire delineation" of their science without "ever saying a single word about the reality of their object,"

---

[46] Longuenesse (1998) has noted this as well, hence the title of her second chapter "The 'Threefold synthesis' and the Mathematical Model". She claims, however, that Kant moved away from this model in the B edition. In Chap. 4, I argue that Cohen's Marburg School and Frege continued to insist on Kant's idea of constructing mathematical concepts, as modeled on the traditional methods of Euclidean geometry, as the necessary pre-requisite for obtaining concepts with the appropriate cognitive content.

citing the fact that Apollonius' derivation of the concept *parabola* does not contain instructions on how to draw one (Allison (1973), 109–110). Kant responds with his own step-by-step description of Apollonius' derivation to clarify his intended meaning and defend his claim that pure intuitions are considered indispensable in geometric proofs:

> Apollonius first constructs the concept of a cone, i.e., he exhibits it a priori in intuition (this is the first operation by means of which the geometer presents in advance the objective reality of this concept).[47] He cuts it according to a certain rule, e.g., parallel with a side of the triangle which cuts the base of the cone (*conus rectus*) at right angles by its summit, and establishes a priori in intuition the attributes of the curved line produced by this cut on the surface of the cone…Consequently the objective reality of this concept, i.e., the possibility of the existence of a thing with these properties [a thing with the properties of a parabola], can be proven in no other way *than by providing the corresponding intuition* (Kant (1973), 110)

Kant's point is that Apollonius presents a pure intuition corresponding to his concepts insofar as he articulates a procedure whereby a curve, having the requisite properties, can be imaginatively constructed and then intuited as the cut on the surface of a cone. The mathematician obtains the pure intuition by demonstrating "the reality of the rule itself, and with it of this concept for the use of the imagination" (Ibid., 127).[48] The fact that Apollonius does not explain how to draw the figure, using compass and ruler or other technical instruments, is irrelevant (Ibid., 111). The fact that Kant feels the need to justify himself relative to Apollonius' derivation shows just how much he depends upon the practices of the working mathematician for legitimizing various aspects of his C&I distinction.

To see that Kant is not simply overlaying his own views on the practices of ancient geometers, let us compare his description to that of Morris Kline. Apollonius begins with defining a double cone as follows: "Given a circle BC and any point A outside the plane of the circle, a double cone is generated by a line through A and moving around the circumference of the circle" (Kline (1972), 90). Apollonius then describes how to cut the cone such that a line (PP') can be found that bisects a chord (QQ') on the curve and all chords parallel to QQ.' He then shows that there is

---

[47] Although Kant uses the expression 'objective reality,' what he means, given my fixing of his terminology, is objective validity.

[48] This passage contains further textual evidence that Kant identifies mathematical entities qua pure intuitions with the rule or procedure for generating spatio-temporal configurations and not with the configurations themselves. The Kant-Eberhard exchange makes it clear that, for Kant, the sphere of pure intuitions and objectively-valid mathematical concepts is circumscribed by the possibility of constructability in Euclidean 3-space, not by our ability to actually construct or visualize particular figures perfectly corresponding to the concept. For instance, the psychological fact that I cannot imagine a perfectly round circular figure would seem to have no bearing on the objective validity of the Euclidean notion of a circle. However, given Kant's account of the transcendental synthesis grounding that notion, it seems that Kant is making some assumptions about my imaginative capacities, namely, that I do not consistently imagine an ellipse. It is important to note the role that Kant ascribes to the imagination and visualization in geometric proofs, since this is one of the features of Kant's philosophy of mathematics that Helmholtz will hone in on when relating Kant's claims to his own research in perceptual psychology.

a point (L) on the cutting plane such that the following equality holds: $(QV)^2 = PV \cdot PL$, where V is the point of bisection and PL is perpendicular to PV. QV is called the ordinate and PL is called the parameter. PP' is the diameter.[49] Kline also explains how Apollonius' equality between geometric magnitudes can be transcribed into the analytic expression for the parabola.[50] As Kline notes, however, "no algebra appears in Apollonius' treatment" (Ibid., 92).

Notice that Apollonius' definition of a cone really is a procedure enabling us to apprehend possible instantiations of the concept; one grasps the definition of the concept 'double cone' if and only if one conceives that some sort of hour-glass figure is being generated thereby. This holds for practically all of the definitions provided by ancient Greek geometers. Euclid's definitions for 'point', 'line' and 'plane' are infamously vague; otherwise, geometric definitions tend to look like disguised postulates or rules for producing something using these basic elements. In other words, Euclid's definitions are almost meaningless without recourse to his postulates. The postulates guarantee that lines and circles can be constructed. The definitions of more complex figures involve stating a procedure whereby they can be produced using only lines and circular arcs, i.e., straight-edge and compass (Ibid., 86–7).

Kant also stressed the role of the postulates in ascribing meaning to Euclid's concepts:

> Now in mathematics a postulate is the practical proposition that contains nothing except the synthesis through which we first give ourselves an object and generate its concept, e.g., to describe a circle with a given line from a given point on a plane; a proposition of this sort cannot be proved, since the procedure that it demands is precisely that through which we first generate the concept of such a figure (A234/B287).

It is the postulated procedure that lends to the concept its general applicability. Consider once again Apollonius' definition/construction of the concept 'double cone.' Depending on one's choice for the location of point A relative to a circle BC, one can produce/imagine an infinite variety of hour-glass type figures. Kant appears then to be correct with respect to traditional Euclidean geometry; articulating a procedure whereby the geometric entity can be constructed using only the basic elements and properties of Euclidean 3-space was logically prior to obtaining a suitably general, cognitively meaningful geometric concept.[51]

---

[49] That is, PP' is a diameter or the axis of the parabola if the cone is right circular, and Kant's analysis seems to presuppose that it is.

[50] Notice that the length of PV equals the distance of a chord (QQ') from point P, so let x = the distance from P to a parallel chord and let y = the corresponding length of the ordinate (QV). Finally denote the length of PL as $2p$. Appollonius equality is then transcribed into $y^2 = 2px$. The lines PL and PP' are perpendicular and so can determine a Cartesian coordinate system. However, Appollonius's proof is slightly more general than Kant acknowledges. He does not restrict the construction to right circular cones. Therefore, it is not necessarily the case that the system of parallel chords is perpendicular to PP'. Thus, Kline's analytic equation references an oblique coordinate system (Kline, 90–92).

[51] By referring to the basic properties of Euclidean 3-space, I mean that Kant and the ancient geometers assumed that the space in which construction occurred was both flat and continuous.

While Kant lauds the constructive methods of the ancients and looks to them for support, he does not endorse whatever a mathematician does as methodologically sound. More specifically, he does not endorse Descartes' analytic approach to geometry. Immediately after arguing that Apollonius' non-algebraic derivation of the concept *parabola* is a paradigm case of what he meant by saying that mathematical knowledge depends upon the construction of concepts, Kant chastises "modern geometers" who "arbitrarily think for themselves such a line (e.g., the parabola through the formula $ax=y^2$)" (Kant (1973), 111).

Providing some historical context, can help us better unpack Kant's remark. Descartes had been critical of the synthetic method of the ancients, the constructability criterion in particular: "[Descartes] explicitly criticized the geometry of the ancients as being too abstract, and so much tied to figures 'that it can exercise the understanding only on the condition of greatly fatiguing the imagination'" (Kline (1972), 308). Descartes was willing to admit anything uniquely defined by an algebraic equation in x and y as an acceptable geometric curve. Leibniz was critical of even this limitation and extended the notion of curve to include transcendental equations. Commenting on this extension and the subsequent elimination of constructability as the criteria granting geometric legitimacy, Newton (1707) writes, "But the Moderns advancing yet much further [than the plane, solid and linear loci of the Greeks] have received into Geometry all Lines that can be expressed by Equations" (cited in Kline (1972), 312). For Kant, this modern trend is becoming too untethered. He thus urges geometers "not to forsake the synthetic methods of the ancients for the analytic method which is so rich in inventions" (*Kant-Eberhard Controversy*, 111).

We can now see what's not so good about Kant's philosophy of mathematics and the Achilles' heel of his epistemological system, namely, its indebtedness to the traditional synthetic method and the objective truth of Euclidian geometry. Kant's immediate predecessors and contemporaries were just beginning to enlarge the domain of geometric inquiry to include any structure that could be defined in algebraic terms.[52] Kant, on the other hand, restricts the scope of geometry to whatever is constructible using the basic elements and primitive operations countenanced in traditional Euclidean geometry. According to Tappenden (1995), this debate over

---

[52] I say that Kant's contemporaries were just beginning to move in this direction because Descartes actually waffles on whether stating a constructive procedure that occurs within a spatio-temporal domain with certain characteristics (continuity and successively ordered) is a necessary prerequisite for introducing some entity as a proper geometric object. For instance, when arguing for recognizing more complex curves, e.g., spirals, as legitimate geometric entities, Descartes begins by citing the various assumptions regarding constructability that are made within traditional geometry. Besides the postulates guaranteeing the possibility of drawing a straight line between any two points and the possibility of constructing circles with any center and diameter, the ancients also introduce "the assumption that any given cone can be cut by a given plane". To this set of primitive constructive operations, Descartes wants to add "only one additional assumption…namely, two or more lines can be moved, one upon the other, determining by their intersection other curves." He then claims that spirals qualify as legitimate geometric entities because they too "can be conceived as described by continuous motion or by several successive motions…." (Descartes (1637b), 43).

the proper subject matter and methods of geometry would continue to be "a topic of discussion" among German mathematicians throughout the nineteenth century (Ibid., 324). Helmholtz was one of those who fully embraced analytic methods, along with the idea that the subject matter of pure geometry can be any collection of entities, e.g. an n-tuple of numbers, so long as they exhibited well-defined algebraic properties and relations. Helmholtz will cite the fact that analytic geometry allows for the representation of structures manifesting properties incompatible with the Euclidean axioms to refute Kant's thesis concerning these axioms.[53]

In conclusion, one of the features that made Kant's philosophy such a triumph in its day, namely, its sensitivity and relevance to the sciences as they were actually conducted, also rendered it vulnerable to developmental changes within those sciences. But before we move to discuss how nineteenth century German mathematicians and philosophers tried to salvage Kant's system in spite of an altered scientific landscape, let me briefly mention one question that plagued Kant's philosophy of mathematics from the outset: how does his picture of mathematics as necessarily involving the construction of concepts capture what goes on in the fields of arithmetic and algebra? The fact that Kant relies so much on geometric examples for motivating and underwriting his account of what characterizes synthetic a priori knowledge makes it difficult to tease out a coherent view on the nature of number theoretic truths.

One possible line of interpretation is that arithmetic concepts involve constructions within the pure temporal manifold in a manner analogous to the pure spatial constructions giving rise to geometric concepts.[54] The objection to this view of arithmetic, both as a reading of Kant and on its own merits, is that while geometry is considered a science *of* space and thus responsible for articulating the properties and relations of spatial configurations, arithmetic is not similarly considered to be a science *of* time. In response to a letter in which this criticism was posed, Kant writes, "Time, as you rightly remark, has no influence on the properties of numbers" (cited in Friedman (1992), 106).[55]

Another possibility is that Kant thinks our knowledge of algebraic properties and relations involves the construction of concepts insofar as it makes an essential use of a symbolic notation:

> But mathematics does not merely construct magnitudes (*quanta*) as in geometry; it also mere magnitude (*quantitatem*), as in algebra, where it entirely abstracts from the constitution of the object that is to be thought in accordance with such a concept of magnitude. In this case it chooses a certain notation for all construction of magnitudes in general (numbers), as well as addition, subtraction, extraction of roots, etc., and, after it has also

---

[53] See Sect. 3.1.1.

[54] The relevant texts for developing this line of interpretation are: A142–3/B182; A241/B300 and the *Prolegomena*, 4:283.

[55] In his "Arithmetic and the Categories", Charles Parsons (1992b) also resists an interpretation of Kantian arithmetic that treats time as its subject matter: "the identification of a number as a schema would have its difficulties, for it attributes a temporal content to the notion of number itself" (Ibid., 148).

designated the general concept of quantities in accordance with their different relations, it then exhibits all the procedures through which magnitude is generated and accordance with certain rules in intuition...(A717/B745).

Pursuing this line, Kant maintains that by tracking and manipulating arithmetic and algebraic notation we obtain knowledge that would not be available via concepts alone. Unlike geometric construction, algebraists are not constructing the objects themselves, but rather constructing a notation. The problem here is that Kant's claim that that arithmetic or algebra is a body of synthetic a priori truths dependent on the construction of concepts only follows by blurring the distinction between a number and its symbolic denotation. In any case, Kant's views on number theory are notoriously obscure, so much so that Philip Kitcher suggests that this obscurity partially motivated Frege's logicism (Kitcher (1979), 252). And it is certainly true that few nineteenth century neo-Kantians defended Kant's views on number theory with the same tenacity and vigor with which they defended his views on geometry and other aspects of his epistemological framework.

References

Allison, H. E. (Ed. and Trans.). (1973). *The Kant-Eberhard controversy*. Baltimore/London: The Johns Hopkins University Press.

Allison, H. E. (Ed. and Trans.). (1983). *Kant's transcendental idealism*. New Haven/London: Yale University Press.

Berkeley, G. (1710/1988). *Principles of human knowledge and three dialgues between Hylas and Philonous* (Ed., with an introduction, by R. Woolhouse). London: The Penguin Group.

Berkeley, G. (1713/1988). *Principles of human knowledge and three dialogues between Hylas and Philonous* (Ed., with an introduction, by R. Woolhouse). London: The Penguin Group.

Chignell, A. (2008, Summer). Introduction: On going back to Kant. *The Philosophical Forum, xxxix*(2), 109–124.

Descartes, R. (1637b/1954). *The geometry of René Descartes* (D. E. Smith & M. L. Latham, Trans.). New York: Dover Publications.

Frege, G. (1953). *Die Grundlagen der Arithmetic* (J. L. Austin, Trans. as *The foundations of arithmetic (FA)*). Evanston: Northwestern University Press.

Frege, G. (1997). *The Frege Reader (FR)* (Ed. and introduction by M. Beaney). Oxford: Blackwell.

Friedman, M. (1992). *Kant and the exact sciences*. Cambridge: Harvard University Press.

Friedman, M. (2000a). *A parting of the ways*. Chicago/La Salle: Open Court.

Friedman, M. (2000b). Geometry, construction, and intuition in Kant and His successors. In G. Sher & R. Tieszen (Eds.), *Between logic and intuition*. Cambridge: Cambridge University Press.

Friedman, M. (2008, Summer). Ernst Cassirer and Thomas Kuhn: The Neo-Kantian tradition in history and philosophy of science. *The Philosophical Forum, xxxix*(2), 239–252.

Hanna, R. (2018). Kant's theory of judgment. *The Stanford encyclopedia of philosophy* (Winter 2018 Edition, E. N. Zalta (Ed.)). https://plato.stanford.edu/archives/win2018/entries/kant-judgment/

Hatfield, G. (1990). *The natural and the normative*. Cambridge, MA: MIT Press.

Hintikka, J. (1967). Kant on the Mathematical Method. Reprinted in Posy (1992), pp. 21–42.

Kant, I. (1973). *On a discovery according to which any new critique of pure reason has been made superfluous by an earlier one in Allison 1973* (pp. 107–160).

Kant, I. (1992). *Lectures on logic* (J. M. Young, Ed. and Trans.). Cambridge: Cambridge University.

Kitcher, P. (1979). Frege's epistemology. *Philosophical Review, 88*(2), 235–262.

Kitcher, P. (1990). *Kant's transcendental psychology*. New York: Oxford University Press.

Kline, M. (1972). *Mathematical thought from ancient to modern times* (Vols. 1–3). New York: Oxford University Press.

Kremer, M. (2000). Judgment and truth in Frege. *Journal of the History of Philosophy, 38*(4), 549–581.

Locke, J. (1690/1978) *An essay concerning human understanding* (Ed. with introduction by J. W. Yolton). London: J. M. Dent.

Longuenesse, B. (1998) *Kant and the capacity to Judge* (C. T. Wolfe, Trans.). Princeton/Oxford: Princeton University Press.

Newton, I. (1707). Universal Arithmetick: Or, a treatise of arithmetical composition and resolution. Written in Latin by Sir Isaac Newton, and Trans. by the late Mr. Ralphson, and Revised and Corrected by Mr. Cunn. (London 1728), pages i–iv, 1–257. Reprinted in *The mathematical works of Isaac Newton* (Vol. 2, Dr. D. T. Whiteside, Ed., (1967), pp. 3–136).

Parsons, C. (1992b). Arithmetic and the categories. In Posy (1992), pp. 135–158.

POSY, CARL J. (1992), Kant's Philosophy of Mathematics (Dodrecht: Kluwer Academic Publishers).

Poma, A. (1997). *The critical philosophy of Hermann Cohen* (J. Denton, Trans.). Albany: State University of New York Press.

Sluga, H. D. (1980). *Gottlob Frege*. London: Routledge & Kegan Paul, Inc.

Tappenden, J. (1995). Geometry and generality in Frege's philosophy of arithmetic. *Synthese, 102*, 319–336.

Tolley, C. (2014). Kant on the content of cognition. *European Journal of Philosophy, 22*(2), 200–228.

Young, J. M. (1992a). *Functions of thought and the synthesis of intuitions*. Reprinted in Guyer (1992), pp. 101–122.

Young, J. M. (1992b). *Lectures on Logic*. Cambridge: Cambridge University Press.

Chapter 3
Disambiguating Kant for the Sake of Science

Abstract This chapter analyzes the differences between Helmholtz's and Cohen's readings of Kant, both of which aimed at constructing an epistemological framework suitable for housing progressive developments in scientific practice. Section 3.1 explains how Helmholtz interprets and then evaluates Kantian terms and dicta in light of research advances in perceptual and cognitive psychology and the study of non-Euclidean geometry. I conclude that the neo-Kantian framework proffered by Helmholtz anticipates Quine's (1969) proposed "Epistemology Naturalized." In Sect. 3.2, I argue that Cohen's school of Kantian interpretation was largely an effort to undermine Helmholtzian naturalism, demonstrating that it could not account for the normativity and objectivity accorded to science proper. I further conclude that Cohen's objections to Helmholtz's research on non-Euclidean geometry are not the mere ramblings of someone standing in the way of progress. Rather, these objections indicate that Cohen presaged Kuhn in worrying that radical changes in terminology and methodology, if not managed carefully, threaten the very notion of scientific progress.

Keywords Hermann Helmholtz · Hermann Cohen · Neo-Kantianism, · Philosophy of science · Philosophy of geometry · Scientific naturalism · Scientific progress

Since the 1980s, Frege scholars have been embroiled in a dispute as to whether Frege should be interpreted as a Platonic realist or a Kantian idealist. Philip Kitcher (1979), Hans Sluga (1980), Joan Weiner (1982) and Gottfried Gabriel (2002) have all argued for reading Frege as significantly committed to Kantian epistemology. Michael Dummett (1991) and Michael Resnik (1980) counter that if there are any Kantian influences in Frege's thought, they are minimal at best and dropped by the wayside as his philosophical views matured. As this dispute has made exceedingly clear, however, referring to Frege as 'a Kantian' is trivially true, on the one hand, and extremely problematic on the other.

© Springer Nature Switzerland AG 2020   63
T. Merrick, *Helmholtz, Cohen, and Frege on Progress and Fidelity*,
Philosophical Studies in Contemporary Culture 27,
https://doi.org/10.1007/978-3-030-57299-0_3

A cursory glance at Frege's historical setting reveals a whole slough of self-proclaimed Kantians.[1] Frege entered the University of Jena in the spring of 1869 and stayed until the winter of 1870. He then attended Gottingen University from 1871 until he received his Ph.D on December 12, 1873. He returned to Jena as a faculty member for the next 44 years. The period of 1865–1881 was the third phase in the evolution of Neo-Kantianism, marking its rapid dissemination throughout Germany. Kant became the most read philosopher in German universities. There was a threefold increase in the number of courses on Kant in the 1870's as compared to the 1860's. Gottingen and Marburg were "growing into centers of Kant instruction" (Kohnke (1987), 206). Although Neo-Kantianism was pervasive, it was hardly cohesive:

> The connection of the 'Kantians' one with another was so loose, their interest in Kant so variously grounded, that the meaning of the concept 'Neo-Kantianism' might seem to be reducible to saying that a 'neo-Kantian' was a philosopher the focus of whose endeavors lay in a compounding with at least some parts of the philosophy of Kant (Ibid.)

By the end of this period, Neo-Kantianism could be divided into no less than seven "tendencies":

> 1) the physiological tendency (Helmholtz, Lange); 2) the metaphysical tendency (Liebmann, Volkelt); 3) the realist tendency (Riehl); 4) the logicalist tendency (Cohen, Natorp, Cassirer—the Marburg School); 5) value-theoretical criticalism (Windelband, Rickert, Munsterberg—the South-West or Baden school with which Bauch, too, is closely connected); 6) the relativist remodeling of criticalism (Simmel); 7) the psychological remodeling deriving from Fries (neo-Friesian school, Nelson).[2]

Given the philosophical landscape of Frege's day, to say that a significant Fregean thesis has Kantian roots says little unless it can be associated with one of the branches making up the Kantian clan.

Sluga (1980) argues that Frege's doctrine concerning the essentially predicative or unsaturated nature of concepts is evidence of his commitment to Kant's transcendental idealism and that this commitment came via the influence of Hermann Lotze's 1874 *Logik*. Though reluctant to embrace Sluga's thesis that Frege is a transcendental idealist, Gabriel (2002) supports his general contention that Frege is a part of the Neo-Kantianism significantly shaped by Lotze (Ibid., 39). He thus provides additional historical evidence for Sluga's contention by showing the significant parallels in the thought of Frege and Wilhelm Windelband, founder of the Southwest school of Neo-Kantianism, claiming that members of the Marburg school "seem to have been influenced only indirectly" (Ibid, 41). I do not intend to deny Gabriel's claims concerning the influence of the Southwest school on Frege's thought and even want to affirm his assertion that Frege followed Lotze and other nineteenth century German philosophers in using the term 'logic' so as to include

---

[1] See Kohnke (1987) and Peckhaus (2000).

[2] This is T.K. Oesterreich's (1880–1949) survey of the various neo-Kantian camps represented in the latter half the nineteenth century cited in Kohnke (1987), 200.

epistemology (Ibid., 44).[3] Still, I maintain that aligning Frege with Cohen's Neo-Kantianism is exceedingly fruitful in terms of clarifying Frege's position on the necessity of the Euclidean axioms and better understanding his C&O distinction and its role in combating his formalism. My reasons are as follows.

First, Frege specifically identifies Helmholtz as formalist (*TPW*, 213)[4] For Frege, formalism in mathematics is related to both empiricism and psychologism. Formalists are empiricists in the sense that they refuse to grant existence to anything that has not come by means of the senses. Thus, the existence of numbers is thought to be guaranteed by identifying numbers with numerical signs (*TPW*, 124). We will see, in Chap. 4, that the formalists relied on Helmholtz' research in perceptual psychology to argue that number signs possessed the relevant arithmetical properties. Formalism is related to psychologism because it attempts to ground the validity of mathematical propositions on facts about our psychophysiology. In Sect. 3.1, we learn that Helmholtz himself applied this research in addressing Kant's question: how are synthetic a priori judgments possible? As already noted, when Frege introduces his C&O distinction in the *Foundations,* he claims to be recovering the truth of a Kantian doctrine that is lost when one privileges the more subjectivist, psychological connotations in Kant's texts (*FA*, 37). Frege's introduction and use of the C&O distinction to combat formalism is thus thoroughly bound up with rejecting Helmholtz's reading of Kant.

Second, we know that by 1885 Frege had read Cohen's *The Principle of the Method of Infinitesimals and its History (PIM)* and that this work, in conjunction with Cohen's 1871 and 1885 editions of *Kant's Theory of Experience (KTE),* is known for initiating the movement aimed at "extricating Kant from Helmholtz's interpretation" (Patton (2005), 111). In his review of *PIM,* Frege lambastes Cohen for his obscure writing style and for his failure to adequately appreciate recent developments in mathematics. He ends, however, with the following cryptic remark: "…I agree with Cohen that knowledge as a psychic process does not form the object of the theory of knowledge, and hence, psychology is to be sharply distinguished from the theory of knowledge" (*CP*, 111). So, if we are looking for a reading of Kant that Frege himself saw as standing in opposition to Helmoltz' naturalized reading, Cohen's Kantianism seems a natural choice.

There is a third reason why tracking Cohen's line of Neo-Kantianism makes sense, given the story of the C&O distinction that I am interested in telling. I have said that Frege's use and elucidations of the distinction shows it to be a de-psychologized, strictly methodological (not ontological) successor of Kant's C&I distinction. I have also said that establishing this line of succession is part, if not all, of Frege's justification for treating the C&O distinction as a primitive logico-epistemological fact. This raises the question whether Frege and his intended

---

[3] This is why I have used the modifier 'logico-epistemological' when referring to Frege's C&O distinction as a fundamental principle introduced in *Foundations.* See Patricia Blanchette (1994) for a defense of the epistemological import of Frege's logicist project against Paul Benacerraf's 1981 argument to the contrary.

[4] See also *FR,* 132, and *CP,* 208.

audience were familiar with this manner of revising and reintroducing Kantian doctrines. The answer is yes, and here again, we are pointed in the direction of Cohen. Longuenesse (1998) describes Cohen's interpretation as "the neo-Kantian reduction of the *Critique of Pure Reason* to an epistemology of the natural sciences" (Ibid., 4). Cohen and his students, Natorp, and Cassirer, pushed for a de-psychologized reading of the *Critique*:

> Cohen and his neo-Kantian followers favored the orientation of the B edition because of what they perceived to be the more epistemological character of its argument, as opposed to the dubious psychologism of the A edition (Ibid., 59).[5]

They then explained how adherence to Kantian methods and doctrines, suitably interpreted, could ward off the skepticism implicit in psychologism and empiricism, without falling prey to the excesses of dogmatic rationalism. In the course of articulating their version of Neo-Kantianism, they specifically address the methodological errors that were of concern to both Kant and Frege, namely illicit appeals to abstraction and confusion regarding the notions of logical possibility, objective validity and objective reality. In fact, if there is one place where Lotze does have a direct influence on the Marburg school, it is his claim that Lockean abstraction cannot yield concepts suitable for scientific use. As we will see, Cassirer invokes Lotze to reject a Helmholtzian account of concept formation and to support Cohen's insistence that science stands in need of a priori concepts to ground the normativity, generality and necessity accorded to scientific propositions properly so-called. Moreover, Cassirer credits Cohen's books on Kant with first bringing "full methodological clarity" to Lotze's category of 'validity' (Cassirer (2005), 103). Therefore, by reading Frege within the context of the Helmholtz-Cohen debate and aligning Frege' with Cohen's Kantianism, we can see how Frege's elucidations and use of the C&O distinction serve to convince his contemporaries that the distinction warrants acceptance as a fundamental principle in a properly Neo-Kantian epistemology of science.

Fourth, as is well known and seemingly inexplicable, Frege maintained allegiance to a Kantian description of Euclidean geometry as system of synthetic a priori truths, dependent on axioms expressing "basic facts of our intuition" throughout his lifetime (*FG*, 7 and *PW*, 273). However, he never presents an explicit, detailed defense of Kant's theory of geometry. Beginning in 1870, Helmholtz launched an all-out attack on Kant's claim that Euclidean geometry was a body of synthetic a priori truths. In 1885, despite Helmholtz' objections and the advent of non-Euclidean geometries, Cohen defended Kant's theory of geometry. Members of the Marburg school continued their defense of Kant on the nature and primacy of

---

[5] Similarly, Alan Kim (2004) characterizes the Marburg reading of Kant as follows: "The first of their modifications stems from an anti-psychologistic critique of Kant himself, namely of what they see as a confusion in first *Critique* between the task of a transcendental grounding of the sciences and that of a transcendental logic of human cognition. The former is in their view the genuine critical enterprise, for it promises to reveal the autonomous sources of objective knowledge, whereas the latter threatens to trace science back to psychological, and therefore contingent, subjective...wellsprings (Kim (2004), 3).

Euclidean geometry until 1921, after which Cassirer acknowledged that the empirical confirmation Einstein's theory of relativity decided against Kant's privileged class of synthetic a priori geometric propositions. By examining Cohen's response to Helmholtz, we can get a clue as to how Frege or anyone could have marshaled a defense of the apriority of the Euclidean axioms in the midst of revolutionary changes in geometry and related sciences.

Finally, the Helmholtz-Cohen debate is worth examining in its own right, irrespective of the fruit it bears in resolving puzzles in Frege scholarship. Both Helmholtz and Cohen recognized Kant as a founding father of a relatively new science, the science of epistemology. Both recognized that advances in modeling non-Euclidean geometric structures called for rethinking and reinterpreting some of Kant's most significant epistemological claims. In other words, Helmholtz and Cohen saw themselves confronting two potential paradigm shifts, one in geometry and another in the epistemological framework housing the sciences generally. As we will see, Helmholtz responded by developing a Neo-Kantian version of naturalized epistemology, where the objectifying and validating conditions generating cognitively significant representations are discovered through an empirical investigation of our cognitive capacities. According to Van Fraassen (2002), no such epistemology can explain how the choice to adopt a new scientific paradigm can viewed as prospectively rational (Ibid., 74–81). In Chap. 5, we reconsider Helmholtz's Neo-Kantianism to see if it serves as counter-example to Van Fraasen's broad thesis.

Cohen, in contrast to Helmholtz, responds to the trans-paradigm epistemic crisis by insisting that the normativity and necessity accorded to cognitively-significant representations cannot be accounted for except by invoking a modified version of the Kantian a priori. Cohen subsequently applied his Neo-Kantian epistemological precepts to the case of religio-ethical knowledge, explaining how the propositions of a religious tradition can retain their requisite necessity and still undergo revision in the light of new knowledge. As noted previously, Friedman (2008) argues that the Marburg school's modification of the Kantian a priori cannot account for the application of pure concepts and propositions to sensible experience. In Chap. 5, we reconsider Cohen's epistemology of science in conjunction with his epistemology of religion to see if, pace Friedman, Cohen can provide adequate resources for responding to our post-Kantian, post-Kuhnian epistemic situation. Let us now turn to the Helmholtz-Cohen exchange.

3.1 Helmholtz's Naturalized Kantianism

Helmholtz grounds his critique of Kant's philosophy on nineteenth century research in the areas of sense perception and analytic geometry. Based on this research, he concludes that certain Kantian propositions are confirmed, but others must be rejected. The proposition that Helmholtz says must be given up, without "surrendering any essential feature of the Kantian system", is the claim that Euclidean geometry is a body of synthetic a priori truths, whose validity derives from our form of intuition (*OMG II,* 686).[6] The "essential" features of Kant's system that Helmholtz retains are a hostility to metaphysical reasoning, a skepticism regarding our knowledge of things-in-themselves, and the belief that some minimal spatial structure must be recognized as a subjective form of intuition (*OMG II,* 686, and *FP,* 696–7).

Helmholtz attacks Kant's characterization of Euclidean geometry from three different angles: (1) demonstrating its incompatibility with current research in analytic geometry; (2) demonstrating its incompatibility with current research into the psychophysiological mechanisms giving rise to sense perception; and (3) illustrating by a series of thought-experiments that, pace Kant, it would be possible for us obtain empirical evidence confirming or denying Euclid's axioms.[7]

3.1.1 The Argument from Analytic Geometry: Euclidean Axioms Are not Necessities of Thought

Helmholtz begins by noting a problem with determining the content and justification of geometric propositions when geometers restrict themselves to traditional synthetic proof techniques:

> The main difficulty in these inquiries is, and always has been, the readiness with which the results of everyday experience become mixed up as apparent necessities of thought with the logical processes, so long as Euclid's method of constructive intuition is exclusively followed. It is in particular extremely difficult, on this method, to be quite sure that in the steps prescribed for the demonstration we have not involuntarily and unconsciously drawn in some most general results of experience, which the power of executing the certain parts of the operation has already taught us practically (*OMG I,* 667).

Specifically, the problem is that "the foundation of all [synthetic] proofs" resides in establishing congruence relations between geometric figures (Ibid.). This assumes that these figures can be moved and superimposed on one another without undergoing any distortion in their shape or size. 'Free mobility' is the term that Helmholtz uses to refer to this property regarding the movement of figures within a space.[8] His

---

[6] I will use the following abbreviation scheme in citing Helmholtz' works: *OMG I – The Origin and Meaning of Geometrical Axioms* (1876); *OMG II – The Origin and Meaning of Geometrical Axioms* (1878a); *FP – The Facts in Perception* (1878b); and *NME – Numbering and Measuring from an Epistemological Viewpoint* (1887).

[7] The three axioms that Helmholtz recognizes as distinctive of Euclidean geometry include the following: (1) if the shortest line drawn between two points is called a straight line, there can only be one such straight line; (2) the parallel postulate, which he considers equivalent to the claim that the sum of the angles of any triangle is equal to two right angles or 180°; and (3) free mobility – the fact that a figure can be moved in any direction and rotated within the space without it bending or distorting (*OMG I,* 672).

[8] What Helmholtz' means by 'free mobility' can also be explained by considering a space in which the distance measurements remain constant and geometric figures retain their shape and rigidity

concern is that one may mistakenly believe that the free mobility of geometric figures is a necessary truth, when it is only an empirical fact that "we all have experienced from our earliest youth" (Ibid.). The implication is that Euclid's method cannot be used to decide which propositions qualify as necessities of thought, since it must already invoke the truth of at least one proposition that cannot necessarily lay claim to that status.

So, Helmholtz continues, to decide if the Euclidean axioms qualify as necessities of thought, we must begin by "applying the analytic method of modern algebraic geometry" (*OMG*, 668). This involves stripping geometric terms of their traditional Euclidean connotations and interpreting them in a manner more conducive to number theoretic calculations. For example, instead of Euclid's vague characterization of a point as a spatial entity having no part, 'point' will now refer to any ordered n-tuple of numbers, where the numbers indicate the distance measured from a fixed coordinate system. A geometric curve or figure will no longer be conceived as something that is produced via the continuous motion of more primitive Euclidean elements, but as a set of ordered n-tuples satisfying an equation. Helmholtz also provides a strictly algebraic definition of free mobility, which anticipated the application of abstract group theory to geometry (Rosenfeld (1988), 333–338). Finally, Helmholtz proposes generalizing the notion of a spatial manifold to include any system where all internal positions, a.k.a. points, can be determined by providing n measurements from some fixed position within the system. Riemann had first developed and investigated such systems, designating them 'n-dimensional extended aggregates.' The dimension of the system is defined as the number of measurements needed to determine a position. The advantage of viewing space(s) as various kinds of Riemannian extended aggregates is that "all operations consist in pure calculations of [numerical] quantities, which quite obviates the danger of habitual perceptions being taken for necessities of thought" (*OMG I*, 673).[9]

The question whether the Euclidean axioms are necessities of thought will now be decided by asking whether the axioms hold true for all 3-dimensional extended aggregates. Helmholtz cites current developments in analytic geometry showing

---

while moving through the space as viewed from the perspective of someone standing outside of the space in which the measuring occurs. I am indebted to Jeff Barrett for helping me to explain this notion more carefully.

[9] It was not uncommon in the nineteenth century to view algebraic calculations as on a par with logical inference in the sense that both were considered truth-preserving regardless of the semantic content expressed by the propositions or equations. Thus, G. Boole writes: "Those who are acquainted with the present state of the theory of symbolic algebra are aware that the validity of the processes of analysis does not depend on the interpretation of the symbols which are employed but solely upon the laws of their combination. Every system of interpretation which does not affect the truth of the relations supposed is equally admissible, and it is thus that the same process may under one scheme of interpretation represent the solution of a question [of] the properties of number, under another that of a geometrical problem, and under a third that of optics" (from *The Mathematical Analysis of Logic* (1847), 3, cited in Enriques' *Historic Development of Logic* (1929), 127).

that the answer to this question is no. Riemann paved the way for proving that any extended aggregate with constant curvature will also exhibit free-mobility. In other words, free-mobility can also be a property of a space with properties violating Euclid's parallel postulate, i.e., a space with non-zero constant curvature. In addition, Helmholtz claims to have developed a consistent geometry for extended aggregates in which free mobility does not hold (*OMG I*, 676). Thus, Helmholtz not only argues that our belief in the free mobility of geometric figures is most likely based on experience, but also that the property of free mobility, algebraically defined, does not uniquely pick out a Euclidean space. He concludes that the Euclidean axioms should not be considered a priori necessities of thought, since "the special characteristics of our own flat space...are not implied in the general notion of an extended quantity of three dimensions and of the free mobility of the bounded figures therein" (*OMG I*, 677).

Let us step back now and see what points Helmholtz has scored against Kant at this stage in his argument. First, he has shown that the Euclidean axioms are not a priori propositions that can be deduced from the meaning of 'a 3-dimensional spatial manifold,' as this term is defined in nineteenth century analytic geometry. Second, Helmholtz has demonstrated that other consistent geometries and their corresponding structures are "conceivable, in the sense that they can described within the resources of analytic geometry" (Hatfield (1990), 219). What does this conceivability amount to? Helmholtz has shown that non-Euclidean structures are not merely conceivable as logical possibilities, i.e., something that can defined without engendering a logical contradiction, but also conceivable as abstract structures manifesting well-defined algebraic properties and relations. However, demonstrating that non-Euclidean geometries and spaces are conceivable in this sense is not enough to show that non-Euclidean concepts and propositions are objectively valid. Remember, to say that a concept or, in this case, an entire system of concepts is objectively valid is to say that it can be applied to experience. For Kant, this entailed that what was expressed in the concept or proposition could be realized in Euclidean space. Helmholtz has not yet shown, nor did he intend to show, that non-Euclidean concepts and propositions are potentially applicable or can be constructed within Euclidean 3-space.[10]

Helmholtz also acknowledges that the findings of analytic geometry alone cannot decide whether beings with our cognitive capacities can obtain an intuitive representation of a non-Euclidean spatial manifold or its geometric properties. For example, using the language of analytic geometry, Riemann was able to mathematically represent an *n*-fold extended aggregate with n > 3, but Helmholtz maintains that we could not intuit such structures: "As all our means of sense-perception extend only to space of three dimensions, ...we find within ourselves by reason of our bodily organization quite unable to represent a fourth dimension" (*OMG I*, 681).

---

[10] Of course, Helmholtz knew that non-Euclidean spaces could be modeled in Euclidean 3-space (*OMG II*, 687). I simply want to emphasize that he does not appeal to that fact in arguing that the Euclidean axioms are not necessities of thought.

Therefore, it could be that the space-relations described in the Euclidean axioms are necessities imposed by the constitution of our faculty of sense perception. To counter this claim, Helmholtz will draw upon research in the psychophysiology of perception.

3.1.2 The Argument from the Science of Perception: Euclidean Axioms Are not Necessities of Intuition

What Helmholtz hopes to gain by reciting current research in perceptual psychology is stated succinctly in the following claims:

> Space can be transcendental without the [Euclidean] axioms' being so (*FP*, 716).

> Space may very well be a form of intuition in the Kantian sense, and yet not necessarily involve the axioms (*OMG II*, 686).

Helmholtz will argue that while it may be true that certain spatial relationships are immediately presented to us, simply in virtue of our sensory apparatus, these spatial relationships and the minimal spatial form they determine are a far cry from the more particularized spatial form described by Euclid's axioms.

By 1870, Helmholtz put the finishing touches on his sign theory (*Zeichentheorie*) of perception. According to the theory, the initial effects on our sense organs should be conceived as uninterpreted signs since they fail to present us with any definitive properties or relations corresponding to a particular external stimulus:

> What physiological investigations now show is that the [difference in the modality of sensations] does not depend, in any manner whatsoever, upon the kind of the external impression whereby the sensation is excited, but is determined alone and exclusively by the sensory nerve upon which the impression impinges (*FP*, 693).

Helmholtz cites experimental results showing that the excitation of the optic nerve produces similar light effects, regardless of whether the nerve is excited by "objective light" (aether vibrations impinging on it) or an electric current being passed through the eye or by applying pressure on the eyeball (Ibid.). The same external stimulus will also produce radically different effects, depending on which sensory nerve is excited: "The same aether vibrations as felt by the eye as light, are felt by the skin as heat." (Ibid., 694).

Uninterpreted sensory qualia can, however, become sensory "symbols," capable of alerting us to the presence of a certain type of stimulus: "Inasmuch as the quality of our sensation gives us a report of what is peculiar to the external influence by which it is excited, it may count as a *symbol* for it" (Ibid., 695). How does this transition from uninterpreted sign to sensory symbol occur? Regularly occurring qualitative distinctions are registered on memory, leaving traces subsequently associated with one another and definitive spatiotemporal characteristics. Helmholtz describes this process as a series of subconscious inductive inferences operating on memory traces similar to the more familiar conscious inductions operating on content expressible in words or propositions:

> There appears to me in reality only a superficial difference between the inferences of logicians and those of inductive inferences whose results we recognize in the intuitions of the outer world we attain through sensations. The chief difference is that the former inferences are capable of expressions in words, while the latter are not, because instead of words they deal only with sensations and memory-images of sensations.[11]

Through a series of "unconscious inferences" operating on a "major premiss" composed of memory traces and a "minor premiss" consisting of "the new sense impression," none of which are "formulated in words," uninterpreted sensory qualia eventually become interpreted signs or "symbols" enabling us to track the "lawfulness of processes of the actual world" (*FP*, 695 and 703).[12]

To get a better picture of how these subconscious inductive inferences work, Helmholtz provides the following example. The perceiving subject experiences brightness, upon moving her eye the brightness fades, moving it back the brightness intensifies. He likens these eye movements to an "experiment through which" we arrive at law-like generalizations regarding the spatial and temporal characteristics of the qualia (*FP*, 706). In this case, the generalization arrived at is 'if I move my eyes to the right, I will experience brightness' and 'if I move my eyes to the left, I will experience darkness.'[13] Through this process of experimentation and unconscious inductive inferences, the alternating sensory signs, brightness/darkness, come to be interpreted as "an enduring existence of different things at the same time one beside another":

> One does not yet need to think of substantial things as what are here supposed to exist one beside another. 'To the right it is bright, to the left it is dark'…could for example be said at this stage of knowledge, with right and left being only names for certain eye movements…. (Ibid., 698)

Once these visual cues become similarly associated with certain tactile cues, etc., the subject infers the sensible image of a luminous object over there in front of me and to the right.

Helmholtz then applied his Zeichentheorie to decide what features of Kant's epistemology should be retained, modified, and rejected. First, he concluded that a roughly Kantian picture of the phenomenal world was correct and sought to displace a naïve form of materialism and empiricism fashionable in Germany since 1850 (Poma (1997), 3). Against the naïve empiricist, Zeichentheorie states that sensations present us with no information about their external stimulus until the

---

[11] Helmholtz as cited in *Hatfield* 1990, *201*.

[12] For a more thorough and hence more charitable account of Helmholtz' Zeichentheorie and a summary of the current debate over where Helmholtz sits relative to Kant and the various nineteenth century neo-Kantian camps, see *Hatfield* 1990, *Friedman* 1997, *DiSalle* 2006, *Lenoir* 2006 and *Patton* 2009 and *2012*. Since our goal is to view Helmholtz' epistemology through Frege's eyes, I am stressing those elements sure to raise his eyebrows.

[13] Remember, however, that these experiments and generalizations are supposed to be sub-linguistic, not expressible in language.

perceiving subject organizes those sensations within a stabilized spatio-temporal manifold. This manifold of sensations is thus the result of sensory input, the nature of the subject's sense organs, the nature of her mind and the series of subconscious inferences performed by that mind. To say that the manifold of sensations functions as a symbolic language of perception is to say that this lawlike ordering of perceptions can acquaint us with the "lawlike order in the realm of the actual," but not with a picture or image of "the real" or "Kant's thing-in-itself" (*FP*, 710). Helmholtz succinctly describes this more sophisticated, Kantian "empiricist theory" of sensible representations: "[O]nly the non-understood material of sensations originates from external influences, while all representations are formed from it in accordance with laws of thought" (Ibid., 705).

Helmholtz attaches at least two different meanings to the term 'transcendental': (1) subjective, dependent on our psychophysiology; and (2) necessary, either causally or as a presupposition of other inferences. The theory supports the idea that there is a spatial form that qualifies as transcendental in both senses of the word. This space, like color, is nothing other than how external stimuli present themselves to us as effects upon our sensory apparatus and should not be construed as real properties of an object (*FP*, 696). It is a necessary pre-condition for distinguishing the inner realm of consciousness from the world of outer appearances: "We comprehend as the world of inner intuition, as the world of self-consciousness, that in which no spatial relation is to be perceived" (Ibid., 697). It is causally necessary, since it is a requisite first stage of cognitive development in our ability to attribute spatial determinations to more complex external phenomena, such as the chair is beside the table. Helmholtz concludes that Kant was correct in thinking that a spatial form is given to us prior to our perception of empirical entities.

Second, the research is also said to confirm Kant's particular brand of skepticism. Helmholtz draws out this Kantian consequence by stressing that meaningful sensations are still only symbols, not images, of their external stimuli (Ibid., 695). Although sensory symbols are capable of tracking law-like regularities in sensory qualia and therewith the corresponding law-like regularities in the parade of external stimuli, they are incapable of expressing the "real" properties or relations of that stimuli (Ibid., 724). Or, as Helmholtz prefers to put it, sensory symbols do not represent "Kant's 'thing-in-itself…'"(Ibid., 710).[14]

---

[14] We will see that Helmholtz waffles on whether the findings of perceptual psychology force us to adopt this skeptical attitude. In proposing his test for empirically confirming the Euclidean axioms, Helmholtz assumes that we can obtain knowledge of the spatial properties of the external world. In his account of Helmholtz's sign theory, Friedman highlights Helmholtz's rejection of nativist theories of perception in favor of the "learned or acquired" process whereby perceptual signs allow for interpretation (Friedman (1997), 31). Helmholtz himself often claims that he is presenting an empiricist theory of perception, in contrast to the nativist theories of his predecessors or colleagues. I agree that Helmholtz clearly rejects nativist perceptual theories which treat the spatio-temporal ordering of perceptions as innate in the sense of being entirely due to the constitution of our sense organs or, as he puts it, arriving "ready-made," without a process of unconscious experimentation and inductions involving memory traces (Helmholtz, *FP*, 705). However, I also believe that it is a fair reading of Helmholtz (1876) and (1878b) to say that the unconscious processes

There are, however, features of Kant's epistemology that must now be modified or rejected. Helmholtz presents his notion of an interpreted perceptual symbol as the scientifically upgraded model of Kant's previous notion of an intuition, understood as the sensible representation of a particular given object:

> I believe the resolution of the concept intuition into the elementary processes of thought to be the most essential advance in the recent period. This resolution is still absent in Kant, which is something that then also conditions his conceptions of the axioms of geometry as transcendental propositions. Here it was especially the physiological investigations on sense perceptions which led us to the ultimate elementary processes of cognition. These processes had to remain still unformulable in words, and unknown and inaccessible to philosophy, as long as the latter investigated only cognitions finding their expression in language. (Ibid., 712)

According to Helmholtz, Kant mistakenly assumed that the initial effects registering on our sensible faculty are immediately invested with certain spatial and temporal characteristics, and furthermore that these spatial characteristics are Euclidean. As the theory of sensible signs has shown, however, the raw inputs of sensibility are not spatial at all. Although Helmholtz grants that these raw inputs cannot become sensory symbols without being invested with some spatial characteristics, he will go on to argue that Kant was wrong in thinking these minimal characteristics yield the spatial form described by the Euclidean axioms.

We have already seen that Helmholtz tends to equate the terms 'a priori', 'pure', and 'transcendental' with innate. As a result, he sets out to settle the question whether the Euclidean axioms are a priori propositions grounded on pure intuition by asking whether the axioms express "initial, original facts of perception" that a subject could become acquainted with independently of receiving any externally-given stimulus (*FP*, 697 and 706). It is consistent with the theory, says Helmholtz, to assume that we are innately equipped to distinguish the changes in sensory qualia that can be induced by "an impulse of our will" and the changes in qualia over which we have no control (*Ibid.*, 697). He then defines 'spatial relation' as a relationship that we alter in an immediate manner by the impulses of our will (*Ibid.*).[15] As previously mentioned, Helmholtz thinks of these alternating state changes as a

---

whereby raw sensations attain their most primitive spatiotemporal characteristics are presented as processes towards which we are innately predisposed. And more importantly for our purposes, we will see that this is how Cohen reads Helmholtz. Besides, it would certainly be a mistake to interpret Helmholtz as claiming that this primitive spatiotemporal form is learned or acquired from experiencing particular, sensibly perceived objects, since he maintains that there are unconscious inferences yielding "initial, original facts" of spatial ordering which are necessary for and prior to perceiving objects in the outer world (Helmholtz, *FP*, 706). The fact that Friedman himself is not ascribing a straightforwardly empiricist view of perception to Helmholtz is especially evident in Friedman (2000b), where he likens Helmholtz's account of unconscious inferences operating on the data received from voluntary bodily motion to Kant's account of the transcendental synthesis of the imagination.

[15] One of the consequences that Helmholtz draws from this definition is that we also have a priori knowledge, an innately pre-disposed idea, that space is our form of outer sense: "In this case, space will also appear to us—imbued with the qualities of movement—in a sensory manner, as that through which we move, through which we gaze forth" (*FP*, 697).

kind of experiment, which can lead us to intuit a particular spatial configuration, e.g., to the left and to the right. He also maintains that these experiments can be initiated voluntarily and independently of an actual encounter with an external object (*FP,* 698). Reflecting on everything that nature has endowed us with, Helmholtz concludes that a subject could become acquainted with the spatial relationship denoted by 'one beside another' prior to, and as a causal condition of, perceiving more complex phenomena, e.g., the chair is beside the table.

Helmholtz thus argues that Kant was not entirely wrong in identifying space as a transcendental form of intuition, just wrong in identifying this space with the space described by Euclidean geometry. Given that the perceiving subject can voluntarily initiate some experiments on sensory input and that these experiments can be conducted independently of and causally prior to any perception of objects in Euclidean 3-space, it is reasonable to identify the resulting space-relation, say 'one beside another', as a constituent part of a transcendentally given space:

> [S]pace would be a *given form* of intuition, possessed *prior to all experience*, to the extent that its perception were connected with the possibility of motor impulses of the will the mental and corporeal capacity for which had been given to us, by our make-up, before we could have spatial intuition (*FP*, 697).

This general form of spatial intuition, however, is a far cry from the more differentiated structure picked out by the Euclidean axioms: "Space may be transcendental without the [Euclidean] axioms' being so (*FP*, 716).[16] He thus rejects Kant's thesis that the Euclidean axioms are synthetic a priori propositions, since it cannot be supported by the reigning theory in perceptual psychology. Finally, by modifying Kant's system in light of new developments in mathematics and the natural sciences, Helmholtz sees himself avoiding metaphysical speculation and hence "more consistently Kantian" than Kant himself (Lenoir 2006, 141).

3.1.3 Helmholtz' Thought Experiments: Euclidean Axioms Are Contingent Truths

Before running through Helmholtz' thought experiments, it is important to see what he did and did not expect them to illustrate. They were not intended to cast doubt on the truth of axioms. In 1878, neither Helmholtz nor his colleagues seriously questioned whether Euclidean geometry was a true, or approximately true, description of physical space: "[T]he founders of 'Non-Euclidean Geometry' have never maintained its objective truth…"(*OMG II*, 685). I also maintain that the most charitable reading of these thought-experiments is to assume that Helmholtz is not expecting them to refute Kant's claim that the axioms are necessities of thought or intuition.

---

[16] See too *OMG II*, 686.

That work was supposed to be accomplished in the previous arguments.[17] Notice Helmholtz' introduction to the thought experiments:

> In discussing the question whether space-relations can be imagined in meta-mathematical [non-Euclidean] spaces, the first thing to settle is the rule by which we shall judge of the imaginability of an object that we have never actually seen. I advanced a definition which was to the effect—that for this we need the power of representing the sense-impressions which the object would excite in us according to known laws of our sense-organs under all conceivable conditions of observation, and by which it would be distinguished from other similar objects. (*OMG II*, 687)

Here he stipulates that the perceptual experiences that we can imagine having will be those experiences that conform to the "known laws of our sense-organs." He is thus already relying on the theory of perceptual signs, according to which it would be possible for us to intuit non-Euclidean spatial relationships.

The arguments from analytic geometry and the theory of sense perception were intended to show that the Euclidean axioms *could be* contingent, empirical truths. Now the goal is to show that *they are* such truths. For Helmholtz, if the axioms qualify as empirical truths, then they must susceptible to testable, observational verification. The point of the thought experiments is indicate what should count as empirically confirming or falsifying the axioms, thereby establishing the a posteriori nature of the Euclidean axioms:

> Let us examine the opposite assumption as to their origin being empirical, and see if they can be inferred from the facts of experience and so, established, or if, when tested by experience, must be rejected. If they are of empirical origin, we must be able to represent to ourselves a series of facts, indicating a different value for the measure of curvature from that of Euclid's flat space (*OMG I*, 677).

Helmholtz goes on to paint a picture of a world exhibiting positive constant curvature, explaining how its non-Euclidean properties could be detected by its inhabitants.

The properties of this non-Euclidean world are most easily visualized by restricting ourselves to two-dimensions and considering the surface of a sphere. Helmholtz asks us to imagine creatures, with perceptual and intellectual capacities similar to our own, inhabiting such a world and engaging in their own geometric investigations. Their straight lines are drawn or described by stretching a string until it is taut against the surface. The shortest line between two points in their world is determined as follows. Consider the great circle(s) on the sphere passing through the two points. These points will divide the circle into two arcs; the arc that is smaller in length is the shortest, straightest line between the two points. He then calls attention to the geometric facts that they would observe contradicting one or more Euclidean axioms. For example, it would not be true that, for any two points, there is only one straightest and shortest line between them. If the two points are the endpoints of a diameter of the sphere, there are an infinitely large number of great circles passing through them and the semi-circular arcs connecting the two points will all qualify as

---

[17] I mention this because some commentators suggest otherwise (e.g., Coffa (1991), 47–54).

the shortest, straightest lines between them. More importantly, in terms of determining the curvature of this space, "the sums of the angles of a triangle would always be greater than two right angles, increasing as the surface of the triangle grew greater" *(OMG I, 669)*.

The moral that Helmholtz expects the reader to draw from this story is that beings, "whose powers of reasoning are quite in conformity to our own," would be able to obtain measurements verifying whether or not the Euclidean axioms truly describe the space of their world *(OMG I, 16)*. Helmholtz then cites astronomical calculations applied to the triangles formed by the lights rays of a distant star observed 6 months apart (the base of the triangle is major axis of the earth's orbit) as confirmation that the curvature of our space is indistinguishable from zero. Since the truth of the axioms is determined not by the structure of our capacities, but by the contingent features of our world, the status of geometric truths is the same as any other scientific facts confirmed on the basis of observational evidence.[18]

However, as Helmholtz notes, the reliability of this observational evidence presupposes that we have accurate measuring instruments:

> [A]ll our geometrical measurements depend on our instruments being really, as we consider them, invariable in form, or at least on their undergoing no other than small changes we know of, as arising from variation of temperature, or from gravity acting differently in different places *(OMG I, 678)*.

He drives this point home with a second thought experiment, this time involving creatures inhabiting a world of negative constant curvature. Here we envision how our world appears in a convex mirror, e.g., the reflecting globe in a garden. Standing directly in front of and at a certain distance from globe, the reflected image is indiscernible from things as they appear in our own flat space. Imagine yourself measuring the angles of a triangle with a protractor and your reflected image doing the same, both coming up with 180°. Now you and your protractor back away from the globe, constructing and measuring triangles. Your reflected image would shrink and contort such that the angles of a triangle in this position would appear to be less than 180 degrees. The state of affairs reflected on the globe is thus indicative of the geometric properties of a space with constant negative curvature. But even though *you* can observe that the reflected triangle is no longer a rectilinear triangle as it appears in Euclidean space, according to Helmholtz, the man in the mirror will not:

> The image of a man measuring with a rule a straight line from the mirror would contract more and more the farther he went, but with his shrunken rule the man in the image would

---

[18] In Helmholtz' day, it was believed that the empirical measurements of stellar parallax triangles confirmed that our space was Euclidean. Today, due in part to technical advances in our measuring instruments, these same measurements are taken to confirm that our space is not Euclidean. Furthermore, as Jeff Barrett pointed out to me upon reading a draft of Chap. 3, this method of detecting the curvature of space presupposes that we must think of light as traveling in straight lines/geodesics and a Kantian might simply not be willing to grant this assumption. As we shall see, this is pretty much what Cohen says in response to Helmholtz' proposed method of confirmation. Apparently, some attenuated form of the Cohen v. Helmholtz debate regarding the status of geometric principles lingers to this day.

count out exactly the same number of centimeters as the real man. And, in general, all geometrical measurements of lines or angles made with regularly varying images of real instruments would yield exactly the same results as in the outer world…In short I do not see how
men in the mirror are to discover that their bodies are not rigid solids and their experiences
good examples of the correctness of Euclid's axioms (*OMG I*, 769).[19]

So, unless physical bodies really are invariable in form such that we can trust them
as reliable measuring devices and capable of constituting rigid geometric figures,
any geometric calculations we would make are not sufficient for distinguishing a
Euclidean world from one exhibiting constant negative curvature.

Helmholtz' argument for the empirical grounding of Euclid's axioms has hit a
snag. To confirm the truth of the axioms, we must obtain certain measurements, but
we cannot trust our measurements without presupposing another empirical-
geometric fact, namely that whatever we use as measuring rods and whatever we
measure retains its rigidity while moving through space. Helmholtz thus anticipates
and parries an objection from someone wanting to defend a more orthodox Kantian
view of the axioms. Suppose, like Kant, we insist that the Euclidean geometry is
necessarily true of space, any calculations indicating otherwise will simply be construed as evidence that bodies, e.g., measuring sticks and empirical triangles, shrink
and contort as the result of some unknown physical force. Helmholtz admits that
this move is available to his opponent, but that it purchases the necessity of the axioms at the cost of their cognitive significance:

> Taking the notion of rigidity thus as a mere ideal, a strict Kantian might certainly look upon
> the geometrical axioms as propositions given, a priori, by transcendental intuition, which
> no experience could either confirm or refute, because it must first be decided by them
> whether any natural bodies can be considered rigid. But then we should have to maintain
> that they are not synthetic propositions, as Kant held them; they would merely define what
> qualities and deportment a body must have to be recognized as rigid (*OMG I*, 682).

To retain their immunity from verification via empirical observation and physical
measurement, the strict Kantian could make the Euclidean axioms constitutive of
the meaning of 'rigid natural bodies'. But, says Helmholz, he would then have to
treat them as trivially true, analytic propositions, since they would be nothing more
than consequences derived from the mere definition of a certain concept.

Helmholtz concludes that the only means of demonstrating that Euclidean geometry has "real import," the capacity for representing relations adhering in "real
things", is to conjoin the axioms with principles of mechanics asserting the existence of rigid bodies, thus insuring that they "can be verified or overturned by
empirical observations…" (*OMG I*, 683). In effect, Helmholtz makes both the

---

[19] The idea is that a non-Euclidean world consisting of physical bodies that do not possess the
property of free mobility could be observationally equivalent, for its inhabitants, to a Euclidean
world where physical entities do possess this property. Similarly, a Euclidean world, with natural
bodies bending and contorting as they moved through space, could yield geometric calculations
convincing its inhabitants that they lived in a space with positive or negative constant curvature: "It
is possible to imagine conditions for bodies apparently solid such that the measurements in Euclid's
space become what they would in spherical or pseudospherical space" (*OMG I*, 678).

objective validity and the objective reality of Euclidean geometry dependent on the natural sciences. He allows that Euclidean geometry can be treated as its own isolated system of propositions, expressing pure geometric relations in a particular mathematical structure.[20] In this case, it is on a par with all other mathematically conceivable geometries, all of which share a certain kind of necessity and none of which can claim supremacy as a body of universal and necessary propositions about experience. Or, we can treat Euclidean geometry as a potentially applied science, a possible description of space-relations obtaining in our universe. In this case, geometric propositions are inseparable from other propositions in the natural sciences, all of which lay claim to cognitive significance insofar as they can receive empirical confirmation and none of which, therefore, are beyond refutation (*OMG I*, 683). Given the cogency of his arguments, Helmholtz has gone a long way in showing that, in spite of Kant's efforts, synthetic a priori propositions are not possible: "geometry is… the first and most perfect of the natural sciences" (*OMG II*, 686).

3.1.4 Summarizing Helmholtzian Neo-Kantianism

Recall that Helmholtz claims to have retained the most significant features of Kantian philosophy. In particular, he has preserved Kant's hostility to metaphysical reasoning. Indeed, Helmholtz thinks the theory of geometry presented in the *Critique* is an example of the sort of metaphysical reasoning that Kant generally dislikes. So, by denying geometric propositions the status of synthetic a priori truths, Helmholtz sees himself as rendering Kant's system more internally consistent (*OMG II*, 686).

There are at least two ways in which Kant's account of the synthetic a priori might appear to Helmholtz as a piece of metaphysical reasoning: (1) it relies upon a method characteristic of traditional metaphysics and (2) it implies that we possess the kind of knowledge that metaphysicians have thought possible. With respect to (1), Helmholtz interprets Kant's argument for the a priori and apodictic nature of the axioms as relying upon the method of "transcendental intuition," which he understands as an infallible, inner inspection of the contents of consciousness in order to uncover the original facts of perception (*FP*, 721). He also ascribes to Kant the view that a priori intuitive representations do not presuppose a synthesis whereby sensibility is affected. In other words, as Helmholtz reads him, Kant is appealing to an infallible, non-sensible intuition directed toward uncovering innate ideas as the method whereby the content and nature of geometric propositions can be discovered, and such a method is certainly characteristic of rationalistic metaphysics.

With respect to (2), one of the goals of metaphysical reasoning is to yield indubitable propositions about the nature of reality. On Kant's account, Euclidean

---

[20] By 'pure geometric relations in a mathematical structure', I mean those relationships as they are defined in analytic geometry, not as they are characterized under in the traditional synthetic treatment of geometry.

geometry is a collection of indubitable propositions about the spatial structure of the physical universe. For Helmholtz, on the other hand, in order for a proposition or system of propositions to have any objective content whatsoever, it must be susceptible to experimental testability (*FP*, 726). Experimental confirmation confers only probable truth, not absolute certainty. Therefore, if Euclid's axioms purport to be "laws of nature," they can only achieve "the approximate provability of all laws of nature through induction" (*FP*, 718). By positing a body of synthetic a priori propositions, Kant is giving way to wishful thinking about the possibility of a metaphysical science.

Having removed from the Kantian project what he takes to be the dross of metaphysics, Helmholtz relocates Kant's epistemological inquiry squarely within the natural and mathematical sciences. A question like what space-relations can be represented independently of experience is either a purely mathematical one, decided by the best available methods within that practice, or a question about what features of sensory experience are imposed by our faculties, decided by the best available research in the psychophysiology of perception. What is characteristic of Helmholtz' Neo-Kantianism is that the notion of transcendental psychology, an investigation into the cognitive functioning of a purported transcendental knowing subject, is reduced to empirical cognitive psychology, an investigation into the cognitive functioning of actual, physical human subjects. Helmholtz thus anticipates Quine's recommendation that epistemology fall into place "as a chapter of psychology and hence of natural science" (Quine 1969), 82).

Helmholtz's approach is clearly a relevant critique of Kant's epistemology.[21] For instance, Kant justifies the move from the objective validity of mathematically constructed concepts to their objective reality as instantiated forms in external appearances by appealing to a particular account of perception, namely that the possession of the schema for a mathematical concept is necessary for perceiving objects manifesting that form. This is a claim about empirical perception and thus subject to counter-evidence from investigations into the psychophysiology of perception. Similarly, Helmholtz is correct in thinking that recent mathematical progress in providing models for non-Euclidean geometries threatens Kant's theory of geometry. Helmholtz' critique takes seriously Kant's criterion that the objective reality of mathematical concepts and judgments depends upon their instantiation in experience. Nineteenth century mathematical research opened up the possibility of con-

---

[21] Richardson (2006) points out that Kant's *Critique* poses a genuine interpretive challenge: "Should we emphasize the generally psychological language of Kant's explanation of the transcendental conditions of knowledge…? Or should we, rather, stress the way in which the transcendental conditions of a priori knowledge are…formal conditions of objectivity revealed through attention to the formal conditions of judgment?" Following the first line of interpretation urges us to recognize the deep connection between psychology and epistemology, since both are concerned with the "subjectivity of conditions of objective knowledge." Following the second line, urges us to see the connection between epistemology and logic, since both are concerned with the conditions of objectivity "expressed in formal conditions of judgment." Richardson reads Cohen's *PIM* as his "most sustained and systematic attempt" at interpreting Kant in the second way (Ibid., 215). It should be obvious that I read Helmholtz interpreting Kant in the first way.

firming this presumed instantiation. Given that possibility, it is hard to see why the axioms of Euclidean geometry should be considered distinct from the axioms characterizing other freely-generated geometric structures and/or ordinary a posteriori claims about the world.

However, Helmholtz's Neo-Kantianism has its limitations, both as an adequate interpretation of the *Critique* and as an extension of the Kantian problematic. In attempting to distance Kantian epistemology from traditional rationalism, Helmholtz pushes it too close to traditional empiricism. Thus, he equates Kant's notion of a transcendental form of intuition with Locke's notion of secondary properties (*FP*, 693). The spatial form that is given to us a priori no more picks out the relations adhering in objects than does an ascription of color:

> Thus Kant considers spatial specifications too as belonging as little to the world of the actual—or to 'the thing in itself'—as the colours which we see as attributes of bodies themselves, but which are introduced by our eye into them (*FP*, 696).

Here we see that even though Helmholtz believes his investigation confirms Kant's thesis about the unknowability of 'things-in-themselves,' the skepticism that Helmholtz actually endorses is that which arises as a result of a Lockean-type of investigation into how sense perception hooks onto an alien, pre-Copernican Revolution, thing-in-itself world of external objects. Thus when Helmholtz explains how this link is established and the objectivity of sense perception presumably accounted for, his explanation is strikingly similar to that offered by Hume and Berkeley.

In addition, Helmholtz's claim that a sensible sign is simply a nineteenth century upgrade of Kant's previous notion of an intuitive representation rests on a serious misunderstanding of the latter. Helmholtz traces Kant's mistaken view of the Euclidean axioms to the influence of nativistic perceptual theories of his time, which assumed that "ready-made representations of objects are elicited through our organic mechanism" (*FP*, 705).[22] Helmholtz believed that Kant equated full-fledged intuitive representations with the raw givens of sensibility:

> Whereas, [Helmholtz] believed, for Kant spatial intuitions were unanalyzable into more primitive components and arose without the activity of the understanding, he had shown how they are in fact psychologically derivative, produced by processes that underlie all judgment and belief (Hatfield (1990), 203).

But Helmholtz is mistaken. For Kant, full-fledged intuitive representations are the result of a cognitive synthesis, a process whereby the raw inputs of sensibility are combined in accordance with elements and functions that must be contributed by the understanding.

Besides, even though Helmholtz' theory of perception describes a synthesis of sensations guided by "the elementary processes of cognition" that is responsible for generating representational sensory symbols, these symbols should not be considered identical to Kant's intuitions. What Helmholtz means by the elementary

---

[22] See also *FP*, 701 and 712.

processes of cognition are "unconscious processes of the association of ideas going on in the dark background of memory."[23] The result of these elementary processes, at least at the initial stage, is purportedly distinct qualia mapped to a spatial location internal to the perceiving subject. Thus, Kant would not include Helmholtz' sensory symbols in the class of cognitions, i.e., objective perceptions. For Kant, a representation counts as a cognition only if it refers to something inter-subjectively ascertainable.

Finally, because Helmholtz substitutes an empirical investigation into the conditions of human knowledge for Kant's transcendental investigation, the normative vocabulary in which Kant cast epistemological questions is lost. Concepts are no longer rules legislated by the understanding, no longer the standard by which the objectivity or inter-subjective validity of the sensory given is judged. Rather, concepts and judgments are formed in a manner analogous to his previous account of the formation of sensory symbols. The process of unconscious inference yielding perceptual representations is "an elementary process lying at the foundation of everything properly called thought…" (*FP*, 703). Both the formation of concepts and the inferring of judgments consist in accumulating particular sense experiences, the propensity of memory to eradicate incidental fluctuations and retain traces of repeated regularities, and the association of these memory traces according to psychological laws. Helmholtz considered the reduction of judgment and conceptualization to sense perception and then to the psychological association of ideas as "one of the chief advantages of his theory of unconscious inferences":

> The solvent performing these reductions was the notion that underlying the supposedly distinct operations that result in sensory perception on the one hand and judgments on the other is one kind of psychological operation—the association of ideas. Mental life is unified. Having something look some way to you is not a different kind of achievement, and does not rely on different psychological operations, than does subsuming an object under a concept or drawing an inference (Hatfield (1990), 204).

Helmholtz' Neo-Kantianism thus collapses Kant's all-important distinction between objective and subjective perceptions. The epistemic cache of 'Fido is a dog' becomes equivalent to 'It looks to me as if Fido is a dog'. It is the attempt to recover these distinctions and restore the normative language in which Kant described his inquiry that marks Cohen's Neo-Kantianism and motivates his critique of Helmholtz's interpretation.

---

[23] From Helmholtz' *Handbuch der physiologischen Optick* 3rd ed., vol. 3, p. 24, translated and cited in Hatfield (1990), 201.

3.2 Cohen's Contra-Helmholtz Kantianism

According to Cohen, Helmholtz' reading of Kantian philosophy ignores, misinterprets, and rejects its most essential features: the introduction of transcendental reasoning in epistemology; the post-Copernican meanings of 'a priori' and 'experience'; and the unique role that the synthetic a priori concepts and propositions play in accounting for our knowledge of the natural world. Like Helmholtz, Cohen presents his quintessential Kant as a slight deviation from the original. How slight remains to be seen. I begin with Cohen's argument that Helmholtz has misunderstood the task of Kantian epistemology and its most fundamental notions. I then present his defense of Kant's thesis concerning the special status of the Euclidean axioms. Finally, I review the extent to which Cohen's Marburg School modified Kant's notions and doctrines in the course of defending them against Helmholtz' critique.

3.2.1 The Transcendental Method, Kant's 'A Priori' and the Copernican Turn

The fact that Helmholtz interpreted Kant's thesis that Euclidean space is our pure form of intuition as the claim that Euclidean space is the spatial form of external perception for which we are innately pre-disposed is not surprising, given the prevailing hermeneutics applied to Kant's *Critique* at the time. Prior to Cohen's *Kant's Theory of Experience* (1871), it was common to see Kant's arguments that space and time are a priori subjective forms of intuition as his siding with the rationalists on the debate over innate ideas. This, in turn, encouraged the idea that Kant's theses could be bolstered or refuted by drawing from empirical psychology (Hatfield (1990), 111). With the publication of this book, Cohen began what would become a career long project of presenting a thoroughly de-psychologized version of Kantian epistemology.[24] The first step was to prove that "Kant had overcome the pre-critical disjunction: Innate or acquired?", and thus, the terms 'apriori' and 'innate' should not be treated as synonyms in an appropriate lexicon of Kantian terminology (*KTE*[1], 87).[25]

For Cohen, the truly Kantian epistemologist does not concern himself with a Lockean dissection of the "soul-apparatus" of knowledge or a description of the

---

[24] Because Cohen's book marked a turning point in Kant scholarship and rejuvenated Kant's distinctive argument that psychologism cannot account for the normativity ascribed to logical laws, Anderson (2005) refers to it as "the founding text of orthodox neo-Kantianism" (Ibid., 302).

[25] In first edition of *Kant's Theorie der Erfahrung*, Cohen is explicitly confronting J.F. Fries and especially J.F. Herbart. Helmholtz' *The Origin and Meaning of Geometrical Axioms* and *Facts of Perception* did not appear until 1878. However, given the nature of Cohen's criticisms, it is clear they are also applicable to Helmholtz' reading of Kant, and in 1885 edition Helmholtz is explicitly and similarly accused.

mode and manner of consciousness, but rather with cognition understood as a "fact" that is "objectively given" in and upon science (*PIM*, 4–5):

> [The transcendental method] does not investigate the principles of human reason, but it seeks the scientific validation of the foundation of the sciences…What makes them sciences? What is the source of the character of their universality and necessity? From which concept do we derive their worth as knowledge, valid within their area? What characteristics and methods of knowledge clarify those historical facts of knowledge? This is the question of the sciences, when they are compelled to consider their own principles. This, and nothing else, is the transcendental question (Cohen's 1881 Preface to Lange's *History of Materialism,* cited in Kluback (1987), 11).

This insistence on the proper explanandum of Kantian epistemology is distinctive of Cohen's Neo-Kantianism and essential to his critique of Helmholtz's naturalized Kantianism. According to Ernst Cassirer, Cohen located "the fundamental error" of "naturalistic" interpretations of Kant in a misunderstanding about what ought to count as the given for the epistemologist (Cassirer (2005), 97) The Helmholtzian epistemologist begins with a given domain of objects with specifiable causal properties, a domain of "existing actualities and their causal interactions" (Ibid.).[26] The epistemologist's aim is then seen as articulating the causal relationship between these existing actualities and the "representations and processes in the thinking individual (Ibid., 96)." The objectivity of these representations and processes thus depends on telling a causal story tracking sensibly-given actualities to the actualities of a subject's cognitive representations. Within this class of cognitive representations is a "particular class of psychophysiological 'actualities" identified as "apriori truths" (Ibid.).[27]

By contrast, Cohen maintained that the consequence of Kant's Copernican revolution was the introduction of a "new concept of experience," experience "as it occurs in unbroken progress in genuine science:"[28]

> Experience itself becomes concept, which we must build up in pure intuition and thought…We build up a concept of experience as a synthetic unity of experiences, according to transcendental principles (*KTE*[1], 104).

The given towards which the epistemologist must continually orient himself is the objective content expressed in scientific propositions and theories:

> [I]f I take knowledge not as a type or method of consciousness, but as a *fact*, which came about in *science* and which continues to take place *from a given grounding,* then the investigation no longer refers to a subjective fact, but to a state of affairs given objectively and

---

[26] Cassirer's use of the term 'actualities' is significant since Frege also insists on a sharp distinction between actuality [*Wirklichkeit*] and objectivity [*Objektivität*].

[27] *Cassirer* 2005, 96. Given Cassirer's and Cohen's account of the Helmholtzian epistemologist it is easy to why Frege believed that "physiological psychology" provided the "most striking case" of a supposed logical inquiry beginning with "a realistic point of departure" and leading to "an extreme form" of idealistic epistemology (Frege 1997, 244–5).

[28] This description of the Marburg school's reconstruction of Kant's Copernican turn is taken from Paul Natorp, as cited in Kohnke (1987), 181.

founded on principles, not on the process and apparatus of cognition, but on the result of these, science (Cohen (1883), cited in Patton (2004), 90).

So just as Cohen argued that Kant introduced a new concept of experience, so too he argued for a more "fundamental" analysis of "the concept of object itself" and the view that a sufficiently "certain" and "objective" reference to the actual entities studied in the natural sciences can only be established by means of the "language of mathematical physics" (Cassirer (2005), 97). Thus, the paradigmatic objective representation of a single object is no longer the empirical perception of sensed particulars, but particular entities referenced in a proper scientific discourse. In this way, Cohen's Kantianism opens the door for introducing abstract entities as singular objects and allowing the syntax of a suitably policed scientific discourse to decide when a cognitively significant reference to such objects is established.

Recognizing that Kant is operating with a new concept of experience also explains, according Cohen, why the post-Copernican notions of 'a priori' and 'pure' must be seen as having nothing to do with the debate over innate ideas. Experience is a synthetic unity built up according to "transcendental principles," and so what we should "call a priori" are the "constructive elements" required for generating the subject matter of a science, specifically the subject matter of mathematical physics (*KTE*[1], 104). These a priori constructive elements guarantee the objective validity of the 'facts' of natural science by establishing that these facts are or are derivable from "secure and exact *judgments*" (Cassirer 2005, 97).

Given Cohen's contra-Helmholtzian Kantianism, the epistemologist's task is articulating the "ideal" structure of inferentially related propositional truths whereby it is clear how the "principles and axioms of mathematics" confer objective "content and sense to any statement of the natural science about actuality" (Ibid.). This structure includes the propositions of the natural sciences and the a priori mathematical and logical propositions from which the surety and objectivity of the former is lawfully derived (Ibid., 97–8).[29] The objectivity of propositions and the entities referenced therein is secured by displaying their role and place within this structure. In *PIM*, Cohen drives this point home by substituting 'Erkenntniskritik' [critique of cognition] for 'Erkenntnislehre' [theory of cognition], with the explicit purpose of distinguishing the objective from "psychological representations," a distinction Kant himself still struggled with (*PIM*, 6). This title change is intended to remind us of Kant's "original discovery" that the "critique of reason" just is "the critique of

---

[29] According to Cassirer, Cohen recognized that this structure would undergo constant revision as science progressed. He also intended for this structure to include ethical and aesthetic propositions. The 'possibility of objective experience' and the 'unity of consciousness' refers to the entire network of propositions and relations of validity obtaining between logic, mathematics, the natural sciences, ethics and aesthetics. The given of epistemology is most accurately described as "the givenness of the *problem,*" the ongoing problem of articulating an idealized structure of interrelated truths as the bodies of scientific, ethical and aesthetic knowledge expand (Cassirer 2005, 100). In Chap. 5, we will examine Cohen's *Ethics of Maimonides* to identify the methodological and hermeneutical principles that he proposed for faithfully maintaining while continually revising this structure of interrelated truths.

cognition or science" and that this critique discovers the "pure in reason" by discovering the "conditions of certainty" upon which "cognition as science" is founded (Ibid., 6).

3.2.2   Helmholtz Is Wrong: The Euclidean Axioms Are Synthetic A Priori

Given Cohen's de-psychologized reading of Kant's *Critique,* the thesis that Euclidean geometry is a body of synthetic a priori propositions is rendered entirely immune to any new developments in the area of perceptual psychology. In the Transcendental Aesthetic, Kant poses the question: "What our representation of space be for such a cognition [the synthetic a priori cognition in Euclidean geometry] of it to be possible?" (A25/B41) And he answers: "Space is nothing other than merely the form of all appearances of outer sense, i.e., the subjective condition of sensibility…" (A26/B42). For Cohen, this answer is misleading because it renders Kant's thesis susceptible to Helmholtz' attack. Neither the apriority of Euclidean space itself nor our justification for treating it as the space in which objects of experience must be located can be explained by identifying space with our form of sensibility: "In its final analysis, therefore, the apriority of space does not depend on the formal nature of our sensuality…" (*KTE¹*, 93).

Instead, Euclidean space is an a priori or pure form of space because everything that can be known about this spatial structure is obtained from a set of fundamental principles laid down by the practitioners of geometry, without having to import any empirically verified, contingent matter of fact:

> Here the form is a priori because it is definable a priori in accordance with general principles (*KTE¹*, 97)

> These general principles, however, are the axioms of geometry. The common bond of the axioms makes up the meaning and content of the spatial form (*KTE²*, 238).

According to Cohen and Natorp, at the core of each science is a collection of "basic concepts and propositions" expressing the "primary objective content assumptions" for that discipline (Natorp (1887), 252). One of Kant's most significant insights was recognizing that the generality, necessity, and cognitive content ascribed to scientific concepts and propositions can only be secured when practitioners of a science construct their respective domains of "*objective unities*" in accordance with a priori principles (Natorp, 263).

When it comes to geometry, these "objective content assumptions" are expressed in the Euclidean axioms. The axioms express the objective and cognitively significant meaning of the basic concepts—point, line and straight line--from which all other properly geometric notions are derived. We have seen that within the context of traditional synthetic Euclidean geometry, the axioms include postulates asserting the possibility of constructing the basic elements, e.g., an infinitely extended straight line, from which all other Euclidean constructions are obtained. So, given this tradition, it is accurate to say that the axioms contain "the meaning and content" from which the entire spatial form of Euclidean geometry is built up.[30]

Cohen and Natorp also insisted that the basic propositions lying at the foundation of a discipline were not themselves subject to any additional, external justification:

> We become certain of the truth within the proper internal network of the science, developed from primary objective content assumptions as they are formulated in the basic concepts and propositions of that science....The mathematician or physicist who truly grasps the nature of his science will find it superfluous to seek the grounds for the laws of truth for his knowledge in psychology. He will in principle deny such a search; he recognizes only the laws of his own science, not an alien science, as the judge of truth (Natorp, 252).

The fundamental concepts and propositions at the core of each science serve as the final arbiters of truth and fix the objective meaning of the primitive terms it employs. One cannot ask for further justification of these basic constitutive elements without running the risk of slipping into psychologism, substituting "the laws of psychic life" for "laws of truth" and trading away the objective subject-matter of the sciences for the subjective content of consciousness (Ibid., 249). Deciding which propositions express the basic objective content assumptions of a science is a task internal to that science: "Which axioms the geometry has to accept and how to formulate the same is a matter for geometry" (KTE^2, 228). Helmholtz' attempt to base the objective meaning and truth of Euclidean geometry on either empirical observation or perceptual psychology thus violates the autonomy and sphere of governance accorded to geometry as a science.

So far, however, nothing that has been said warrants the claim that the spatial form derived from Euclid's axioms and postulates is the unique pure spatial form. By 1870, it was well-known that non-Euclidean geometries had been developed and apparently engendered no logical contradictions. It was also known that non-Euclidean geometries could be modeled in Euclidean 3-space. What right, then, did Cohen have for conferring special status on Euclidean geometry as the geometry containing propositions that are necessarily and universally valid concerning the domain of spatial entities? For any axiomatized geometric system would have its own corresponding sphere of objective content as circumscribed relative to its basic concepts and propositions. Since the meaning and validity of geometric notions is, on the Marburg account, determined solely in reference to a particular set of axioms, what sense does it make to ask which geometry, Euclidean or non-Euclidean, qualifies as the true description of anything spatial? Euclidean geometry will be universally and necessarily true of its pure spatial form, and non-Euclidean geometries will be universally and necessarily true of theirs as well.

It is at this stage in his defense of Kant's thesis that Cohen draws on the crucial fact that Newtonian physics was the governing paradigm of the natural sciences. Remember, experience in Cohen's post-Copernican sense is a "synthetic unity of

---

[30] For more on the nineteenth-century debate over the use of analytic versus synthetic proof techniques in geometry, see Tappenden (1995) and Wilson (1992).

experiences" constructed according to a priori transcendental principles (*KTE¹*, 104). For a pure spatial form to be designated by the expression 'pure intuition,' it is not enough to show that this form is definable in accordance with a set of general principles stipulated by the practitioners of geometry. It must also be shown that this spatial form and its constituting elements are required for constructing experience as it presents itself in the continual progress of science since the time of Galileo and Newton:

> Since Newton, there is a science built upon principles, which is conscious of its foundations and presuppositions and produces itself continually according to the mathematical method. It was first then that the object was given about which the transcendental question of the possibility of a priori cognition could prove to be legitimate (*PIM*, 7)

> Newton made geometry a part of mechanics, not in the raw empirical sense but in the transcendental sense. The concept of pure intuition designates this inner condition as a condition of pure experience (*KTE²*, 230).

Galileo and Newton were the first to transcribe "the rhapsody of perception" into the law-like unity of inter-subjectively valid representations of objects captured in the propositions of natural science (*KTE¹*, 101). This transcription was made possible by rejecting the representation of entities given to us in "direct sensual appearance" in favor of a representation of natural phenomena conforming to mathematical and logical forms (Natorp, 255). Because the natural phenomena of Newtonian physics are represented in conformity to the axioms, concepts and constructive method of traditional Euclidean geometry, Euclidean space qualifies as the spatial form designated by 'pure intuition.' To say that Euclidean space is the form of pure intuition is to say that "the form of [Euclidean] geometry is essential for the natural sciences" (*KTE²*, 238).

It is irrelevant whether or not non-Euclidean geometries are logically consistent systems or even if perceptual psychology indicates that non-Euclidean spaces can be visualized:

> He [Helmholtz] says: 'If spaces of a different kind are imaginable then one could disprove that the axioms of geometry are necessary consequences of an *a priori* given transcendental form of our intuitions in Kant's sense.'… If those spaces are now imaginable and even intuitable, then a lot is still missing until they can be considered products of pure intuition as well; for that, their scientific fertility has to be proven or, if not yet ripened, they have to be anticipated and designated as leading prospects (*KTE²*, 231).

To challenge the claim that Euclidean geometry is the body of universal and necessary truths about anything scientifically recognized as having spatial characteristics, Helmholtz and others must show "that these [non-Euclidean] spaces are necessary and fruitful to deepen" our understanding of experience as it occurs in the sciences as a whole (Ibid., 230).

To appreciate the full force of Cohen's response to Helmholtz, we need to remember that it was not until 1919 that Einstein's theory of relativity was empirically confirmed and hence that non-Euclidean geometry demonstrated its fruitfulness for physics. Subsequently, Ernst Cassirer (1921) published a supplement to his

Substance and Function acknowledging that this revolution in mathematical physics rendered the Kantian thesis untenable:

> The *factum* of geometry has lost its unambiguous definiteness; instead of the one geometry of Euclid, we find ourselves facing a plurality of equally justified geometrical systems, which all claim the same intellectual necessity, and which, as the example of the general theory of relativity seems to show, can rival the system of classical geometry in their applications, in their fruitfulness for physics. (Ibid., 353).

So until 1919, both Cohen's and Frege's insistence on the unique status of Euclidean geometry given its indispensable use in natural science was warranted. Moreover, as late as the 1960s, Kuhn and Feyerabend maintained the propositions of Newtonian mechanics could not be translated into the language of Einsteinian physics such that the latter could be reasonably conceived as the genuine progressive successor of the former. Besides, as van Fraassen (2002) and Friedman (2008) explain, even if such a translation is possible, the reasonability it affords is only retrospective. Therefore, it is fair to say that Cohen's response to Helmholtz is tantamount to the claim that the unique epistemic status of classical Euclidean geometry cannot be denied unless and until it is prospectively reasonable to reconceive the Euclidean axioms as part of a more encompassing theory capable of delivering additional insight into the geometric properties and relations presupposed or instantiated in the subject matter of the natural sciences.

In addition to rebutting Helmholtz' arguments from analytic geometry and perceptual psychology, Cohen challenges the proposal to empirically confirm the Euclidean axioms. According to Cohen, the mere fact that Helmholtz would propose such a test demonstrates that he is still operating with the old, pre-Copernican notions of experience and nature:

> He [Helmholtz] does not proceed from the inner connection of mathematics and natural science, in order to determine this as a unit of experience and the conditions of the latter. For him, too, nature exists in and for itself…Furthermore, experience is accepted in the un-Kant-like sense in which it is not inherently mathematical, but since considered a source of consciousness, it could at least apply to mathematics. (*KTE²*, 227)

The mistake incorporated in Helmholtz' proposal is treating nature as existing "in and for itself," something that exists independently of our scientific methods and can be experienced through mathematically indeterminate sense impressions. To obtain observational evidence confirming the Euclidean axioms, Helmholtz must assume that our experience of the external world is mathematically neutral and thus capable of adjudicating between various geometric systems.

As we saw, Helmholtz tends to conflate the world of external objects with Kant's thing-in-itself. He thus likens space as a pure form of intuition to one of Locke's secondary properties, which ultimately encourages him to adopt a skeptical position reminiscent of David Hume. Cohen now points out this error in Helmholtz' understanding of Kant:

> Helmholtz says that Kant considers spatial designations 'just as little pertaining to the world of realism, or to the thing-in-itself, as the colors we see, belong to substances.' Here almost every word requires a question mark. (*KTE²*, 233)

On Cohen's reading of Kant, the world of the objectively given is "life as it is represented in scientific reality" (Ibid., 235). The spatial designations dictated by Euclidean geometry, therefore, *do* represent real properties and relations of the objects to which we refer subjective sensory qualities: "[Euclidean] space...constructs the external objects from which experiential impressions proceed" (Cohen (1871) in Kohnke (1987), 180).

Given Cohen's understanding of what it means to follow Kant in making the Copernican turn, it is hard to see how Helmholtz' criteria for establishing the contingency of the Euclidean axioms could ever be met. Helmholtz bases his argument for the contingency of the axioms on the possibility of measuring the angles of empirical triangles, e.g. star parallax triangles. Such a procedure counts as an unbiased test of the axioms only if we can identify empirical instances of triangles without presupposing one of the competing geometric systems. For the Marburg Kantians, this is simply at odds with how the natural sciences must function if they are to produce objectively-valid judgments. The mere accumulation of empirical observations cannot ground the generality, necessity, and normativity implicit in objective judgments, even when those judgments concern an empirical matter of fact (cf. Cassirer (1910), 242, 246–7).[31] Rather, cognitive access to empirical entities must be mediated by a network of a priori mathematical and logical principles and concepts:

Mathematics brings forth the foundations, by means of which physics is capable of investigating the nature of existence (Cohen (1881) in Kluback (1987), 15).[32]

The objectification of appearance is carried out in the reduction to law; there is no other way. In this, the autonomy which science claims is guaranteed. When the appearance is reduced to law then the appearance is reduced to the object appearing in it...The given appearance is explained by its connection with other known laws. On the other hand, that

---

[31] Although I sometimes reference works post-dating Cohen's debate with Helmholtz, whatever I ascribe to Marburg Neo-Kantianism is already present in Cohen's earlier works and constitutes his refutation of Helmholtz' argument for the contingency of the axioms. My choice in citation is determined by the clarity and succinctness with which its point is made. By 1877, Cohen had already identified the objective experience of the stars, planetary motions, etc., with "the facts of scientific reality" represented in an astronomical theory, rather than "the star-things" presented to us in ordinary sense perception. As Kim (2004) explains: "[Cohen's] point is that the scientific or epistemic value of, say, astronomy, is not to be found in what is given and observable by the senses, but rather in the mathematical exactness of its equations. These alone constitute and underwrite the truth-value of astronomy's propositions....[T]he essential characteristic of science lies in its objectivity, and that objectivity is rooted in its lawfulness" (Ibid., 3). The Marburg School insists on distinguishing three questions: (1) how is the objective experience of astronomy possible? (a philosophical/epistemological question); (2) what laws account for our ability to observe stars, etc.? (a question in psychophysiology); and (3) what laws govern the objects, their properties and relations, treated in astronomy? (an question internal to astronomy) (Ibid.). Cohen sees Helmholtz' Kantianism conflating these questions.

[32] This quote is from Cohen's Introduction to the second edition to Lange's *History of Materialism*. Besides this introduction and Cohen's *PIM*, historically-informed Frege scholars claim that Frege would have been very familiar with Cohen's *Kant's Theorie der Erfahrung* (1871).

representation which accepts the object according to direct sensual appearance frustrates the reduction to law and thus cannot have objective validity (Natorp (1887), 225).

Operating within these methodological constraints, Helmholtz' proposed confirmation can not get off the ground, for there is no epistemic warrant for picking out empirical straight lines or triangles without invoking a particular geometric system from which those notions derive their objective meaning.

Cohen thus denies that the truth of the axioms can be verified in the manner that Helmholtz needs for his contingency argument to work:

> The axioms are not 'mere forms of representation' that may be confronted with the 'conditions of the real world.' Instead, the axioms validate the forms of representation and it is this representation that constitutes the conditions of the real world. (*KTE*[2], 234–5, the quotation marks refer to Helmholtz' *Facts in Perception*).

There can be no testing of pure geometric hypotheses against the spatial structure of the real world as envisioned by Helmholtz, because the spatial form described by Euclid's axioms is already part of that world as one of its a priori, constituting, objectifying elements.

Before moving beyond Cohen's specific objections to Helmholtz's view on the nature of the Euclidean axioms, we should note the difference between what Cohen and Helmholtz mean by 'pure intuition.' For Helmholtz, a full-fledged intuition is a completely interpreted sensory sign, resulting in an image of a particular sensed entity with spatial and temporal characteristics, e.g. a book presented now and to the left of me. A *pure* intuition of a spatial configuration would be the perception of a spatial relationship that a subject could obtain simply in virtue of the various capacities and dispositions she was born with, e.g. one beside another. For Cohen, the notion of pure intuition has nothing to do with the cognitive functioning capabilities of a knowing or sensibly perceiving subject:

> Intuition like thought are abbreviations for scientific methods, that is, methods that are so autonomous from the particular content of research to which they relate themselves that they form, much more, the general pre-conditions of all scientific research (*PIM*, 3)

'Pure intuition' in Cohen's sense refers to a method, i.e., the method of generating and cognizing the spatial structure(s) studied within geometry and, more importantly, the spatial structure presupposed by the natural sciences. This method includes the basic concepts, axioms, and procedure of traditional Euclidean geometry. On Cohen's de-psycholgized reading of Kantian epistemology, the notion of space as a pure intuition and as a source of a priori knowledge is completely severed from the notion of a sensible faculty or the form of that faculty. The a priori source of knowledge identified as pure intuition consists of the axioms, basic spatial configurations and method of traditional Euclidean geometry.

We also saw that the Marburg School resists the idea that the Euclidean axioms stand in need of any external justification. Our recognition of the truth of the axioms does not derive from the purported fact that they describe an innate form of sensible intuition nor that they that they describe the spatial properties and relations of the external world understood in a pre-Kantian, pre-Copernican revolution sense.

Rather, to the extent that the truth or validity of the axioms stands in any need of justification, this need can only be met indirectly. In order to account for body of knowledge represented in the practice of classical Euclidean geometry and the body of knowledge represented in the practice of the natural sciences since the time of Galileo and Newton, nineteenth century scientists must affirm the Euclidean axioms as universally and necessarily true concerning everything spatial. For a Marburg Neo-Kantian, to say that the Euclidean axioms express basic facts of pure intuition is not to say that they *describe* the spatial relationships of either a subject's form of external perception or things-in-themselves. Rather, it is to say that they *prescribe* the space-relations comprising the objective, cognitive content and traditional sense of the terms 'point,', 'line,' and 'straight line' of a properly functioning scientific discourse.

This is not to say that Cohen would denounce any research involving non-Euclidean geometries. When criticizing Helmholtz, Cohen demands that developing new geometric structures should, at the very least, expand and deepen our understanding of geometry (KTE^2, 231). He also suggests that the a priori elements comprising the foundation of geometry will be modified as science progresses; the "foundation that forms the treasure house for the content of knowledge" can be "eternally increased" (Cohen (1876), cited in Kluback (1987), 13). Hermann Hankel (1867) claimed that the synthetic treatment of projective geometry contributed to our knowledge of Euclidean relationships. It would seem, then, that Cohen could endorse projective geometry as a legitimate trajectory for geometric research, so long as a tight connection with traditional Euclidean geometry is maintained. Cohen's main concern is that mathematicians adopt methods preserving the continuity of the subject matter of geometry and add to our knowledge of that subject matter: "all new foundations ought to be presented as deepening the old ones" (cited in Kluback (1987), 13). However, negating Euclid's parallel postulate, emptying geometric terms of their traditional meanings, and investigating any logically consistent system of geometric axioms threatens the status of geometry as a science. From our vantage point, one can see Cohen urging mathematicians to engage in practices that could have forestalled Kuhnian concerns that the actual practice of scientists undermines the very notion of science as cumulative process of gaining knowledge about a well-circumscribed domain of entities.[33]

---

[33] Throughout my exposition of Cohen's critique of Helmholtz's position on the Euclidean axioms, I have been guided by the thought as to how their debate would have looked to Fregean eyes. In consequence, Helmholtz' empiricism and psychologism have been brought to the fore. For a less Fregean perspective of the Helmholtz-Cohen debate, see Coffa (1991), 54–61 and Hatfield (1990), 218–226.

3.2.3 Summarizing Cohen's Kant: What Is Retained, Tossed and Modified?

Not surprisingly, the most drastic modifications to Kantian epistemology come from trying to expunge it of all things psychological. Talk of faculties, whose distinct modes of representation must be processed and combined by the knowing subject so as to produce objective representations, gives way to talk about sources of knowledge [*Quellen des Erkennens*]. It has been said of Cohen's transcendental idealism that it is idealism without the subject, and this is accurate in the sense that Kant's references to the activities and faculties of the knowing subject are downplayed to the point of extinction.[34] In 1871, Cohen begins chipping away at the idea of sensibility as a subjective capacity through which we are affected by objects, preferring to characterize it as a ground of pure knowledge:

> Sensibility—a real source of knowledge. Sensibility—the ground of apriority. (*KTE¹*, 17)

> One can either say that our sensibility is merely subjective, i.e., that it counts as nothing objective. Or, one can honor it as the first source of knowledge and thus acknowledge its products as objective products of knowledge (Ibid., 59).

With the publication of his *Principle of the Infinitisimal Method* (1883), Cohen replaces the terms 'sensibility' and 'understanding' altogether with 'intuition' and 'thought,' both of which are explicitly identified as abbreviations for scientific methods that "constitute [*bilden*] the general pre-conditions of all scientific research" (*PIM*, 3). Intuition is an abbreviation for space qua method of Euclidean geometry and time qua method of arithmetic. Thought is an abbreviation for the method(s) by which the content of the mathematical and physical sciences are determined according to logical forms or pure categories.

Because Frege also speaks of pure thought and pure intuition as sources of knowledge and because his appeal to three sources of knowledge—sense perception, the logical source, and the geometric source—has been cited as evidence of his commitment to Kantian epistemology,[35] it is especially important to review precisely what Cohen intends by referring to pure intuition and pure thought as sources of knowledge or methods. Poma (1997) explains what Cohen intends to rule out by identifying intuition as a method:

> Intuition does not signal a vague faculty of knowledge, nor an unjustified given; rather, it signals the act, that is, the constructive method by which mathematics reaches its knowledge: 'A priori space is not physical space, neither is it geometrical space, in the exact sense, but merely the process of production and formation of the latter. This is the meaning of space as pure intuition' (Poma (1997), 50–1)

---

[34] Natorp (1887) makes this explicit in his investigation as to the objective grounds of knowledge: "Each appeal to the subject of knowledge and the way in which consciousness participates in knowledge must ...appear to us from the start as a category mistake" (249).

[35] See Kitcher (1979), 252–256.

Neither 'pure intuition' nor 'pure thought' refer to things existing within the knowing subject or independently of that subject, which must be called upon to validate the basic laws or elements of geometry or logic.

For Cohen, pure intuition is "the constructive method by which mathematics reaches its knowledge," the method of "production and formation" of the geometric space that constitutes the subject matter of geometry and also serves as the spatial form presupposed throughout the natural sciences (Poma (1997), 51). In *PIM,* Cohen notes the ambiguities in Kant's own use of the term 'intuition' and redefines it so as to more sharply fix its meaning as characterizing a "general mathematical lawfulness" (*PIM,* 19). Thus disambiguated, Cohen introduces 'intuition' as a technical term within the science of Erkenntniskritik that designates the mathematically lawful methods of generating both spatially extended and successively given magnitudes (*PIM,* 18–21). The a priori source of knowledge designated by 'pure intuition' therefore includes the axioms, basic spatial configurations and synthetic method of Euclidean geometry. Similarly, 'pure thought' no longer signals the faculty of the understanding, but a network of a priori logical elements and laws, through which "the originary relationality of knowledge and its object are grounded…where [epistemologica] reflection encounters an ultimate ground" (Manfred Brelage (1965) cited in Poma (1997), 63).

As Brelage explains, by identifying a network of principles as the ultimate ground of knowledge, Cohen sought to retain the Kantian theme that a knowledge claim carries with it the connotation of legislative authority concerning its subject matter, that a knowledge claim expresses what is objectively valid and intersubjectively binding concerning the objects falling within its scope. Furthermore, by insisting on the autonomous and spontaneous nature of knowledge, Cohen's transcendental idealism blocks any move to reduce the process of cognition to a psychological process. This also explains why Cohen treats 'pure thought' as a method or a source of fundamental logico-epistemological principles:

> It highlighted, on the one hand, the independence of this pure thought from any presupposition of a being in itself, on the other that of pure thought from concrete subjectivity. The assurance of the one was realized with its defense of the independence of transcendental logic from metaphysics, and that of the other with the defense of its independence from psychology (cited in Poma (1997), 64)

Natorp thus follows his teacher Cohen when he insists that a methodologically sound psychological inquiry must recognize its own dependency on the necessary conditions for objective experience and rationality:

> [T]he psychological inquiry is, Natorp argues, necessarily posterior to the philosophical one, since there is 'no immediate access to the immediacy of psychic experience; it can only be reached by a [methodical] regression from its objectivations [object-formations], which must therefore [first] be secured in their own purely objective justification (Kim 2004).

Discovering the conditions supplying the necessary objective validity of a properly functioning scientific inquiry into our psychological experience requires pursuing the a priori philosophical method that Cohen advanced, which as Cassirer describes, involves laying bare the ideal logico-mathematical inferential structure that secures the objective content and validity of empirical propositions.

Cohen's substitution of sources of knowledge for the faculties of sensibility and the understanding induces further modifications. Pure sources of knowledge, in contradistinction to Kant's sensibility and the understanding, possess within themselves adequate resources for producing objective ideas and judgments. The methods of intuition and of thought each deal with their own "particular content," content general enough to account for their respective spheres of applicability and validity (*PIM*, 3).[36] Cohen thus dismisses the idea that logic is a purely "formal science" lacking in cognitive content, rather its task is "to secure" the "entire [objective] content of knowledge" by the "method of thought" (Ibid.). Pure thought "in itself and only from itself" is capable of yielding pure knowledge (cf. Poma (1997), 79). Not only is the method of pure thought, i.e. logic, capable of producing its own objective content, but it also lays the foundation upon which the more particular sciences, including geometry, build (Natorp (1887), 251). Cohen's Neo-Kantianism, therefore, enlarges the domain of full-fledged knowable entities to include non-sensible, pure geometric and logical objects.

Notice, however, that Cohen's insistence on two distinct and fairly autonomous a priori sources of knowledge constitutes a significant departure from Kant himself. For Kant, a network of pure logical laws and concepts cannot be a self-contained source of knowledge from which cognitively significant a priori propositions could be derived. Recall his description of knowledge as a synthesis in which pure forms of thought are combined with the manifold of sensible intuition. Only schematized logical forms and conceptualized sensible manifolds could qualify as sources of pure knowledge. For Cohen and his followers, this dependence of thought upon sensibility for obtaining its content is symptomatic of the psychological, subjectivist tendencies within Kant's system that needed to go. Reinterpreting Kantian epistemology, sans sensibility, was part of an effort to keep logic distinct from psychology and to recognize the former as a science in its own right, with its own distinct, objective subject matter, and as the source of pure knowledge with the widest possible range of legislative authority and applicability.[37]

Despite this tinkering to Kant's system, Cohen's reading of Kant goes a long way in recapturing his claims about the alleged fallacious practices characteristic of the pre-critical empiricist and rationalist traditions. By insisting that autonomously legislated, foundational principles and their constituent concepts first fix the objective content of thinking and perceiving within the sciences, Cohen's Kantianism easily recovers and reinforces Kant's distinction between subjective and objective representations. Like Kant, Cohen and company maintain that concepts are rules

---

[36] In *PIM*, vacillates on whether pure intuition and pure thought denote self-contained sources of knowledge yielding their own objective products or whether they must be combine their resources. In his review, Frege cites the latter, but would affirm the former in the interest of maintaining a strict delineation of number theory/logic and geometry (see "Review of H. Cohen" and "On Formal Theories of Arithmetic" in *CP*, 108–144, and "Sources of Knowledge" in *PW,* 267–274).

[37] See *PIM,* §§7–8, 21–25, Natorp (1887), 250–25, Friedman (2000a), 28–33 and Richardson (2006).

stipulating a list of must-have properties and relations that make it possible to render, normatively necessary, truth-valued claims about the particular givens of sensory experience. In other words, they share Kant's notion of a concept as a rule that confers the objective unity on particulars enabling them to function as the singular entities referenced in the context of an objectively-valid judgment. Thus, Natorp writes that the particular givens of cognition "can only be grasped…in concepts" and "when it is grasped in concepts it is no longer absolutely immediate and subjective, but has always been objectified" (Natorp (1887), 262).[38]

Cohen's Marburg School thus reaffirms Kant's model of human cognition as necessarily judgmental, along with its implication that any cognitively significant reference to a given object must be conceptually mediated:

No object is given to us in any other way than in knowledge. (Natorp (1887), 253)

What is here called the 'object' of knowledge acquires determinate meaning only by being referred to a certain form or function of knowledge. (Cassirer (1929), 321)

Directly apprehended sense data or innate ideas are merely subjective, psychological phenomena incapable of grounding objectively valid judgments.

Because the Marburg Neo-Kantians retain what Hans Sluga describes as Kant's doctrine of "the priority of judgments over concepts," they can also recapture Kant's claim that an empiricist appeal to abstraction cannot ground scientifically fruitful, objectively valid concepts (Sluga (1980), 90–1). As we have seen, Cohen and Natorp routinely state that a priori concepts and principles are necessary for ensuring the objective content of concepts in the natural sciences. Moreover, by their lights, Helmholtz's psychophysiological description of concept formation is simply incompatible with the notion of a concept as a knowledge-generating rule. Recall that, for Helmholtz, general ideas are formed by repeated presentations of sensibly-given particulars and the memory traces that these leave behind. Memory traces wash out irregularities, retaining the shared features ranging over a group of particulars. In order for the general ideas thus produced to be truth-tracking, one must suppose that particular entities, taken as things-in-themselves, break up into natural kinds and that our psychophysiological apparatus is such that the traces left behind form general ideas mapping onto these kinds. Presumably, Helmholtz would have believed that such a supposition was entirely reasonable, at least it does not seem entirely unreasonable to me. However, this is a supposition that Cohen et al. are unwilling to grant, given their orthodox Kantian position that even repeated

---

[38] See also Cassirer's notion of a concept as a necessary rule for judging objects and his endorsement of Frege's notion of a concept as a function in *The Philosophy of Symbolic Forms* (1929), 287–288 and 291–293. Natorp does not draw a sharp distinction between a particular idea as a component part of the objective content of a proposition and the singular entity that is referred to via that content. Prior to Frege's distinction between sense and reference distinction, it was common to conflate the objective meaning of a term with the reference of that term. Even when presented with Frege's distinction between sense and reference, Russell continued to hold that Frege's referents were themselves component parts of the asserted objective content (*PMC*, 169). What Cohen and Natorp are clear on, however, is that objective particular representations and objective particular represented must be distinguished subjective mental states and their content.

experiences of how things are can never deliver the normative necessity implied when asserting 'This is an *F*'.[39]

Cassirer presents the clearest statement of the Marburg complaint against an empiricist abstractionist account of concept formation. According to Cassirer, abstracting from sensed particulars is simply too arbitrary to yield scientifically fruitful and cognitively significant concepts:[40]

> There is nothing to assure us that the common properties, which we select from any arbitrary collection of objects include the truly typical features, which characterize and determine the total structures of the members of the collection. We may borrow a drastic example from Lotze: If we group cherries, and meat together under the attributes red, juicy and edible, we do not thereby attain a valid logical concept but a meaningless combination of words, quite useless for the comprehension of the particular cases. (Cassirer (1910), 7)[41]

Cassirer's point is that any set of common properties is not an automatically cognitively informing set, and so abstraction of these properties must be guided by some rational purpose or "intellectual criterion" (Ibid., 7). The concept itself is the criteria for grouping particulars in a cognitively significant manner. The objects making up the extension of a concept are not "disconnected particularities." They are related to one another according to a "specific rule", namely, the concept. Hence, in order to abstract from particulars just those features that enable us to comprehend that sort of thing, we are not deducing a concept, but presupposing it (Ibid., 17).

Thirdly, Cassirer argues that even if we assume that abstraction alone could yield cognitively significant empirical concepts, it could never account for the concepts employed in the mathematical sciences. In his *Logik*, Benno Erdmann takes up Helmholtz' theory of memory traces and unconscious inferences and uses it to explain how we come into contact with abstract entities corresponding to mathematical concepts. As Cassirer tells it, the fundamental idea behind Erdmann's account is that memory traces eventually "exert a special and independent influence on the act of perception itself" such that "newly occurring content is apprehended and transformed according to them" (Ibid., 11). Erdmann believes that this process not

---

[39] Cassirer also maintains that the supposition of natural kinds is tantamount to an unwarranted metaphysical judgment about the nature of being (Cassirer (1910), 7–8).

[40] Cassirer does not lodge this critique of the abstractionist account of concept formation against Helmholtz's naturalized Kantianism. Indeed, he speaks favorably of Helmholtz' Zeichentheorie of perception, seeing it as implicitly endorsing the notion of a concept as a function yielding an objective output. In other words, Cassirer ignores Helmholtz' references to memory traces and psychological associationism. Cassirer's critique is explicitly targeted at Mill and Benno Erdmann. Helmholtz himself, however, credits Mill's *System of Logic* for inspiring his own views on the formation of perceptual signs through a process of unconsciously performed inductive inferences (see Hatfield (1990), 199–200, 204 and De Kock (2018), 51–52. It is also important to note that, for Frege, Erdmann is the poster child for all that goes wrong when logical and psychological investigations are conflated (cf. *BLA*, 13–25).

[41] The reference to Lotze here is significant since it shows that these debates about abstraction, what it is and what it accomplishes, were going on at a time and by parties that Frege could have been aware of prior to developing his C&O distinction. It also shows that Cohen's Marburg School was directly influenced by Lotze with respect to this issue at the very least.

only affords us sensibly detectible evidence for the existence of abstract entities, but also explains the application of mathematical truths to experience. For example, objectively real geometric abstracta must trace their lineage to common spatial properties of actually perceived empirical entities, whose variable and non-geometric features have been appropriately washed away. The abstract entities, e.g. point, line, triangle, then exist as mnemonic images, "nothing more than the contents of perception" (Ibid., 11). These images causally interact with subsequent perceptions of sufficiently similar empirical entities, modifying how we perceive them such that they become imbued with stable geometric properties and thus subject to geometric reasoning.

Disregarding the obvious worry that Erdmann guarantees the existence of mathematical entities only by turning them into ideas internal to the perceiving subject, Cassirer goes after the description of abstraction as a transformative process, licensing our attributing to physical and psychic entities certain mathematical features that they do not actually possess:

> The concept of the point, or of the line, or of a surface cannot be pointed as an immediate *part* of physically present bodies and separated from by simple 'abstraction'...But psychical existence is denied no less than physical to the objects of geometrical definitions. For in our mind we never find the presentation of a mathematical point, but always, the smallest possible sensible extension.... 'Abstraction,' as it has hitherto been understood, does not *change* the constitution of consciousness and objective reality, but merely institutes certain limits and divisions in it; it merely divides the parts of sense-impression but adds no new datum. (Ibid., 12–14)

Cassirer, like Kant, stresses that abstraction does not underwrite the existence of objects falling under non-empirical concepts nor does it account for the applicability of these concepts to empirical particulars. 'Abstraction', insofar as we mean the removal of characteristics, is a process that legitimately operates only on concepts. We remove a conceptual mark and thereby obtain a concept under which more objects fall. We do not take away or add properties so as to present ourselves with an abstract entity instantiating our a priori concepts (Ibid., 6).

For Cassirer and his fellow Marburg Neo-Kantians, the objective validity of mathematical concepts is not established by appealing to any presumed psychological capacity for transforming sensible entities, mental or physical, into entities possessing the right mathematical properties. And, *contra* Kant, it is also not established by showing that they function as necessary rules, determining the sensible manifold and making it possible for us to unite and intuit pure or empirical sensibly-given instantiations. Instead, the objective validity of mathematical and logical concepts is established via their indispensability in rendering empirical phenomena as possible subjects of judgments within the context of the sciences. Here the "sensuous manifold" is replaced altogether by "another manifold, which agrees with certain theoretical conditions" (Ibid., 14).

As we have seen, Cohen replaced Kant's pure sensory manifold and the unity of consciousness with the "unity of synthetic principles" which grounds the "validity" of experience and the "possibility of objectivity" (Cassirer (2005), 98). As a consequence, for Cohen's Kantianism, the paradigmatically given object is not, as it was for Kant, the object as it is presented to us in empirical perception, but rather "the fully determinate and individualized" entities that would present themselves to us at the end of scientific inquiry (Friedman (2000a), 78).[42] Furthermore, since the objects comprising the subject matter of Euclidean geometry, and presumably the logical objects treated in arithmetic, are necessary posits of Newtonian physics and fully determined according to the foundational principles of their respective sciences, these objects are just as given, actually more so, than empirical entities at this stage of development in the physical sciences. There is little appreciable difference between the pure and empirical particular givens of cognition, for both are presented to us as "thought-contents" of the sciences and neither is bound to the "here and now given" of sense perception (cf. Natorp (1887), 259, and Kohnke (1987), 183).

Since the givenness of objects is no longer tied to sensible affectation or empirical perception, it is not surprising that though the Marburg School retains a distinction between logical possibility and objective validity, it tends to collapse the distinction between the objectively valid and the objectively real. Evidence of their commitment to the former distinction is seen in Cohen's unwillingness to grant epistemic significance to non-Euclidean geometries simply on the basis that they form an internally consistent system. For Natorp too, "simple lack of inner contradiction" and "the consistent connection of thoughts" is not enough to ensure that a concept or system of thoughts is cognitively significant or potentially applicable. Here he explicitly recalls Kant's distinction between logically consistent, empty forms of knowledge and forms having cognitive content (Natorp (1887), 246). There is also textual evidence that Cohen and Natorp agree with Kant that the objective validity of pure mathematical concepts requires constructing a corresponding mathematical object, where the construction indicates how the mathematical concept could be applied to empirical phenomena.[43]

But since Cohen and company do not link the cognitive value and objectivity of representations to sensibility, they do not follow Kant in distinguishing a concept's objective validity from its objective reality. In Sect. 3.1.2, I argued that though Kant takes the mathematical construction of a concept to be sufficient for ascribing objective validity to that concept, he does not take this to suffice for treating the concept as objectively real, i.e., as having a full-fledged, existing object instantiating the concept. To grant real existence to an object and thus objective reality to the concepts it instantiates, Kant demands that the object be sensibly perceived or causally related to something perceptible. Geometric concepts are objectively real if and only if the procedure by which the geometer constructs the concept is the same

---

[42] As Friedman notes, Natorp and Cassirer conceive empirical science, in contrast to the exact sciences of pure mathematics and logic, as a never-ending process of further determining its objects via more refined concepts as science progresses. Therefore, they consider the "real object of empirical cognition" to be "a never completed 'X' towards which the methodological progress of science is converging" (Friedman (2000a), 79).

[43] Both Cohen and Natorp stress the importance of construction as the means whereby the link from mathematical concepts to mathematical objects to natural phenomena is established (See Cohen (1871), 11, 23–24, 104–5; Natorp (1887), 263).

procedure that we perform in order to perceive empirical entities manifesting that geometric shape. For Cohen, by contrast, neither sensible perceptibility nor causal connection to the sensibly perceptible is necessary for granting full-fledged object-hood. If the geometer articulates a procedure for constructing a mathematical entity in accordance with fundament laws of her discipline and if that constructed concept has proven itself to be universally applicable, say by enabling Galileo and Newton to mathematize nature, then there is no further, legitimate question about the reality of the concept or it corresponding mathematical object:

> When I construct the triangle, and depict it in pure Intuition this is a priori. Whoever wants to ask further, is it really? He evidently debases the a priori. He hears that the triangle is strictly necessary and universally applicable and he asks if it is real! (*KTE¹*, 35)

Cohen thus refuses to acknowledge any significant difference between objective validity and objective reality.

In summarizing Helmholtz' Neo-Kantianism, I complained that he pushed Kantian epistemology too close to traditional empiricism. One might similarly complain that Cohen's Neo-Kantianism looks too much like traditional rationalism. Ordinary perceptual experience certainly does not play as big a role in harnessing our theoretical speculations as it did for Kant. The immediate givens of sense perception, the keyboard in front of me, are not treated as the paradigmatic objects and seem to pale in comparison to the pure objects of mathematics and the fully determinate objects posited at the hypothetical end of scientific inquiry. Moreover, Cohen does see Kant following in the footsteps of Plato and Descartes:

> Idealism in its classic forms, in Plato, in Descartes, in Kant, is a method not a doctrine, and its products, positive and negative, are the outcome of epistemological critique (Cohen in Kohnke (1987), 191).

It is important, however, to understand that in making this association Cohen does not mean that the task of critical idealism is to articulate the metaphysical underpinnings of proper scientific inquiry. This would be to read the *Critique* primarily onto-logically, as articulating the existence and nature of various entities that make knowledge possible, and Cohen insists that the *Critique* should not be read in this manner. Rather the "given" task of the properly oriented Neo-Kantian epistemologist is to reveal, maintain and suitably revise the a priori logico-epistemological principles, a priori mathematical concepts and propositions and the inferentially-related structure of propositions that comprise the ever accumulating knowledge from the traditions of the natural sciences, aesthetics, and religious ethics (Cassirer (2005), 98; Poma (1997), 103–30).

We have discussed the fact that a primary motivation for Cohen's interpretation of pure intuition and pure thought as methods was to avoid any suggestion that they were things or faculties for perceiving things and that it was these things which guaranteed the truth of mathematics and logic. Helmholtz' naturalized reading of Kant's *Critique* is thus one example of the kind of ontological reading of Kantian epistemology that Cohen opposes. Helmholtz interprets Kant as positing the form of our sensible faculty as the *thing* which guarantees the a priori truth of the Euclidean axioms. Cohen defends Kant's thesis on the status of the axioms by arguing that Kant is simply articulating the methodological constraints imposed on geometry and mathematical physics insofar as these disciplines are entitled to the title of an ongoing scientific enterprise.

On Cohen's reading of Kant, critical idealism overcomes both the skepticism associated with British empiricism and the speciousness of dogmatic rationalism precisely because it is first and foremost an inquiry into the methodology, not the ontology, underwriting objectively valid knowledge claims:

> Critical idealism …does not consist, like 'any other kind of idealism', in simply identifying being with idea, but in considering every given as a problem, and in positing thought as methodical foundation of science, rather than metaphysical ground of being: 'Idealism in general resolves things in phenomena and ideas. On the other hand, the critique of knowledge analyzes sciences according to *presuppositions and grounds*, that are admitted in and for its *laws*. Thus, idealism as critique of knowledge has *scientific facts* as its objects, not things and events, not even those of simple consciousness.' (Poma (1997), 57–8, quoting from Cohen's *PIM*, 6)

When Cohen aligns Kant with Plato and Descartes, he means that Kant is committed to a "methodically grounded rationalism" and that he shares with Plato and Descartes an appreciation of mathematics and logic as laying down the initial conditions for objective validity, truth and falsity, necessary for developing a science of nature (Ibid., 57). The Marburg School retains Kant's transcendental method and thus preserves the autonomy and authority of philosophical reasoning, while keeping it from "ascending into the aether of [metaphysical] speculation" (Kim 2004). This is accomplished by anchoring the transcendental method "in facts (eminently the fact of mathematical physics)" and insisting that the sole task of the critical philosopher is "to establish the conditions of possibility or justification" of those facts (Ibid.).

Finally, we also saw, in Cohen's critique of Helmholtz, that the entities or subject matter investigated by geometers must similarly prove their worth in supplying the conditions for asserting objectively-valid propositions in the natural sciences. So, it is fair to say that, for a Marburg Neo-Kantian or for someone sympathetic with Cohen's read of Kant, the entities or subject matter of the pure sciences, e.g., mathematics and logic, will be determined and justified by how well they function as an a priori reservoir supplying the necessary normative principles and concepts securing the objectivity and cognitive content for the less general, empirical sciences.

If Cohen's Kant looks like less of an empiricist than Kant himself, this is because pure thought and pure sensibility have here been reduced to methods or sources of knowledge and any trace of transcendental psychology has been wiped away. Thus, we see in Cohen's Neo-Kantianism exactly what we hoped to find. There was historical precedent for reworking Kantian principles and distinctions, ridding them of any psychological, subjectivist elements and presenting them strictly as safeguards against various methodological errors threatening the sciences, especially the errors threatening the exact sciences: "And one may assert that generally difficulties arise in the exact sciences, when researchers deviate from the principles which Kant certified concerning these sciences" (Cohen *Kant's Einfluss auf die deutsche Kultur*, speech honoring the birthday of the Kaiser Wilhelm I, March 17 1883, 8).

References

Anderson, L. (2005). Neo-Kantianism and the roots of anti-Psychologism. *British Journal for the History of Philosophy, 13*(2), 287–323.

Benacerraf, P. (1981). Frege: The last logicist, repr. in Demopoulos (1997), 41–67.

Blanchette, P. (1994). Frege's reduction. *History and Philosophy of Logic, 15*, 85–103.

Brelage, M. (1965). *Studien zur Transzendentalphilosophie*. Berlin: de Gruyter.

Cassirer, E. (1910). *Substanzbegriff und Funktionsbegriff*, trans. and repr. In W. C. Swabey & M. C. Swabey (Eds.), *Substance and function and Einstein's theory of relativity* (pp. 1–346). New York: Dover Publications, 1923.

Cassirer, E. (1921). *Zur Einsteinschen Relativitatstheorie*, trans. and repr. In W. C. Swabey & M. C. Swabey (Eds.), *Substance and function and Einstein's theory of relativity* (pp. 347–456). New York: Dover Publications, 1923.

Cassirer, E. (1929). *The philosophy of symbolic forms* (Vol. 3, R. Manheim, Trans.). New Haven: Yale University Press, 1957.

Cassirer, E. (2005 [1912]). Hermann Cohen and the renewal of Kantian philosophy (L. Patton, Trans.). *Angelaki* (Abingdon: Routledge), *10*(1), 95–108, repr. in Luft (2015), 221–235.

Coffa, J. A. (1991). *The semantic tradition from Kant to Carnap* (L. Wessels, Ed.). Cambridge: Cambridge University Press.

Cohen, H. (1871). *Kant's Theorie der Erfahrung* (1st ed., selected portions J. Prasse & K. G. T. Merrick, Trans.). Berlin: F. Dummler.

Cohen, H. (1876). Introduction to 1902 edition Friedrich Lange's *Geschichte des Materialismus*. Leipzig: J. Baedeker.

Cohen, H. (1881). Preface to the 1887 edition of Friedrich Lange's *Geschichte des Materialismus*. Iserlohn: J. Baedeker.

Cohen, H. (1883). *Das Princip der Infinitesimal-Methode und seine Geschichte* (selected portions K. Hildebrand & T. Merrick, Trans.). Berlin: F. Dummler.

Cohen, H. (1885). *Kant's Theorie der Erfahrung* (2nd ed., selected portions J. Prasse, K. Gallagher, & T. Merrick, Trans.). Berlin: F. Dummler.

Cohen, H. (2004 [1908]). *Ethics of Maimonides* translated with commentary by Almut Sh. Bruckstein. Madison: University of Wisconsin Press.

De Kock, L. (2018). Historicizing Hermann von Helmholtz's psychology of differentiation. *Journal for the History of Analytical Philosophy, 6*(3), 43–62.

Demopoulos, W. (Ed.). (1997). *Frege's philosophy of mathematics*. Cambridge, MA: Harvard University Press.

Disalle, R. (2006). Kant, Helmholtz and the meaning of empiricism, in Friedman and Nordmann (2006), 123–139.

Dummett, M. (1991). *Frege and other philosophers*. Oxford: Clarendon Press.

Enriques, F. (1929). *The historic development of logic* (J. Rosenthal, Trans.). New York: Henry Holt.

Ewald, W. (Ed.). (1996). *From Kant to Hilbert: A source book in the foundations of mathematics*. Oxford: Clarendon Press.

Frege, G. (1952). *Translations from the philosophical writings of Gottlob Frege (TPW)* (P. Geach & M. Black, Ed.). Oxford: Basil Blackwell.

Frege, G. (1964). *Basic laws of arithmetic: Exposition of a the system (BLA)* (ed. and trans. with introduction by M. Furth). Berkeley: University of California Press.

Frege, G. (1979). *Posthumous writings (PW)* (H. Hermes, F. Kambartel, F. Kaulbach, Ed., & P. Long & R. White, Trans.). Oxford: Basil Blackwell.

Frege, G. (1980). *Philosophical and mathematical correspondence (PMC)* (B. McGuinness, Ed. & H. Kaal, Trans.). Chicago: University Chicago Press.

Frege, G. (1984). *Collected papers on mathematics, logic and philosophy (CP)* (B. McGuinness, Ed. & M. Black, Trans.). New York: Basil Blackwell.

Frege. G. (1997). *The Frege Reader (FR)* (ed. and introduction by M. Beaney). Oxford: Blackwell.

Friedman, M. (1997). Helmholtz's *Zeichentheorie* and Schlick's *Allgemeine Erkenntnislehre*: Early logical empiricism and its nineteenth-century background. *Philosophical Topics, 25*(2), 19–50.

Friedman, M. (2000a). *A parting of the ways*. Chicago/La Salle: Open Court.

Friedman, M. (2000b). Geometry, construction, and intuition in Kant and his successors. In G. Sher & R. Tieszen (Eds.), *Between logic and intuition*. Cambridge: Cambridge University Press.

Friedman, M. (2008). Ernst Cassirer and Thomas Kuhn: The Neo-Kantian tradition in history and philosophy of science. *The Philosophical Forum, xxxix*(2 Summer), 239–252.

Gabriel, G. (2002). Frege, Lotze, and the continental roots of early analytic philosophy. In E. H. Reck (Ed.), *From Frege to Wittgenstein: Perspectives on early analytic philosophy* (pp. 39–51). Oxford: Oxford University Press.

Hankel, H. (1867). *Vorselung uber die Complexen Zahlen und ihre Funktionen*. Leipzig: L. Voss.

Hatfield, G. (1990). *The natural and the normative*. Cambridge, MA: MIT.

Helmholtz, H. (1876). The origin and meaning of geometrical axioms (*OMG I*), in Ewald (1996), 663–685.

Helmholtz, H. (1878a). The origin and meaning of geometrical axioms (*OMG II*), in Ewald (1996), 685–698.

Helmholtz, H. (1878b). The facts in perception (*FP*), in Ewald (1996), 689–727.

Helmholtz, H. (1887). Numbering and measuring from an epistemological point of view (*NME*), in Ewald (1996), 727–752.

Kim, A. (2004, Summer). Paul Natorp. In E. N. Zalta (Ed.), *The Stanford encyclopedia of philosophy*. http://plato.stanford.edu/archives/sum2004/ entries/natorp/.

Kitcher, P. (1979). Frege's epistemology. *Philosophical Review, 88*(2), 235–262.

Kluback, W. (1987). *The idea of humanity*. Maryland: University Press of Merica.

Kohnke, K. C. (1987). *The rise of Neo-Kantianism: German academic philosophy between idealism and positivism* (R. J. Hollingdale, Trans.). Cambridge: Cambridge University Press, 1991.

Lenoir, T. (2006). Operationalizing Kant: Manifolds, models, and mathematics in Helmholtz's theories of perception, in Friedman and Nordmann (2006), 141–210.

Longuenesse, B. (1998). *Kant and the capacity to judge* (C. T. Wolfe, Trans.). Princeton/Oxford: Princeton University Press.

Luft, S. (Ed.). (2015). *The neo-Kantian reader*. New York: Routledge.

Natorp, P. (1887, October 1981). On the objective and subjective grounding of knowledge (David Kolb, Ed. & Trans.). *Journal of the British Society for Phenomenology, 12*(3), 246–266, repr. in Luft (2015), 164–179.

Patton, L. (2004). *Hermann Cohen's history and philosophy of science*. Ph.D dissertation, McGill University, Montreal.

Patton, L. (2005). The critical philosophy renewed: The bridge between Herman Cohen's early work on Kant and later philosophy of science. *Angelaki, 10*(1), 109–118.

Patton, L. (2009). Signs, toy models, and the a priori: From Helmholtz to Wittgenstein. *Studies in the History and Philosophy of Science, 40*(3), 281–289.

Patton, L. (2012). Hermann von Helmholtz. In E. N. Zalta (Ed.), *The Stanford encyclopedia of philosophy* (Winter 2012 ed.). http://plato.stanford.edu/archives/win2012/entries/hermann-helmholtz/.

Peckhaus, V. (2000). Frege: Kantianer oder Neukantianer? Uber die Schwierigkeiten, Frege der Philosophie seiner Zeit zuzuordnen. In G. Gabriel & U. Dathe (Eds.), *Gottlob Frege – Werk und Wirkung. Mit den unveroffentlichten Vorschoagen fur ein Wahlgesetz von Gottlob Frege* (pp. 191–209). Paderborn: Mentis.

Poma, A. (1997). *The critical philosophy of Hermann Cohen* (J. Denton, Trans). Albany: State University of New York Press.

Quine, W. V. O. (1969). Epistemology naturalized. In *Ontological relativity and other essays* (pp. 69–90). New York: Columbia University Press.

Resnik, M. D. (1980). *Frege and the philosophy of mathematics*. Ithaca/London: Cornell University Press.

Richardson, A. W. (2006). "The fact of science" and critique of knowledge: Exact science as problem and resource in Marburg Neo-Kantianism, in Friedmann and Nordmann (2006), 211–226.

Rosenfeld, B. A. (1988). *A History of Non-Euclidean Geometry, trans. Abe Shenitzer with ed. assistance of Hardy Grant*. New York: Springer-Verlag New York, Inc.

Sluga, H. D. (1980). *Gottlob Frege*. London: Routledge & Kegan Paul.

Tappenden, J. (1995). Geometry and generality in Frege's philosophy of arithmetic. *Synthese, 102*, 319–336.

Van Fraassen, B. C. (2002). *The empirical stance*. New Haven: Yale University Press.

Weiner, J. (1982). Putting Frege in perspective, repr. in Haaparanta and Hintikka (1986).

Wilson, M. (1992). Frege: Royal road from geometry, reprinted in Demopoulos (1997).

Chapter 4
Disambiguating Frege for the Sake of Charity

Abstract This chapter argues that Frege's notorious claims about the nature of geometric propositions and the distinction between concepts and objects become clearer and more defensible if we interpret them through the lens of Cohen's Kantianism. Section 4.1 shows that Frege's life-long endorsement of Kant's thesis that Euclidean geometry is a body of synthetic a priori propositions appears more reasonable once we recognize that Frege follows Cohen in rejecting Helmholtz's interpretation of this thesis. Section 4.2 paints a more general picture of Frege's Kantianism and where it sits relative to the Kantianisms of Helmholtz and Cohen. A main take-away is that Frege shares Cohen's view on the methodological constraints necessary for ensuring the requisite semantic continuity and objective content of scientific terms and assertions. In Sect. 4.3, I argue that Frege's attempts to elucidate the nature of the distinction between concepts and objects should not be dismissed as the unwarranted claims of a platonic realist. Read against the backdrop of the Helmholtz-Cohen exchange, we can see these elucidations and, more importantly, Frege's use of the distinction as intended justification for reinstituting a norm of scientific thought, one capable of blocking the inferences licensed by Helmholtz's naturalized epistemology. I conclude that Frege's distinction between concepts and objects is a Cohen-like reworking of Kant's previous distinction between concepts and intuitions.

Keywords Gottlob Frege · Neo-Kantianism · Synthetic a priori · Mathematical formalism · Mathematical naturalism · Fregean objects · Fregean concepts

Beginning then in 1871 and culminating in his 1883 public address, Hermann Cohen urged those wanting to prevent the methodological mistakes capable of derailing the sciences to dust off their copy of Kant's *Critique* and attend to principles contained therein. According to Frege, such mistakes were prevalent in mathematics at the time. He accuses his fellow mathematicians of engaging in practices that render their concepts and propositions devoid of cognitive content and potential applicability, of failing to appreciate the appropriate foundation for their respective

© Springer Nature Switzerland AG 2020   105
T. Merrick, *Helmholtz, Cohen, and Frege on Progress and Fidelity*,
Philosophical Studies in Contemporary Culture 27,
https://doi.org/10.1007/978-3-030-57299-0_4

disciplines and of misconstruing the nature of the objects under investigation.[1] He sums up the situation as follows:

> There is little cause for satisfaction with the state in which mathematics finds itself at present, if you have regard not to the outside, to the amount of it, but to the degree of perfection and clarity within. In this respect it leaves almost everything to be desired if you compare it with the ideal you may reasonably propose for this discipline...If you ask what constitutes the value of mathematical knowledge, the answer must be: not so much what is known but how it is known, not so much its subject-matter as the degree to which it is intellectually perspicuous and affords insight into its logical interrelations. And it is just this which is lacking (*PW*, 157).

There is also evidence that Frege looked to Kant for guidance in straightening out this sad state of affairs. In his review of Cohen's *PIM*, Frege insists that the person best qualified to mine mathematics for its epistemological and logical import must be someone with "facility in both mathematical and philosophical thought" (*CP*, 108).[2] Kant is then identified as one of these rare people. This may explain, says Frege, why "so little has been done" along the lines of subjecting the rich structure of mathematical thoughts to a logico-epistemological investigation "except by Kant" (Ibid.).[3]

Frege scholars have warned against reading too much into Frege's pro-Kant remarks, suggesting that they are little more than politically correct gestures towards a cultural icon. Consider Coffa's (1991) dismissal of the footnote to *Foundations* §27, where Frege implies that his characterization of concept and object as distinct categories of objective ideas is borrowed from Kant and captures his "true view":

> Extending an undeserved olive branch to the past, Frege added, 'It is because Kant associated both meanings with the word that his doctrine assumed such a subjective, idealist complexion, and his true view [!] was made so difficult to discover' (Ibid., 66–67, the exclamation mark is Coffa's).

---

[1] See for example, "On Formal Theories of Arithmetic" (1885) in *FG*, 142–146; "Frege against the Formalists" (1903) in *TPW*, 167 and 213; and "Logic in Mathematics" (1914) in *PW*, 203–250. Throughout my exposition of Frege's views, I will try to support my points with references dated prior to and after the collapse of Frege's logicist program. This emphasis on dates is important in order to bolster my contention that his interest in the C&O distinction is detachable from his particular aim of showing that number theory reduces to logic. Furthermore, those skeptical of any attempt to link Fregean doctrines to Kantian or neo-Kantian epistemology often cite the fact that Frege's pro-Kant remarks occur either early in his career, before his philosophical thought had reached full maturity, or after he had been crushed by Russell's discovery of the paradox. Therefore, they claim, such remarks are not to be taken seriously and Kantian epistemology played little role in framing Frege's strongly-held, fully developed philosophical theses (e.g., Dummett (1991b), 125–157). To block this argument, one must show that the Kant-Frege link in question persisted throughout his writing career.

[2] By a facility in philosophical thought, Frege means a facility in epistemology and logic.

[3] See also "On Formal Theories of Arithmetic" (1885) in *FG*, 142, and "Sources of Knowledge" (1924/5) in *PW*, 273, where Frege also claims that their lack of facility in both fields is to be blamed for the fact that his contemporaries commit the aforementioned mistakes.

Since I intend to make much of this footnote and other pro-Kant remarks for the purposes of pinning down Frege's Neo-Kantian commitments, I need to respond to those skeptical of granting them too much significance.

First, as evidenced in the Helmholtz-Cohen debate over Kantian exegesis, Frege's contemporaries felt no obligation to pay homage to Kant's theses if they were considered unworthy of such praise. In the latter half of the nineteenth century, opinion regarding Kant's system was simply too fragmented to impose the kind of political correctness presumed to motivate Frege's endorsements. Second, Frege showed little reluctance in rejecting Kantian doctrines which he believed to be in error:

> From all the preceding it thus emerged as a very probable conclusion that the truths of arithmetic are analytic and a priori; and we achieved an improvement on the view of Kant (*FA*, 118).

> I must also protest against the generality of Kant's dictum: without sensibility no object would be given to us. Nought and one are objects which cannot be given in sensation (*FA*, 101).

Finally, the Kantian thesis that Frege most unabashedly and repeatedly championed, that geometric truths are synthetic a priori and grounded on pure intuition, was the very one which Helmholtz and others agreed must be jettisoned. Coffa himself characterizes Frege's intellectual milieu as follows:

> By the second half of the nineteenth century, people began to wonder first about the exclusive necessity of Euclidean geometry and then about the role of intuition in *any* geometry, Euclidean or otherwise. As neo-Kantians were forced to address this issue more extensively, it slowly emerged that the master's silence was not a sign of unspoken wisdom. Revealingly, the cleverest among neo-Kantians silently pushed Kant's pure intuition to the corner of their doctrine of geometry...(Ibid., 42–3).

By Coffa's own account then, Frege's persistently published claim that geometry is founded on intuition seems to align him with only the most orthodox schools of Neo-Kantianism. For these reasons, I maintain that Frege's explicit endorsements of Kant's philosophy can be significant and illuminating.

I agree, however, with skeptics of the Frege-is-a-Kantian position that just because Frege employs Kantian terms or asserts a Kantian thesis does not imply that he means what Kant meant or that he ascribes to the doctrine in its original form (see Dummett (1991b), 130). As we have seen, Kant's terminology and theses were variously interpreted, and there was no consensus on which interpretation represented the real Kant. So, before we can state with any accuracy the extent to which Frege is indebted to Kant or to a particular Neo-Kantian camp, we must first determine the interpretation of Kantian terminology with which Frege is operating.

Often Frege states precisely how he is using a term and this would appear to be the case with 'intuition':

> What is objective...is subject to laws, what can be conceived and judged, what is expressed in words. What is purely intuitable is not communicable (*FA*, 35).

It is in this way that I understand objective to mean what is independent of our sensation, intuition, and imagination, and of all construction of mental pictures out of memories of earlier sensations…(*FA*, 36).

Here intuition is placed in the category of non-objective, psychological entities. These passages are immediately followed with Frege's definition of a subjective idea: "what is governed by psychological laws of association" and "is of a sensible, pictorial character" (*FA, 37*). For Frege then, intuitions are subjective ideas, "which are often demonstrably different in different men", and distinct from objective ideas, which are "the same for all" (Ibid.).

However, it is unlikely that Frege consistently adheres to this usage of the term 'intuition' throughout his writings. In his critique of David Hilbert's axiomatization of geometry, Frege complains that it neglects the fact that geometric axioms express "basic facts of our intuition" (*FG*, 7, 25 and 27). Surely it would be a mistake to read Frege as faulting Hilbert for failing to appreciate that geometric axioms are laws describing essentially private, incommunicable mental states. Anyone familiar with Frege's mantra that discovering the grounds of truth for a proposition is not a matter for psychological investigation should balk at the idea that he considered the whole of geometry to be an exception to the rule. This suggests that figuring out Frege's intended meaning of 'intuition' and any other suspected Kantian expressions will require more than just looking at his explicit stipulations. To decide the sense that Frege attaches to 'intuition', we must look at the specific textual and polemical context in which it occurs and determine the role it plays in that context.[4] This is the interpretative strategy that I will also pursue in deciding what Frege intends by his use of the terms 'concept' and 'object'.

In this chapter, I explain how Frege wields his C&O distinction against his opponents, the formalists, with the aim of showing that the C&O distinction plays the same role within this polemical context as the C&I distinction played for Kant within his. The first two sections are devoted to stage setting and character development. In Sect. 4.1, "What Frege meant by 'Kant is Right about Geometry,'" I articulate Frege's views on intuition, geometry and the conditions for asserting objectively valid claims. My goal here is twofold: (1) explain why Frege uses 'object', not 'intuition', to denote the class of particular objective ideas picked out by the singular terms occurring in a proposition and (2) begin to align Frege with Cohen's understanding of the Kantian problematic. I begin by focusing on Frege's remarks on the grounds of geometric knowledge because it is there that we find the most substantial textual evidence that Frege endorses Cohen's reading of Kant and because the pay-off for have recognized this endorsement is fairly immediate.

Section 4.2 "Frege's Kantianism vis-à-vis Helmholtz and Cohen" extends my argument that Frege would credit Cohen's 1871 *Kant's Theorie der Erfahrung* and his 1883 *PIM* with discovering Kant's "true view" insofar as reading Kant through Cohen's eyes is crucial in achieving Frege's goal of convincing his peers to forsake

---

[4]Although the expressions 'sense' and 'meaning' appear throughout this chapter, I do not intend for the reader to assume that I am using them in Frege's technical sense unless specifically noted.

the misguided epistemology of science and logical errors promulgated by Helmholtz and his fellow formalists. Let me be clear on what I am and am not asserting concerning Frege's endorsement of Cohen's Kantianism. To my knowledge, Frege's awareness of Marburgian Neo-Kantianism was limited to Cohen's 1871 intervention in the Adolf Trendelenburg and Kuno Fischer debate over Kant's views on the nature of space and time, the 1871 edition of *Kant's Theorie der Erfahrung,* and of course the 1883 *PIM.* There is no evidence that Frege's work influenced the thought of Cohen or that of his students, Paul Natorp and Ernst Cassirer, until well after 1903 and the discovery of Russell's paradox (Heis (2010), 383–6). Once Cassirer did become aware of Russell's and Frege's logicism, he credited it with providing the technical proofs necessary to support several distinctive Marburgian theses on mathematics and so-called 'formal logic' (Heis (2010), 391–3). My claim that Frege endorsed Cohen's Kantianism is limited strictly to the claim that Frege relied on Cohen's Kantianism to support his own critique of Helmholtz's naturalized Kantianism and, most importantly, for establishing the precedent whereby Kantian terminology and doctrines could be reworked and reintroduced as sound methodological precepts for the sciences.

Section 4.3, "Deploying the Distinction Between Concepts and Intuitions" focuses on those passages where Frege specifically states what he means by 'concept' and 'object'. The characterization of Fregean concepts and objects that emerges from this exegesis will then be compared and contrasted with Kant's original notions of concepts and intuitions as presented in Chap. 2. I then show that the methodological errors that Frege associates with formalism are identical to those Kant associated with empiricism and rationalism. Just as Kant traced these mistakes to blurring the distinction between concepts and intuitions, so too Frege traces them to blurring the distinction between concepts and objects. I conclude that Frege's introduction of the C&O distinction in the footnote to §27 of the *Foundations,* his use of the distinction to highlight the methodological errors committed by the formalists, and the fact that his use of the C&O distinction mirrors Kant's use of the C&I distinction should be recognized as crucial to his elucidation of and justification for treating the C&O distinction as a primitive logico-epistemological fact.

4.1 What Frege Meant by 'Kant Is Right About Geometry'

As is well known though Frege explicitly rejected Kant's account of arithmetic, he consistently affirmed his account of geometry:

> I regard it as one of Kant's great merits to have recognized the propositions of geometry as synthetic judgments, but I cannot allow him the same in the case of arithmetic. The two cases are anyway quite different. The field of geometry is the field of possible spatial intuition; arithmetic recognizes no such limitation ("Letter to Anton Marty" in *PMC,* 100).

> To touch only upon what is immediately necessary, I consider Kant did great service in drawing the distinction between synthetic and analytic judgments. In calling the truths of geometry synthetic and a priori, he revealed their true nature (*FA*, 101–2).

Understanding exactly what Frege meant when he claimed that Kant was essentially right about geometry provides an important clue in deciphering the extent and type of Frege's alleged Kantianism because, as Coffa indicates, few of Frege's colleagues defended this particular aspect of Kant's philosophy. From 1873 to 1924, however, Frege continually asserted that Euclidean geometry was a species of a priori knowledge, containing propositions that are universally and necessarily applicable to the natural world and whose ultimate source of justification lies in pure intuition. Frege's position is puzzling, since he was surely aware of the advances in mathematics and perceptual psychology that seemed to challenge Kant's thesis.[5]

We can begin to make sense of Frege's views on geometry by examining what he means by 'intuition.' I will present textual evidence that Frege operates with two different meanings for the term 'intuition'. The first is the notion of an intuition as a sensory representation of a particular; the second is the notion of intuition as the pure source of geometric knowledge. I will argue that what Frege means when using 'intuition' in the first sense is one of Helmholtz's sensory symbols and what he means when using 'intuition' in the second sense is precisely what Cohen intended by designating Euclidean space as the method of pure intuition, the a priori source of geometric knowledge.[6]

---

[5] See Tappenden (1995) and Wilson (1992) for evidence that Frege was not only familiar with Riemann's analytic (algebraic) representation of non-Euclidean manifolds, but also with von Staudt's and Plucker's purely geometric (no numbers) representations of spatial manifolds that violated one or more of the properties of Euclidean 3-space. Furthermore, Frege's life-long mentor and supporter Ernst Abbe was deeply engaged with research in the theory of ophthalmic instrument construction. Helmholtz had invented the ophthalmoscope in 1850. Surely, Abbe would have been aware of Helmholtz' *Facts in Perception*, so it is safe to assume that Frege also knew about Helmholtz' contention that current research in sense perception undermined Kant's claim that geometry was synthetic a priori.

[6] Dummett also notes Frege's vacillating use of the term 'intuition' to challenge Sluga's claim that Frege held Kant's view of geometry and so his view on the nature of space. Dummett does not, however, attempt to explain Frege's use of the term 'intuition' to denote the source of geometric knowledge (Dummett (1991b), 128–130). I maintain that understanding Frege's use of 'intuition' to refer to a source of knowledge holds the key to understanding his commitment to Cohen's Neo-Kantian account of geometry. So, while I agree with Dummett that Frege's views on geometry do not commit him to the thesis that space is a subjective form of intuition, I would argue that Frege's expressed views on geometry do count as evidence against reading him as a metaphysical Platonic realist.

4.1.1 Intuitions as Subjective Ideas

Frege's use of the term 'intuition' to denote the sensory representation of a particular occurs almost exclusively in the *Foundations* and here he self-consciously adheres to the characterization of an intuitive representation as it was presented in Kant's *Critique*. Consider his objection to Hermann Hankel's use of the expression 'a pure intuition of magnitude', where the magnitude that Hankel is envisioning is something "valid for magnitudes in every field" (*FA*, 18). Frege responds that while it is perfectly reasonable to talk about a "*concept* of magnitude" representing "all the different things that are called magnitudes," it is incoherent to talk about an *intuition* of a magnitude in general:

> I cannot even allow an intuition of 100,000, far less of number in the general, not to mention magnitude in general. We are all too ready to invoke inner intuition, whenever we cannot produce any other ground of knowledge. But we have no business in doing so, to lose sight altogether of the sense of the word 'intuition.' (Ibid., 18–19).

According to Frege, the meaning of the word 'intuition' was first fixed by Kant: "Kant in his *Logic* defines it as follows: 'An intuition is an *individual* idea (repraesentatio singularis), a concept is a *general* idea (repraesentatio per notas communes) or an idea of *reflexion* (repraesentatio discursive)" (Ibid.).[7] Frege acknowledges that, in the *Logic*, there is "no mention of any connexion" between sensibility and intuition, but that Kant explicitly made this connection in the *Critique*. He then cites the passage from *Transcendental Aesthetic* (A19/B34). Here sensibility is defined as our capacity for receiving representations through being affected by some indeterminate something, and Kant insists that sensibility "alone provides us with *intuitions*" (*FA*, 19). Given that this detour into Kant scholarship is supposed to undermine Hankel's claim, we can take Frege's suppressed conclusion to be the following: an intuitive representation of a magnitude is simply too particular and too bound up with what is sensibly perceptible to do the work that Hankel expects of it. An intuition in this sense cannot serve as a general idea capable of representing all particular magnitudes from every scientific field.

Notice too that when Frege uses the term 'intuitions' to designate sensory representations, he is thinking of them as the images presenting themselves to a subject through sense perception or the imagination, with all the particularity that that entails. The fact that Frege is thinking of intuitions in this manner is hinted at in his rebuttal to Hankel. It comes out even more starkly, however, in his critique of Kant's view of arithmetic:

> Kant thinks he can call on our intuition of fingers or points for support, thus running the risk of making these propositions appear to be empirical, contrary to his own expressed opinion; for whatever our intuition of 37863 fingers may be, it is at least certainly not pure. Moreover, the term 'intuition' seems hardly appropriate, since even 10 fingers can, in different

---

[7] Frege is quoting here from the *Jäsche Logic*. As noted previously, Frege and his contemporaries referred to this text as *Kant's Logic* and understood it as representing Kant's thoughts on the subject.

arrangements, give rise to very different intuitions. And have we, in fact, an intuition of 135664 fingers or points at all? (*FA,* 6).

Frege's point is that the intuition of four strokes (or fingers) arranged like so '1 1 1 1' is different from the intuition of '11 11.' Furthermore, our capacity for forming images delimits the scope of entities that can be intuited. If we cannot form the image of $1000^{1000\ 1000}$ strokes, dots, etc., we cannot have an intuition of $1000^{1000\ 1000}$ (Ibid., 101).[8]

So, the first meaning that Frege attaches to the term 'intuition' is one that he takes to be equivalent to the Kantian notion of an intuition as a sensory representation of a particular, which, in turn, is interpreted as the image presented to us via our sensory apparatus. Furthermore, we know that Helmholtz had introduced his notion of a sensory symbol as the nineteenth century model of what Kant originally intended by the term 'intuition.'[9] So let us look again at Frege's distinction between objective and subjective ideas, and his rationale for identifying intuitions as subjective ideas:

What is objective…is subject to laws, what can be conceived and judged, what is expressible in words. What is purely intuitable is not communicable (*FA,* 35).

It is in this way that I understand objective to mean what is independent of our sensation, intuition and imagination, and of all construction of mental pictures out of memories of earlier sensations…(Ibid., 36).

An idea in the subjective sense is what is governed by the psychological laws of association; it is of a sensible, pictorial character…Subjective ideas are often demonstrably different in different men, objective ideas are the same for all…It is because Kant associated both meanings with the word ['idea'] that his doctrine assumed such a very subjective, idealist complexion, and his true view was made so difficult to discover (Ibid, 37).

Remember too that, according to Helmholtz's Zeichentheorie, the represented content of sensory perceptions is the construction of mental image by means of the subject's unconscious inductive processing of memories of earlier sensations. Moreover, this process is purportedly not expressible in words and neither is the represented content, at least not at the initial stages. This explains why Frege

---

[8] Michael Resnik (1980) notes that the argument that Frege provides here against the claim that arithmetic is based on intuition would seem to undermine a Kantian view of geometry as well, since we presumably cannot form the image of a chiliagon, a polygon with a 1000 sides. It was remarks such as Resnik's that first drew my attention to fact that we could not begin to make sense of Frege's presumed Kantian commitments unless we first decided what he meant by 'intuition' and that it was extremely unlikely that he was operating with a univocal notion of intuition throughout his writings. As explained in Sects. 2.1.2 and 2.3, Kant does not unequivocally identify the pure intuitions grounding the objectivity validity of mathematical concepts and propositions with the sensible *images* of mathematical entities. Instead, I maintain that the most textually supported and charitable reading would suggest that pure intuitions are the *schemas* or *constructive procedures* employed in the synthetic method of traditional Euclidean geometry.

[9] See Sect. 3.1.4.

classified intuitions as subjective ideas. Intuitions are sensory symbols, and sensory symbols do not meet Frege's criteria for objective ideas.

This also explains why Frege must now use the term 'object,' rather than 'intuition', to denote the objective ideas or meanings corresponding to the singular terms employed in a propositional context.[10] The term 'intuition' had been forever corrupted by those, like Helmholtz, who privileged the psychological reading of the Kant's epistemology. By tossing Helmholtz's sensory symbols into the bin of subjective ideas and arguing that an intuition in this sense cannot be the objective meaning of the term 'point' or 'line',[11] Frege shows himself to be in agreement with Cohen that we ought not look to psychology for supplying the objectively valid meaning of geometric terms. It also suggests that he agrees with the Marburg School that cognitive significant representations of entities are not identical to the "direct sensual appearance" of entities (Natorp (1887), 252). Rather, as we will see, for Frege as for a Marburg Neo-Kantian, cognitively significant representations of entities are identified with the method or mode by which entities are given as the objective content in a properly scientific proposition.[12]

---

[10] See the footnote to *FA*, 37.

[11] See *FA,* 35–6.

[12] Eva Picardi (1996) also claims that Frege's complaint against philosophers who adhered to "a naturalistic reading of Kant" is that they cannot account for the objective meanings of linguistic expressions (Ibid., 313–314). She then argues that, for Frege, providing such an account requires an appeal to a "classical" or "realistic" conception of truth (Ibid., 309 and 314). Picardi only briefly describes the semantic theory that she attributes to Frege, indicating that she follows Dummett's reading of Frege as a semantic realist (Ibid., 312). In *Frege: Philosophy of Language*, Dummett characterizes the difference between a realistic and an anti-realistic semantics as follows: "The fundamental tenet of realism is that any sentence on which a specific sense has been conferred has a determinate truth-value independently of our actual capacity to decide what that truth-value is.... [Non-realism] holds that the sense which we confer on the sentences of our language can be only to the means of recognition of truth-value that we actually possess" (Dummett (1981), 466–467). I have argued that, within Cohen's Kantianism, the objective meanings of the primitive terms in a properly scientific discourse are expressed in the basic propositions of that discipline. For example, the Euclidean axioms determine the objective meanings of the term 'point' and 'straight line.' Furthermore, these basic propositions are the final arbiter of truth or falsity for the assertions made concerning the subject matter with which that discipline is concerned (see Sect. 3.2.2). So, for a Marburg Neo-Kantian, the objective meaning of a sentence uttered with a scientific discursive context will be determined by appealing to on-going scientific tradition with a proven track record in delivering knowledge of the natural world. Appealing to the fundamental propositions or axioms of that traditional practice then provides us with the means for recognizing its truth-value. Since I will argue that Frege shared Cohen's view on what determines the objective meaning and justificatory source for the truth of geometric propositions, I must disagree with Picardi that Frege's anti-psychologism pushed him to adopt semantic realism.

4.1.2 Intuition as the Source of Geometric Knowledge

I have argued that one of the ways in which Frege employs the term 'intuition' corresponds to what Helmholtz meant by this term. Here I will argue that when Frege uses the term 'intuition' to refer to a source of knowledge, what he means is essentially what Cohen meant. In other words, Frege does not use the expression 'pure intuition' to refer to anything intrinsic to the thinking and perceiving subject, not to Helmholtz' innately determined facts of perception and not to Kant's pure form of sensibility. I will show that for Frege, as for Cohen 'pure intuition' refers to the basic concepts, axioms and constructive procedures of traditional Euclidean geometry, which the geometer must employ in the "production and formation" of his subject matter, namely the geometric space presupposed throughout the natural sciences (Poma (1997) 50–1).[13]

Frege's use of the term 'intuition' to denote the ultimate ground of geometric knowledge is scattered throughout his writings:

> [T]he truths of geometry govern all that is spatially intuitable, whether actual or product of our fancy...[Euclidean space is] the only one whose structures we can intuit...(*FA*, 20).

> I call axioms propositions that are true but that are not proved because our understanding of them derives from that nonlogical basis which may be called an intuition of space ("Frege-Hilbert Correspondence" in *FG*, 9).

> From the geometrical source of knowledge flow the axioms of geometry...Yet here one has to understand the word 'axiom' in precisely its Euclidean sense...I cannot emphasize strongly enough that I only mean axioms in their original Euclidean sense, when I recognize a geometrical source of knowledge in them ("Sources of Knowledge" in *PW*, 273).

> I have had to abandon the view that arithmetic does not need to appeal to intuition either in its proofs, understanding by intuition the geometrical source of knowledge, that is, the source from which flow the axioms of geometry (Ibid., 278).

It would be a mistake to read Frege's claim that we can only intuit the structures of Euclidean space as an assertion about our perceptual capabilities and then to assume that this psychological fact is what he presents as the ultimate justification for the truth of Euclidean geometry. First of all, Frege was certainly aware that such a claim was not supported by perceptual psychology.[14] Secondly, Frege did not think the space(s) constituting the subject matter of Euclidean geometry was sensibly intuited at all: "Even the objects of geometry, points, straight lines, surfaces, etc. cannot really be perceived by the senses" (Ibid., 265–6). Thirdly, there is no textual

---

[13] See Sect. 3.2.2.

[14] Based on his references to memory traces of sensations and his close relationship with Ernst Abbe, who was deeply engaged with research in the theory of ophthalmic instrument construction, we can assume that Frege was well aware of current developments in perceptual psychology.

evidence that Frege thought we were capable of a non-sensory, purely intellectual intuition of abstract geometric structures.[15]

Given that Frege identifies the Euclidean axioms and only those axioms as intuition, i.e., the source of geometric knowledge, and given his familiarity with Cohen's notion of pure intuition as a method,[16] I propose interpreting his claim '[Euclidean space] is the only one whose structures we can intuit' as follows. Because the Euclidean axioms are the source of geometric knowledge and because the truths of Euclidean geometry are necessarily valid concerning all that is spatially intuitable, we cannot obtain an immediate, objectively-valid, cognitively significant representation of something as a *spatial* something unless it is represented it in conformity with the axioms, basic elements and synthetic method of traditional Euclidean geometry. If this interpretation holds, then Frege is not looking to psychology to provide the ground for geometric knowledge nor for any external justification of the Euclidean axioms. Instead, like Cohen, Frege is claiming that the axioms articulate the basic objective content assumptions for the science of geometry and serve as general pre-conditions for what is considered spatial within the natural sciences. Thus, they function as prescriptive norms, governing and correcting, our sense perception of spatial relationships.

---

[15] Even Matthias Schirn, who strongly supports reading Frege as a mathematical platonic realist, rejects the idea that Frege would have relied upon intellectual intuition to account for our cognitive access to abstract mathematical entities: "If Frege had been confronted with the postulation of a special faculty of mathematical intuition a la Godel, which is supposed to allow us direct cognitive access to the remote realm of abstract objects, he would probably have stigmatized it as a devastating 'irruption of psychology into logic'" (Schirn (1996b), 118).

[16] We know Frege was aware of Cohen's de-psychologized rendering of Kant's 'pure intuition' as a method by 1885 at the very latest, since this is when he published a review of Cohen's (1883) *Das Princip Infinitisimal Method*, which contains this reading. Noting the date when Frege read Cohen's *Princip* may also explain why, after 1884, Frege becomes increasingly careful in distinguishing his use of 'intuition' to denote the psychological imaging of a particular from his use of 'intuition' to denote the objective (non-psychological) source of geometric knowledge. In the 1879 preface to *Begriffsschrift*, for instance, Frege explains that his logicist project required developing a symbolism capable of representing proofs insuring that nothing "intuitive" could slip in unawares. However, it is unclear whether Frege's concern is with the particularity of intuitions, given that logical laws must "transcend all particulars", or that appeals to intuition import something from the geometrical source of knowledge (*BGS*, 5). §23 of *Begriffsschrift* suggests that his primary interest in excluding intuition is to show that arithmetical objects and concepts can be represented without having to borrow anything from sense perception or geometry, the non-logical sources of knowledge. This is made explicit in an article written sometime between 1924 and 1925: "I have had to abandon the view that arithmetic does not need to appeal to intuition either in its proofs, understanding by intuition the geometrical source of knowledge, that is, the source from which flow the axioms of geometry" (*PW*, 278). Here Frege stresses that intuition, in the second sense, has nothing do with anything psychological and everything to do with the traditional Euclidean axioms (Ibid., 273). Frege does refer to the source of geometric knowledge as a faculty in his 1873 doctoral dissertation. However, to my knowledge, all subsequent references to intuition as a source of knowledge are completely de-psychologized. The fact that Frege never explicitly states that he is appropriating Cohen's use of 'intuition' is not surprising, given that Frege is notorious for failing to acknowledge any debts to his contemporaries.

To see if this reading can be sustained, we first need to ask whether Frege did, as a matter of fact, think the Euclidean axioms expressed the objective meaning and content from which the entire subject matter of geometry is derived. The answer is yes:

> We cannot very well define an angle without presupposing knowledge of what constitutes a straight line. To be sure, that on which we base our definitions may itself have been defined previously; however, when we retrace our steps further, we shall always come upon something which, being a simple, is indefinable, and must be admitted to be incapable of further analysis. And the properties belonging to these ultimate building blocks of a discipline contain, as it were *in nuce*, its whole contents. In geometry, these properties are expressed in the axioms insofar as they are independent of one another ("On Formal Theories of Arithmetic" in *FG*, 143).

According to Frege, there are certain geometric notions, e.g. straight line, that are essentially defined by the set of Euclidean axioms: "Their sense is indissolubly bound up with the axiom of the parallels" (*PW*, 247). All other properly defined, cognitively meaningful geometric terms, e.g. 'angle' or 'triangle' must be composed from these basic constituents. We know, therefore, that the Euclidean axioms are universally and necessarily true concerning spatial entities and relations, since they constitute part of the meaning of geometric terms:

> When a straight line intersects one of two parallel lines, does it always intersect the other?...I can only say: so long as I understand the words 'straight line', 'parallel', and 'intersect' as I do, I cannot but accept the parallels axiom. If someone does not accept it, I can only assume that he understands these words differently (Ibid.)

For Frege, as for Cohen, to claim that intuition is the a priori source of geometric knowledge is to affirm the special role of Euclid's axioms in legislating the intersubjectively binding and cognitively significant meaning of geometric terms.[17]

---

[17] Frege scholars may shudder at the thought that Frege recognized Euclid's axioms as "defining" basic geometric terms. One worry is that Frege's main charge against Hilbert and his defender, Korselt, is that they confuse axioms with definitions. Another is that this characterization of Euclid's axioms makes them sound awfully similar to what Rudolf Carnap later described as meaning postulates, implying that Frege took geometric axioms to be analytic, which he expressly denied (*FA*, 101–2). In response, let me clarify my position by reminding my reader of Kant's distinction between two types of definitions. The first is "a purely verbal definition" of a concept or what he also describes as the definition of "an arbitrarily invented concept" (A593/B621 and A729/B757). The only restriction on definitions of this type is that the resulting concept exhibit logical consistency. However, this is not enough to insure the possibility of applying the concept to any object which might be given to us. Therefore, while definitions of this sort can ground logically necessary judgments, both concepts and judgments may still lack genuine cognitive content. In contrast, "real definitions" must associate the concept-word with some "clear property" insuring the possibility of a determinative application to the entire domain of spatio-temporal objects (A242). Providing real definitions for mathematical concepts requires presenting a rule-guided procedure (schema) that not only captures the meaning of a purely mathematical notion, but also accounts for the applicability of the concept to empirical phenomena. Kant considered the traditional Euclidean axioms and constructive definitions to be exemplars of real definitions, guaranteeing the application and relatively wide scope of validity for geometric concepts.

My point here is that Frege, like the Marburgians, shares Kant's view that the traditional Euclidean axioms and postulates functioned as real definitions for basic geometric concepts. I

The second question is whether Frege thought that pure intuition, i.e., the axioms, concepts, and spatial configurations of traditional Euclidean geometry, are tools necessarily employed in the natural sciences as well. Do they possess the requisite generality to function as preconditions for all scientific inquiry? Again, the answer is yes. Frege maintains that the domain over which the propositions of a discipline are considered valid is "determined by the nature of its ultimate building blocks" (*FG*, 143). He reminds his mathematical colleagues that the axioms of traditional Euclidean geometry include postulates guaranteeing the possibility of performing the "the simplest procedures", e.g., that a straight line may be drawn from any to point to another, from which all other geometric constructions are obtained (*PW*, 206–7). This postulated possibility of generating a geometric line is not a claim about us, about what we can actually draw or are psychologically capable of:

> But what in actual fact is this drawing a line? It is not, at any rate, a line in the geometrical sense that we are creating when we make a stroke with a pencil. And how in this way are we to connect a point in the interior of Sirius with a point in Rigel? Our postulate cannot refer to any such external procedure. It refers rather to something conceptual. But what is here in question is not a subjective, psychological possibility, but an objective one (Ibid., 207).

A postulate is a fundamental truth asserting the possibility of an objective procedure whereby a geometric entity can be constructed. He then argues that postulates assert "the existence of something with certain properties" (Ibid.). The ultimate building blocks of Euclidean geometry are, therefore, the appropriately idealized spatial configurations described in the postulates and conforming to the other axioms. Since the nature of these building blocks "ultimately are spatial configurations", the boundary of Euclidean geometry "will be restricted to what is spatial" (*FG*, 143). This includes anything considered to be a conceptually idealized spatial configuration within the natural sciences as well, such as the line connecting a point in the interior of Sirius with a point in Rigel.

So, according to Frege, the basic concepts and propositions of Euclidean geometry serve as the final arbiters of truth and fix the objective meaning of the terms employed in geometry as a science. Furthermore, the Euclidean postulates assert the

---

argue below that Frege's charge against Hilbert and Korselt is not that they are treating Euclid's axioms as definitions per se, but that they are depriving the axioms of their status as real definitions in Kant's sense and substituting purely verbal definitions for arbitrarily invented concepts. Given this interpretation, is Frege obliged to recognize the Euclidean axioms as analytic? No, not if we apply Frege's criteria for distinguishing analytic from synthetic propositions. Frege never characterizes analytic propositions as those which are true by virtue of the meaning of their constituent terms. In fact, as Michael Beaney points out, Frege could not endorse this notion of analyticity without threatening his logicist agenda: "[I]t is by no means clear that his logicist definitions and axioms embody the sameness of sense that would be seem to be a condition of their 'analyticity" (Beaney (1997), 25). Instead, analytic propositions are those whose scope of validity is maximally general and whose truth can be established by drawing solely from logical laws and definitions (*FA*, 4). Since my reading does not force Frege to claim that the Euclidean axioms have a scope of validity equivalent to arithmetic propositions or that they are derivable from logic alone, it similarly does not force him to claim that the axioms are analytic.

objective possibility of constructing the basic spatial configurations, with specific properties, which the geometer relies on to develop her subject matter, i.e., the space(s) investigated in geometry proper. Finally, the spatial properties and relations expressed by Euclid's axioms and postulates are of such a nature that they can serve as pre-conditions for the investigation of spatial entities in other scientific disciplines. It is for these reasons, I contend, that Frege insists that only "axioms in their original Euclidean sense" should be identified with intuition, conceived of as "the geometrical source of knowledge" (*PW*, 273).

For Frege then, as for a Marburg Neo-Kantian, the notion of intuition as the source of geometric knowledge is not bound up with the subject's capacity for sense perception. Instead, 'intuition,' in this second sense, refers to a method – the axioms, fundamental concepts, and constructive procedures of traditional Euclidean geometry. These, in turn, validate the perceptual observations made within a scientific context:

> In order to know the laws of nature we need perceptions that are free from illusion. And so, on its own, sense perception can be of little use to us, since to know the laws of nature we also need the other sources of knowledge: the logical and the geometrical. Thus we can only advance step by step—each extension in our knowledge of the laws of nature providing us with a further safeguard against being deceived by the senses and the purification of our perceptions helping us to a better knowledge of the laws of nature...We need perceptions, but to make use of them, we also need the other sources of knowledge. Only all taken in conjunction make it possible for us to penetrate ever deeper into mathematical physics ("Logic in Mathematics" in *PW*, 268).

On Frege's account, Kant's thesis that Euclidean geometry is a body of a priori truths grounded in pure intuition is not a thesis about which truths describe the spatial form imposed by our sensory apparatus, but rather a thesis about which geometric propositions and concepts serve as prescriptions for picking out empirically observable points, straight lines, etc. To say that the source of geometric knowledge is intuition is simply to say that these prescriptive notions are Euclidean ones.

Rejecting Kant's thesis is thus tantamount to claiming that the accumulation of knowledge represented in the long tradition of Euclidean geometry and the knowledge of nature that we thought we had achieved since Galileo and Newton is an illusion:

> The question at the present time is whether Euclidean or non-Euclidean geometry should be struck off the role of the sciences and made to line up as a museum piece alongside alchemy and astrology...That is the question. Do we dare to treat Euclid's elements, which have exercised unquestioned sway for 2000 years, as we have treated astrology? It is only if we do not dare to do this that we can put Euclid's axioms forward as propositions that are neither false nor doubtful ("On Euclidean Geometry" in *PW*, 169).

Thus, we see that Frege follows Cohen in refusing to appeal to a faculty of sensible intuition to validate the Euclidean axioms. Instead, when pushed to give a justification for the truth of the axioms, Frege offers the kind of indirect justification that I previously ascribed to Cohen and company. If geometry as it has been practiced to date counts as a science and is understood as contributing to our knowledge of the world, scientific practitioners must acknowledge the truth of these axioms.

I initially motivated the idea that Frege must not be operating with a univocal notion of intuition throughout his writings by citing his complaint that Hilbert neglected the fact that geometric axioms express basic facts of our intuition. We can now better understand the nature of that complaint. I have argued that when Frege says that Euclidean axioms assert basic facts of our intuition, he means that they prescribe the inter-subjectively binding, cognitively significant, objective content of geometric terms and that they postulate the nature of the basic spatial elements from which all properly geometric entities are constructed. Frege is pleased, therefore, whenever Hilbert characterizes the axioms of his system as "certain basic and inter-connected facts of our intuition" that "geometry requires…for its consequential construction" (*FG*, 25).[18] He objects, however, whenever Hilbert seems to deviate from this characterization of the axioms, as evidenced by the latter's ambiguous use of the terms 'point' or 'straight line':

> [I]t also is unclear what you [Hilbert] call a point. One first thinks of points in the sense of Euclidean geometry, and is confirmed in this by the proposition that the axioms express basic facts of our intuition. Later on, however (p. 20), you conceive of a pair of numbers as a point (Ibid., 6–7).

Hilbert responds to Frege's charge of equivocation by stating that he does not want to saddle his primitive geometric notions with any previously understood meaning or reference: "I do not want to presuppose anything as known" (Ibid., 11). The task of the geometer, claims Hilbert, is simply to develop a formal system of inter-related concepts, held together by necessary logical relations. Whether the term 'point' refers to Euclid's spatial simple, an ordered pair of numbers, or a chimney sweep is irrelevant: "the basic elements can be construed as one pleases" (Ibid., 13).

Hilbert thus intentionally divests geometric terms of the meaning and reference which they have within context of traditional Euclidean geometry, and this is the crux of Frege's complaint:

> We are easily misled by the fact that the words 'point,' 'straight line,' etc. have already been in use for a long time. But just imagine the old words completely replaced by new ones especially invented for this purpose, so that no sense is as yet associated with them. And now ask whether everyone would understand the Hilbertian axioms and definitions in this form (*FG*, 60).

For Frege and for Cohen, the axioms and postulates of Euclid's *Elements* express the objective content assumptions for geometry insofar as it qualifies as a science, a body of cumulative knowledge. Hilbert willfully deviates from the traditional usage and appears uninterested in explaining how his schema of formal concepts can be applied: "I do not know how, given your definitions, I could decide the question of whether my pocket watch is a point" (*FG*, 18). Frege concludes that Hilbert's is a system of pseudo-concepts, pseudo-propositions and pseudo-axioms, lacking in "thought-content," "sense," and "knowledge" (Ibid. 27, 85). To the extent that developers of a geometric system are successful in detaching their basic concepts

---

[18] Here Frege is quoting from the description of the axioms that Hilbert presents in the introduction and §1 of his *Foundations*.

and propositions from the ultimate source of validity, namely, the method of Euclidean geometry, they are essentially detaching their system from intuition, the source of geometric knowledge. We can safely assume, then, that Frege is taking a swipe at Hilbert, when he subsequently writes of those whose "recent works have muddied the waters," potentially contaminating the geometric source of knowledge, by attaching "a different sense to the sentences in which the axioms have been handed down to us" (*PW*, 273). And we know for certain that it is this practice of stripping a word of its "traditional sense [*Sinne*]" with no indication of a suitable replacement that Frege calls out as "a sin against science" (*FG*, 58).

4.2   Frege's Kantianism vis-à-vis Helmholtz and Cohen

Having shown how reading Frege against the backdrop of the Helmholtz-Cohen debate over rightly interpreting Kant for the sciences makes better sense of Frege's puzzling stance on the Euclidean axioms, I now turn to more fully developing my argument that Frege's Neo-Kantian commitments were motivated by his interest in warding off the practices that he attributed to Helmholtz's naturalized Kantianism. First, consider Frege's assessment of Helmholtz's proposed epistemology for arithmetic and number theory. In the *Foundations* and the 1885 "On Formal Theories of Arithmetic," Frege aligns formalist accounts of number with a cluster of interrelated logical mistakes. In the 1891 "Function and Concept," Frege describes the epistemological framework housing these errors, citing Helmholtz as its lead architect:

> There is at present a very widespread tendency not to recognize as an object anything that cannot be perceived by means of the senses; this leads here to numerals' being taken to be numbers…(*CP*, 138–9).

Here and elsewhere Helmholtz' work is accused of perpetuating this tendency:

> H v. Helmholtz apparently adheres to a formal theory in his essay 'Zählen und Messen erkenntnisttheoretisch betracthet'…when he says, e.g. 'I regard arithmetic, or the theory of pure numbers, as a method founded on purely psychological facts, which teaches us the consistent application of a system of signs (namely numbers) of unlimited extension and unlimited possibilities of refinement. Arithmetic investigates which combinations of these signs (arithmetical operation) lead to the same end product.' Here too, the signs are endowed with magical powers because their meaning has disappeared. To add to the confusion, psychology and empiricism are dragged in (*TPW*, 213).

In the essay Frege alludes to, Helmholtz applies his *Zeichentheorie* to show that just as one must reject Kant's thesis on the apriority of the Euclidean axioms so too one must reject the thesis that the "axioms of arithmetic" are a priori and grounded exclusively on a supposed "transcendental intuition of time" (Helmholtz (1887), 727).[19] As Frege notes, Helmholtz contends that both the objective application

---

[19] For Helmholtz, these "axioms" include the communativity and associativity of addition and the transitivity of equality [*gleichkeit*].

[*objective Anwendung*] and the objective sense [*objective Sinn*] of arithmetic propositions depends on deriving them from a foundation of psychological and physical facts. This collection of facts must explain how and under what conditions we can correlate a rule-governed system for combining numerical signs with physical relations [*physiche Verknüpfung*]. Establishing this correlation makes it possible to treat these relations as magnitudes comprised of denumerable units that can satisfy the axiom (Ibid., 730).

For Frege, attempting to ground the objectivity of arithmetic in its empirical applications or, worse, in the perceptibility of numerical signs and the psychological facts allegedly justifying their combination is indicative of a widening confusion about the nature of objectivity and the structure of justification existing among the sciences:

> Now objectivity [*der Grund der Objectivität*] cannot, of course, be based on any sense-impression [*Sinnes-eindrucke*], which as an affection of our mind is entirely subjective, but only, so far as I can see, on the reason [*der Vernunft*]. It would be strange if the most exact of all sciences had to seek support from psychology, which is still feeling its way none too surely (*FA*, 38).[20]

On Frege and Cohen's view longstanding a priori mathematical and logical propositions ground the objective validity of a posteriori scientific propositions, not the converse. By Frege's lights, Helmholtz's strategy for ensuring the objectivity of arithmetic reveals a deep misunderstanding about the kinds of questions an epistemology for science ought to pose: "[H]ardly ever has the sense of the epistemological problem been more misunderstood than here" (*TPW*, 213).

Though Frege's explicit references to Helmholtz's epistemological enterprise post-date the *Foundations,* there is textual evidence that Frege wrote the *Foundations* with an eye toward shutting it down. In the preface of his *Basic Laws of Arithmetic* [*Grundgesetze der Arithmetik*], Frege laments the fact that Helmholtz seems unaware of his work as evidenced by the lack of reviews following the publication of his *Foundations* (*BLA*, 8). More significantly for our purposes, Helmholtz commends Ernst Schröder's 1873 *Lehrbuch der Arithmetick und Algebra* for first recognizing that psychological and physical facts must be adduced to establish the objectivity of arithmetic (Helmholtz (1887), 729). In the *Foundations,* Schröder's book is especially targeted as depicting the formalist practices that the C&O distinction is aimed to curtail (*FA*, viii–ix, 54–5, 62–3).

Against the background of Helmholtz's Kantianism and particularly his characterization of intuitive representations as perceptual representations resulting from sub-conscious, sub-linguistic inductive inferences operating on memory traces, let us reconsider Frege's footnote to §27 of the *Foundations* in full:

> An idea [*Vorstellung*] in the subjective sense is what is governed by psychological associations; it is of a sensible pictorial character. An idea in the objective sense belongs to logic

---

[20] In faithfulness to Austin's translation of this passage, I have rendered 'der Vernuft' as 'the reason.' However, I believe that a more faithful translation to Frege's intent would have rendered it simply as 'Reason.'

and is in principle non-sensible, although the word which means an objective idea is often accompanied by a subjective idea, which nevertheless is not its meaning. Subjective ideas are often demonstrably different in different men, objective ideas are the same for all. Objective ideas [*objectiven Vorstellungen*] can be divided into objects and concepts [*Gegenstande und Begriffe*]. I shall myself, to avoid confusion, use 'idea' only in the subjective sense. It is because Kant associated both meanings with the word that his doctrine assumed such a very subjective, idealist complexion, and his true view was made so difficult to discover. The distinction here drawn stands or falls with that between psychology and logic.

I contend that Frege's elucidatory remarks in the *Foundations* are intended to accomplish the following: (1) to show that his use of these terms is already part of a well-recognized lexicon of logico-epistemological terminology; and (2) to signal his intent to recover those features of Kant's C&I distinction that are especially useful in reconstructing an epistemological framework capable of supplanting Helmholtz's naturalized Kantianism.

Frege's initial step towards recovering the relevant features of Kant's C&I distinction is to distinguish his use of 'object' and 'concept' to designate two types of objective perceptions from the use of 'intuition' and 'concept.' This move is necessitated by Helmholtz's rendering of Kantian terminology. Recall that after citing Kant's C&I distinction as it appears in the *Jäsche Logic,* Frege discusses those passages in the *Critique* where intuitive representations are linked with our capacity to obtain sensory representations of given objects. Remember too that Helmholtz's Zeichentheorie provides the historical context for unpacking Frege's argument that Kantian intuitions understood as sensory representations are too irredeemably idiosyncratic and private to play a role in accounting for the objective content ascribed to words and sentences occurring within scientific discourse.[21] A color word used in a sentence asserting a judgment based on "a physical experiment" cannot "signify our subjective sensation, which we cannot know to agree with anyone else's" (*FA*, 26). "Sensation", "intuition," and "all construction of mental pictures out of memories of earlier sensations" lack the inter-subjective accessibility and communicability demanded of objective ideas (Ibid).

Even after the *Foundations'* description of objective ideas and judgeable content is refined so as to distinguish between the objective sense [*Sinn*] and meaning [*Bedeutung*] of proper names and predicates, Frege still insists that a crucial distinction is the one demarcating the objective sense and meaning of an expression from any associated subjective idea. Here again, intuitions [*Anschauungen*] understood as sense impressions, memories of sense impressions and the acts performed

---

[21] Helmholtz's research in perceptual psychology, which was intended to account for the objectivity of our perceptual representations, ended up revealing such profound differences in how human subjects process sensory input that scientists themselves started to rethink what objectivity in the sciences really amounted to (Daston and Galison 2007, Chp. 5) Suffice it to say that when Frege writes that "subjective ideas are often demonstrably different in different men" and refers to the "peculiarities of [two people's] respective intuitions," he can look to the experimental results of Helmholtz's own research for support.

by the mind in arriving at a complete perceptual image [*Anschauungsbild*] are relegated to the class of subjective ideas (*FR,* 154).

Frege maintains that at some point in the history of a science it becomes necessary to provide a more "precise and fixed *Bedeuntung*" to the expressions that practitioners have been employing. He describes this shift in linguistic usage as transitioning from trafficking in "ordinary words" with their fluctuating connotations to trafficking in "technical terms" with a fixed use aimed at facilitating communication among fellow investigators and furthering the scientific goals of the discipline (*FR,* 313) When initiating such a transition, one will perforce rely on elucidations trafficking in ordinary words. One's success in fixing the use and meaning of proposed terminology thus depends on "other's guessing what we have in mind" and forging a practical "meeting of the minds" (Ibid.). Moreover, "logic, as in other sciences" is open to the coining of "technical terms" so long as the assigned meaning is best suited for expressing the laws belonging to that science (Ibid., 237)

I maintain that Frege's efforts to explain what he means by 'concept' and 'object' in the *Foundations* should be read as elucidations in the sense just described. Frege's 'concept' and 'object' are proposed technical term substitutes for the previously used Kantian expressions 'concept' and 'intuition,' expressions with a wide currency among nineteenth century logicians and epistemologists but with significant fluctuations in use. Frege will allow 'intuition' to serve as a technical term for perceptual psychologists to designate the psychophysiological processes described in Helmholtz's sign theory of perception. Frege also grants that this Helmholtzian understanding of the term can trace its lineage back to Kant's original texts, as long as his colleagues admit that these texts conflate two Kantian projects that should be kept separate, namely, the cognitive-psychological project of explaining how mental representations manage to reliably track the lawlike features of the external world and the logico-epistemological project of articulating the conditions ensuring the requisite objectivity of a properly functioning science. As we have seen, Frege blames Helmholtz's conflation of these two projects for perpetuating the logical mistakes characteristic of the formalists. Frege's C&O distinction is intended to recover the utility of Kant's C&I distinction in exposing and preventing such mistakes. His elucidations on the meaning of 'object' and 'concept' should be seen as the first step toward forging a practical agreement among those he considers colleagues to adopt his proposed terminology, including the syntactic and semantic rules governing its use.

What then should we make of Frege's remark that adopting his use of 'concept' and 'object' furthers the recovering of Kant's "true view"? Given Frege's own exposition of the relevant Kantian texts, we cannot take him to mean that his C&O distinction recovers a distinction that is univocally and unambiguously voiced by Kant himself, but neither can we follow Coffa's in dismissing Frege's remark as "extending an undeserved olive branch to the past" (Coffa (1991), 66). Frege's elucidations trade on the fact that the proposed terminology can find a foothold in some shared understanding of expressions already employed by practitioners of the relevant field of inquiry. Throughout the nineteenth century, neo-Kantians of all stripes hoped to frame questions relevant to the epistemology of science such that pursuing these questions might garner scientific respectability in its own right (Richardson (2006), 217). Developing an adequate epistemological framework for science was considered especially pressing due to rapid changes in the methodologies being employed in mathematics and the natural sciences. For better or worse, Helmholtz, Hermann Cohen, Frege and others who saw themselves inaugurating a would-be science of the epistemology of science saw themselves playing on a semantic field largely carved out by Kant (Daston and Galison (2007), 205–215).[22] In sum, to say that Frege's introduction and use of C&O distinction in the *Foundations* should be read as elucidations for proposed technical terms to be employed in this emerging scientific enterprise is to say that these elucidations needed to find a foothold in the nineteenth century Kantian lexicon.

Frege's complaint against Helmholtz's Kantianism is not that it falsely represents Kant's view, but that it privileges one aspect of that view, namely, Kant's interest in the faculties, capacities and limitations of the knowing subject. For Frege, the Kantian perspective better equipped to address the issues at hand is one keeping an ever vigilant eye on the boundary between cognitive-psychological concerns and logico-epistemological ones. In his 1885 review of *PIM*, Frege chastises Cohen for his obscurity, for the lack technical proofs needed to establish his claims and for his failure to recognize the significance of recent efforts to arithmetize the infinitesimal calculus.[23] Still, he shares Cohen's conviction that Kant provides the basis from which to launch "epistemological and logical investigations" into the structure of mathematical thoughts (*CP*, 108). Moreover, he explicitly applauds Cohen's insistence that "knowledge as a psychic process does not form the object of the theory of knowledge, and hence that psychology is to be sharply distinguished from the theory of knowledge" (*CP*, 111). That said, I now turn to more fully defending my thesis that Frege would, or at least should, credit Cohen's 1871 *Kant's Theorie der Erfahrung* and his 1883 *PIM* with introducing the "true" Kant, since reading Kant through Cohen's eyes is crucial in achieving Frege's goal of convincing his peers to forsake the misguided epistemology of science and logical errors promulgated by Helmholtz and his fellow formalists.

When pushed to define what he means by 'concept' and 'object,' Frege claims that they are indefinable primitive terms corresponding to the ultimate building blocks or "simple" substances of logic (*FR*, 108 and 111). In a 1902 letter to Russell, he refers to the C&O distinction as "a primitive logical fact" for which "no possible definition in possible" (*PMC*, 142). Joan Weiner (2002) argues that much of what Frege writes about concepts and objects must be read as elucidations. She holds that a Fregean elucidation can be recognized *as* an elucidation and its content correctly

---

[22] See too MacFarlane (2002) on the widespread recognition throughout the nineteenth century that Kant's views on the nature of pure general logic constituted the authoritative base from which any proposed modifications or extensions in the sense and reference of terms 'logic' or 'logical' must seek its legitimacy (Ibid., 44–46).

[23] In fact, Frege suggests that Cohen's "misunderstanding" of the arithmetic foundations of the calculus could be cleared up by reading his *Foundations* (*CP*, 111).

interpreted only by looking at "the role it plays in a science" (Ibid., 163).[24] If Weiner is right, and I believe she is, correctly interpreting Frege's elucidations of the C&O distinction depends not only on understanding the role the distinction plays in fending off formalist mistakes, but also understanding how this role can be conceived as a proper part of practicing logic as a science. I will argue that Frege's insistence that logic *is* a science provides further evidence that he rejects Helmholtz's Kantianism in favor of Cohen's. I thus need to show that logic, for Frege, is intimately connected to, if not identical with, the science that Cohen named 'Erkenntniskritik.'

Recall that in *PIM*, Cohen explicitly states that the term 'Erkenntniskritik' is intended to recapture Kant's "original discovery" that the "critique of reason" just is "the critique of cognition or science" and that this critique discovers the "pure in reason" by discovering the "conditions of certainty" upon which "cognition as science" is founded (Ibid., 6). He concludes:

> The critique of cognition [Erkenntniskritik] is thus equivalent in meaning with transcendental logic; for their task is the discovery of synthetic fundamental propositions or that foundation of knowledge upon which science builds itself and the validity of which science depends (Ibid., 7)

In *Foundations of Arithmetic,* Frege states: "the aim of proof is…not merely to place the truth of a proposition beyond all doubt, but also to afford us insight into the dependence of truths upon one another" (*FA*, 2). The goal of his logicism is to present the basic logical propositions from which the subject matter of arithmetic is built up and upon which its objectivity and validity is derived. In pursuing this goal, he aims to reveal the logical structure that grounds the validity of inferential practices of science as a whole, establishing the extent to which the propositions of natural science are dependent on mathematical propositions for their justification.[25] Thus, Frege is engaging in the very task that Cohen assigned to the properly oriented Neo-Kantian epistemologist.[26]

In *Foundations*, Frege also distinguishes his conception of induction from an empiricist's conception. The former is a mode of "inference" providing "justification" for beliefs in the truth of a proposition; the latter that treats induction as

---

[24] Weiner's interest is not defending a specific thesis about the C&O distinction, but rather examining Frege's elucidations of the distinction to argue that §31 of *Basic Laws of Arithmetic* is an elucidation despite its prima facie appearance as a proof (Weiner (2002), 149).

[25] By describing this structure as composed of objectively valid inferences, I betray my indebtedness to Danielle Macbeth's *Frege's Logic*. Macbeth argues that Frege's conception of logic differs from Russell's and Wittgenstein's in that, for Frege, the correctness of an inference necessarily depends on recognizing the truth of the premises from which it is drawn (Macbeth (2005), 3). This argument accords with my argument that Frege shares Cohen's view that the correctness and objective validity of the inferences drawn in the natural sciences depends essentially on affirming the truth or objective validity of propositions of the more general and pure sciences of logic and mathematics. I hesitate, however, to suggest that Macbeth's reading of Frege's conception of logic is entirely compatible with my own, given that she focuses on Frege's "mature" view of logic, namely, a post-1884 view strongly informed by his distinction between *Sinn* and *Bedeuntung*.

[26] See Sect. 3.2.1 for a reminder on what that orientation is.

"habituation" that merely "induces men to believe" in their truth (*FA*, xi, 4 and 16–7).[27] Induction, as Frege conceives it, derives its justification from a more general law: the proposition that "the inductive method can establish the truth of a law, or at least some probability for it" (Ibid., 4). The justification for this law necessarily presupposes the truth of the general laws articulated in probability theory, which in turn presuppose the truth of the even more general laws of arithmetic. For Frege as for Cohen and in contrast to Helmholtz, a proper logico-epistemological inquiry aims to uncover the "objective standards" of justification and certitude for scientific cognition as such, and this involves demonstrating the necessary use of mathematical and logical propositions as general, justificatory norms of reasoning in the natural sciences (*FA*, 16–7).[28] And just like Cohen's *PIM*, Frege's *Foundations* is written with the explicit purpose of convincing any epistemologist tempted to base the objective validity of mathematics on the psychophysiology of inductive inference to "examine afresh the principles of his theory of knowledge" (*FA*, xi).

An objection to my claim that Frege conceives of logic along the lines of Cohen's Erkenntniskritik is that Cohen himself distinguishes the two. To counter this objection, we need to examine the rationale behind Cohen's distinction. While it is true that Cohen's *PIM* distinguishes general logic ('logic') from Erkenntniskritik or transcendental logic, the distinction is not based on the purported formality of the former. Indeed, the idea of "logic as a *formal* science" is branded an "unintelligent insinuation" (*PIM*, 3). Remember that in keeping with his intent to wrest Kant from the hands of Helmholtz, Cohen insists that the terms 'intuition' and 'thought' must be interpreted as distinct "scientific methods," methods manifesting an independence or autonomy that enables them to function as "general preconditions for all scientific research" (Ibid.). Logic is identified with the method of pure thought and has the task of securing the entire objective content of cognition, whereas Erkenntnistkritik is identified with the method of intuition (Ibid.) Logic thus validates norms of reasoning that apply to domains of inquiry outside the scope of Erkenntniskritik. In virtue of its maximal generality, Cohen argues that logic is a source of validity for "the moral sciences" as well as the natural sciences (Poma (1997), 98). For Frege too, logic is the "science" tasked with discovering rules holding "with the utmost generality for all thinking, whatever its subject matter" (*FR*, 228). These maximally general rules serve as "prescriptions for making judgments" insofar as these judgments aim at truth (Ibid., 246).

---

[27] Given Helmholtz's description of the subconscious inductive inferences yielding sensory symbols, one can reasonably assume that the "empiricist conception" Frege has in mind is Helmholtz's or one just like it.

[28] See Ricketts (2010) on Frege's conception of logic as "a maximally general science that every science implicitly assumes and draws on" and logical laws as "maximally general truths" expressing "topic-universal modes of inference" (Ibid.,151). For Ricketts, a distinctive contrast between Fregean and traditional conceptions of logic, including Boolean logic, is that Frege privileges modes of inference moving from the general to the specific (Ibid.,150–4). It is my contention that this is a distinctive that Frege shares with Cohen and the Marburg school of Neo-Kantianism.

Erkenntniskritik, says Cohen, focuses more narrowly on identifying sources of validity for the natural sciences, which requires explicating concepts that bridge the gap between reasoning as conducted in pure mathematics and in the empirical sciences:

> From the time of the Trendelenburg-Fischer debate [1871], Cohen was dedicated to the idea that mathematics is a science of *sui generis* reasoning that can legislate for itself, according to the laws of thought. The practice of mathematics must not be constrained by any physical limitation of perception, but only by the most general relationship between the laws of thought and the empirical facts of science, the "given" (Patton (2005), 97).

Cohen describes mathematical reasoning so as to affirm its close affinity to the laws of thought, to logic. Mathematical reasoning operates with a self-legislating autonomy and scope of normative authority distinguishing it from the more constrained forms of reasoning operative within natural science. Frege also appeals to the widespread and necessary use of arithmetical reasoning to argue for its kinship with pure logical reasoning and the inadequacy of formalist attempts to derive the concept of number from sensibly perceptible collections or empirical facts about psychophysical 'actualities':

> Empirical propositions hold good of what is physically or psychologically actual, the truths of geometry govern all that is spatially intuitable….The basis of arithmetic lies deeper, it seems, that any of the empirical sciences, and even that of geometry. The truths of arithmetic govern all that is numerable. This is the widest domain of all; for to it belongs not only the actual, not only the intuitable, but everything thinkable. Should not the laws of number then, be connected very intimately with the laws of thought? (*FA*, 20–1).[29]

For both Cohen and Frege, the problem with a Helmholtzian strategy of grounding the justification and objective content of mathematical propositions on the science of perceptual psychology is that it cannot account for the necessary and normative application of mathematical reasoning throughout the whole of science.

Like Cohen, Frege realizes that to say mathematical propositions serve as norms certifying the reasoning and conferring the objectivity on the propositions of natural science incurs an explanatory burden of showing that they do indeed fulfill this role. So, though Cohen and Frege reject the idea that one should look to psychophysiology or any empirical science to determine the nature, ground the truth or insure the objective content of mathematical propositions, they insist that mathematicians cannot engage in practices making the widespread and fruitful application of mathematics throughout science inexplicably mysterious. In fact, this is one of Frege's main charges against the formalists.

In his 1885 essay 'On Formal Theories of Arithmetic,' Frege makes it clear that a primary motive behind his logicist project is showing that there is no scientifically significant boundary dividing logic from arithmetic, that together they "constitute a unified science," and that logic is not the "barren," content-free discipline some take

---

[29] See MacFarlane (2002) for additional textual references showing that Frege argues *from* the general normative applicability of arithmetic and his critique of formalism *to* the plausibility of logicism (Ibid.,38–9).

it to be (*FG,* 142).[30] Here again, Frege argues that arithmetic bears a striking resemblance to logic based on the premise of their similarly "extensive applicability" (Ibid., 141). Whatever grounds the inferences used in a science determines the scope of valid application for the propositions of that science. If observation plays a role in justifying the inferences countenanced in a discipline, the propositions of that science are only applicable to the domain of the physically observable (Ibid., 142–3).[31] The properties of the "ultimate building blocks" expressed in the fundamental propositions of a discipline similarly set the parameters of applicability: "If, as in the case of geometry, these ultimately are spatial configurations, then the science too will be restricted to what is spatial" (Ibid., 143). Frege's complaint is that formalists like Helmholtz, who try to ground the objectivity of arithmetic propositions and the ultimate building blocks of arithmetic – the numbers – on inferences countenanced in psychophysiology or the actuality of numerical signs, end up rendering the application of arithmetic to anything other than physically observable actualities inexplicable.[32] By Frege's lights, the formalists have purchased the actuality of the numbers at the cost of securing the objectivity and normative force and scope of arithmetic propositions.

Where Cohen and Frege differ, at least during the crucial period between Cohen's publication and Frege's reading of *PIM,* is on whether logic alone can supply the resources needed to satisfy the burden of explaining how mathematical and particularly arithmetic propositions are necessarily applied throughout science. Cohen maintains that logic cannot provide these resources and hence the need to invoke another "source of cognition," the method of intuition distinctive of Erkenntniskritik. Frege maintains that logic can supply these resources, his *Foundations* goes a long way in showing that it can and Cohen ought to read it. So though it is true that Cohen distinguishes Erkenntniskritik from logic, identifying the former as the science tasked with discovering the normatively necessary, universally applicable propositions functioning as sound methodological and inferential rules for the natural sciences, it is also true that the line of demarcation that Cohen draws in *PIM* is much fuzzier that it would be under a Helmholtzian reading of Kant. And according to Frege, it is entirely unwarranted. In his review of *PIM,* Frege suggests that Cohen draws this distinction because he falsely assumes that making sense of the semantics and applicability of infinitesimal Calculus requires borrowing from geometry and mechanics, a "misunderstanding" that could be cleared by reading the *Foundations* (*CP*, 111). Presumably, reading the *Foundations* would also convince Cohen that the division of labor between logic and Erkenniskritik has been further effaced (Ibid., 109).

---

[30] Frege's logicism is as much about proving that logic is a science with its own distinctive subject matter as it is about establishing what grounds the truth of arithmetic propositions.

[31] Here my reading of Frege roughly aligns with Brandom (2009); the objective purport or content of a proposition is determined by the role it plays as a premise or conclusion in the inferences that scientific practitioners are entitled to draw (see Brandom (2009), 166–8 and 204–5.

[32] Frege acknowledges that the formalists have a story to tell about why arithmetic is applicable to non-physical entities, but this story makes an appeal to abstraction that Frege rejects.

For Kant himself, the distinction between logic and transcendental logic was based on theses that both Frege and Cohen reject: (1) that the objective content and cognitive significance of a priori concepts rests on showing their possible relation to an object(s) and that this entails showing that concepts are intrinsically related to our faculty of sensibility; and (2) that space and time are forms of sensibility providing a pre-conceptualized, given manifold from which schematized pure logical and mathematical concepts obtain their essential relatedness to objects of possible experience. Because Kant charges logic with the task of identifying the laws of the understanding in isolation from the understanding's relationship to the givens of sensibility, it follows that logic is a purely formal discipline rather than a science with a circumscribed subject matter about which we can extend our knowledge. By contrast, transcendental logic is a distinct science charged with the task of identifying the necessary concepts and laws making cognition of objects possible.

Since Cohen and Frege reject the encroachment of transcendental psychology into the domains of logic and epistemology and since both identify the objectively 'given' not with the sensibly given actualities, but with the propositions given to us in a properly scientific language or, as Frege states in the *Foundations*, with the objective meaning of the singular terms used in that language, the basis for distinguishing logic from transcendental logic starts to erode. MacFarlane (2002) argues that the debate between Frege and Kant over the nature of logic can be seen as substantial and not a mere quarreling over words, given that Kant's thesis on logic's presumed formality is a consequence deriving from premises grounded in his transcendental psychology (Ibid., 49–53). MacFarlane's argument is important for my purposes in that it lends credence to the claim that, for Cohen and Frege, refuting this presumed formality goes hand in hand with refuting Helmholtzian interpretations of Kant.

Heis (2010) argues that Cohen and his students subsequently rejected the distinction altogether (Ibid., 386–8). For Friedman (2000), Cassirer retains the distinction but only insofar as logic is described as an abstraction from transcendental logic, where transcendental logic must consider the "particular role of space and geometry in the constitution of empirical knowledge" (Ibid., 93). Once Cohen's students, Natorp and Cassirer, became aware of the riches of Frege's new logic, they still did not believe these resources sufficed to derive substantive, explanatorily fruitful regulative principles for natural science (Heis (2010), 385). There are problems, however, with allowing Natorp and Cassirer to have the final say on whether Frege's work is compatible with or furthers the epistemological project initially conceived by Cohen. First, to my knowledge, there is no evidence that Cohen, Natorp or Cassirer were aware of Frege's work prior to the revelation of Russell's paradox that derives from Frege's Basic Law V. And it is Frege's introduction of the natural numbers as the extensions of certain concepts licensed by Basic Law V that does the heavy lifting in explaining how the truths of arithmetic are truths about all n-membered concepts from every scientific discipline. So after Frege became aware of the paradox, he abandoned his logicism and ultimately sought to ground arithmetic in the geometric source of knowledge. Second, neither Natorp nor Cassirer pay much attention to Frege's critique of Helmholtz or the other formalists, and so they are not in a position to say whether the characteristically Fregean theses espoused in this critique support or are supported by Cohen's interpretation of the Kantian project.

Given my argument that Frege's notion of logic is remarkably similar to Cohen's Erkenntiskritik, what follows for correctly interpreting Frege's claim that the C&O distinction is a primitive logical fact? MacFarlane (2002) nicely summarizes the first implication: "To ask whether a primitive law is logical or nonlogical is simply to ask whether the norms it provides apply to thought *as such* or only to thought in a particular domain" (Ibid., 40). For Frege, to say that the C&O distinction is a primitive logical fact is to say that the distinction expresses a norm, the violation of which cracks the foundation from which the objectivity and inferential validity of scientific cognition *as such* is derived. Frege's *use* of the distinction to illuminate the ways in which formalists jeopardize the kind of objectivity and scope of validity rightly ascribed to arithmetic propositions is thus constitutive of the *justification* for identifying the distinction as a primitive logical fact.

A second implication, however, is that merely demonstrating the use of the distinction against the alleged errors of the formalists is not sufficient; Frege must also show that his use of the C&O distinction parallels Kant's own use of the C&I distinction. Recall that both Cohen and Helmholtz were interested in showing that the epistemology might be viewed as a science in its own right, and both looked to Kant as its founding father (Richardson (2006), 216–217; Cohen and Elkana (1977), xii–xiv). Recall too that rapid developments in the sciences raised concerns that though practitioners employed the same terms, the equivocal use of terminology could result in practitioners talking past one another. Frege conceives of scientific research as a collaborative enterprise and this assumes the possibility of translocal, transtemporal communication among practitioners. He insists that the use of scientific terms must warrant this assumption (*CP*, 300; *FG*, 36–7 and 59–60; *PW*, 247–50). In other words, Frege joins Cohen in anticipating the Kuhnian concern that revolutionary changes in science can cause semantic variances and discontinuities that threaten notion of its progressing, and this concern extends to the science of epistemology.

In *PIM*, Cohen raises worries about semantic continuity within the tradition of epistemology founded by Kant. He points out the "vacillating linguistic usage" of Kantian terminology that occurs "from author to author and often by the same author" (*PIM*, 17). To address this problem, Cohen insists that one must make "an exact determination" of expressions best representing what the "the modern scientific language as received from Kant" (Ibid.) When proposing a more precise and fixed meaning of Kantian terms, one is not obliged to stick to the letter of Kant's corpus since Kant's own usage is often ambiguous. Rather, one must capture the spirit of Kant's critical philosophy that is expressed in the use of terms to justify or reinforce one of Kant's "scientific doctrines" (Patton (2005), 113). Cohen's *PIM* calls for redefining or elucidating the Kantian expressions to serve as primitive terms for Erkenntniskritik. These elucidations must preserve the requisite semantic continuity entitling one to see himself as a sinless practitioner of the properly *scientific* form of philosophizing initiated by Kant. My thesis, of course, is that Frege saw himself as one such practitioner and attempted to answer Cohen's call.

Just as Frege believes that 'point' is an indefinable primitive term corresponding to one of the ultimate building blocks of geometry, so too he believes that 'concept' and 'object' are indefinable terms corresponding to the ultimate building blocks ("simple substances") of logic:

> The word 'concept' is used in various ways; its sense is sometimes psychological, sometimes logical, and perhaps a confused mixture of the both. Since this license exists, it is natural to restrict it by requiring that when once a usage is adopted it shall be maintained. What I decided was to keep strictly to a purely logical use....Kerry contests what he calls my definition of 'concept'. I would remark, in the first place, that my explanation is not meant as a proper definition. One cannot require that everything be defined, anymore than one can require that a chemist decompose every substance. What is simple cannot be decomposed, and what is logically simple cannot have a proper definition....On the introduction of a name for something logically simple, a definition is not possible. There is nothing for it but to lead the reader or hearer, by means of hints, to understand the words as is intended. ("On Concept and Object" in *FR,* 182*)*

> [T]he question arises what it is that we are here calling an object. I regard a regular definition as impossible, since we have here something too simple to admit of logical analysis. It is only possible to indicate what is meant [*gemeint*]. ("Function and Concept" in *FR,* 140)

> First of all, I must emphasize the radical difference between concepts and objects, which is of such a nature that a concept can never substitute for an object, or an object for a concept. Here it is impossible to give proper definitions. ("Letter to H. Liebmann" in *FG,* 3)[33]

Elucidations like these, in conjunction with Frege's use of the distinction to combat the mistakes of the formalists, are intended to meet Cohen's specifications for providing a more exact, scientifically fruitful determination of Kantian terminology, thereby justifying the recognition of C&O distinction as the legitimate successor of Kant's C&I distinction and hence a primitive logico-epistemological fact.

4.3   Deploying the Distinction Between Concepts and Objects

4.3.1   Fregean Concepts and Objects – What Are they?

I will now compare and contrast the Fregean notion of a proper concept[34] with the Kantian notion presented in Chap. 2. Frege does not maintain that the potential applicability and cognitive significance of pure concepts depends on schematizing

---

[33] See also Frege's 1902 letter to Russell, where he claims that the distinction between objects and concepts is a "primitive logical fact" and so "no proper definition is possible here" (*PMC,* 142).

[34] In discussing Frege's criteria for a proper concept, I am restricting myself to what Frege sometimes describes as first-level, one-place predicates or functions. This is because when Frege is expounding the essential difference between concepts and objects, he is almost always talking about first-level, one-place predicates. Furthermore, the goal of Chap. 4 is to show how Frege employs his C&O distinction to highlight the mistakes of the formalists and, according to Frege, these mistakes involve confusing objects with concepts of this specific type.

them in accordance with our forms of sensibility. He does, however, restate the difference that Kant noted between obtaining a logically consistent set of conceptual marks and presenting a concept with cognitive content. For Frege, freedom from contradiction is neither a necessary nor sufficient criteria of a properly defined concept: "What is required of a concept…is a sharp boundary—not freedom of contradiction, which is in no way required" (*CP*, 133). A concept has a sharp boundary if, for any given object, it yields a determinate and inter-subjectively binding answer to the question: does the object fall under the concept? (*FR*, 259). To satisfy this condition, the concept must be characterized sufficiently to indicate what "property an object must have in order to fall under the concept" (*BLA, 11*). Throughout Frege's writings, he insists that we have not succeeded in obtaining a proper concept unless the constituent marks of the concept can simultaneously be recognized as properties allowing us to state unequivocally whether an object(s) bears them or not[35]: "An object is said to *fall under* or be *subsumed under* a concept if and only if it has those properties that are characteristics of the concept" (Kluge (1971), xvii). Since Frege regards it "as essential for a concept that the question whether something falls under have a sense," he requires that conceptual marks must be related to properties such that anyone "well acquainted with the [object] in question" can agree on whether it bears those properties (*PMC*, 101 and *FG*, 63).

The first thing to notice about a proper Fregean concept then is that it must be introduced by what Kant described as a real definition. Recall, a "real definition" provides a "clear **mark** by means of which the **object** (*definitum*) can be securely cognized" (A241–2). Real definitions ensure that a concept is "usable in application" (Ibid.). Frege also endorses Kant's view that introducing geometric concepts by specifying the correlative constructive procedure are paradigmatic real definitions. For, it is the concepts obtained in this way that Frege treats as full-fledged geometric concepts, concepts that function as the rules for deciding what spatial entities do and do not fall under them.[36] Of course, he does not follow Kant in thinking that accounting for the potential applicability of pure concepts requires "descending to the conditions of sensibility" (A241/B300). This is because, for Frege as for Cohen, the applicability criterion is met so long as the construction of geometric concepts indicates how physicists can employ the concepts in order to present a law-like representation of natural phenomena. There is no additional requirement to show that the concept functions as a rule whereby the imagination

---

[35] In responding to A. Korselt's objection that this criterion for defining concepts is too stringent, Frege goes on to explain that he is not saying that the definition alone, prior to and independently of any investigation of objects, should enable us to decide whether or not they fall under the concept: "The question of whether a given stone is a diamond cannot be answered by the mere explanation of the word 'diamond' itself. But we can demand of the explanation that it settle the question objectively, so that by means of it everyone well acquainted with the stone in question will be able to determine whether or not it is a diamond" (*FG*, 63).

[36] Here I am challenging Matthias Schirn's claim that Frege never relies on "the construction of geometrical concepts a la Kant" to account for the generality of geometric objects (points, lines, triangles) and thereby accounting for relatively wide scope of validity accorded to Euclidean concepts and propositions (cf. Schirn (1996a), 24–5).

unites raw sensory input and how the concept is thereby applicable to the domain of the subject's perceptual experience.

Frege objects to Hilbert's method of defining geometric concepts by embedding them in a network of logically consistent propositions precisely on the grounds that this is not a real definition, one telling us how to apply these concepts to the whole host of objects with recognized spatial properties:

> I do not know how, given your definitions, I could decide the question of whether my watch is a point. Already the first axiom deals with two points; therefore, if I wanted to know whether it held of my pocket watch, I should first of all have to know of another object that it is a point. But even if I knew this, e.g. of my fountain pen, I should be unable to decide whether my watch and my fountain pen together determine a straight line, because I don't know what a straight line is (*FG*, 18).

Hilbert purports to define hitherto unknown expressions for geometric concepts by stating the logical relations that hold between those concepts. For Frege, the most charitable interpretation of what Hilbert is attempting to do would only result in a system of mutually defined concepts whereby we know which concepts are subordinate to one another and which concepts "fall within" others.[37]

To see the crux of Frege's complaint, consider his exposition of Hilbert's definition of 'is a point of a straight line.' First, Hilbert stipulates that all points on a straight line stand in a relationship to one another designated by 'betweenness'. Second, Hilbert presents a list of axioms clarifying what the betweenness relation amounts to. The following presented axioms are sufficient to understand Hilbert's intent, as well as Frege's objection:

> II.1 If *A, B, C*, are points of a straight line, and *B* lies between *A* and *C*, then *B* also lies between *C* and *A*.

> II. 3. For any three points of a straight line there is always one and only one point that lies between the two. (*FG*, 31)

Hilbert also presents it as an axiom that "there are at least two points on every straight line" (*FG*, 31).

Frege likens Hilbert's proposed definition with providing the following definition of the concept 'is a god':

> Explanation: We conceive of objects which we call gods.
> Axiom 1. Every god is omnipotent.

---

[37] For Frege, subordination is a relation that can hold only between concepts of the same level, primarily between concepts of the first-level. Propositions of the form "For all x, if x is an F, then x is a G" or "All F's are G's" assert that a concept *F* is subordinate to a concept *G*, which is equivalent to asserting that the characteristic marks comprising the concept *G* are also characteristic marks of the concept *F* (*FG*, 4). "Falling within" is a relation that can hold only between concepts of two different levels, primarily between first and second level concepts. Such relations are asserted in propositions with the form 'There is an F' or 'There are at least two F's' or 'There are six F's'. These propositions assert facts concerning the instantiation of first-level concept in question, e.g., whether or not any objects fall under the concept *F* and how many there are.

Axiom 2. There is at least one god (*FG*, 32).[38]

What is Frege's point? Even if we allow Hilbert to treat axioms as definitions for concept-words to which we presumably attach no prior meaning, then the information contained in Hilbert's axioms only tells us that the concept denoted by 'standing in a betweenness relation' is a constituent mark of the concept that he denotes by 'is a point on a straight line' and that this first-level concept does and does not fall within certain second-level concepts. For instance, since Frege parses sentences of the form 'There are at least two F's' as asserting that a first-level concept *F* has at least two distinct objects falling under it, Hilbert's axiom 'There are at least two points on a straight line' guarantees that his concept 'is a point of a straight line' falls within the second-level concept 'There is at least two' and not the second-level concept 'There is zero.'[39] Frege claims, however, that the information contained in the whole of Hilbert's axioms is not enough to associate any recognizable properties with these newly introduced concept-words: "They do not provide an answer to the question, 'What property must an object have to be point, a straight line, a plane, etc.?' (*FG*, 19).[40] Since Hilbert has not satisfied this demand, he has provided "neither sense nor reference for the grammatical predicate 'is a point'" and "every proposition with this predicate is senseless" (*FG*, 63).[41]

---

[38] Frege restates this analogy whenever he discusses Hilbert's (1899) *Foundations of Geometry* (see *PMC*, 90, and *FG*, 19).

[39] It should also be noted that Hilbert's definitional criteria is fulfilled both by the Euclidean concept of 'is a point on a straight-line' and certain non-Euclidean concepts associated with this term. This is what Frege means when he says that Hilbert may have succeeded in presenting a system of higher-level geometric concepts and propositions, in relation to which the system of first-level Euclidean geometry "presents itself as a special case" among "innumerable other [first-level] geometries, if that word is still admissible" (*FG*. 37). In spite of this admission, Frege still maintains that Hilbert does not achieve what he set out to. He has not shown the independence of Euclid's parallel axiom from the other Euclidean axioms. All that he has shown is that the Euclidean notion of parallelism is independent of Hilbert's higher-level notion and that the Euclidean notion of parallelism is independent of certain non-Euclidean concepts associated with 'is a point' or 'is a straight line.' Besides, we already know that Frege does not think that 'geometry' is an admissible word for denoting first-level non-Euclidean systems of mutually consistent concepts and propositions, if by 'geometry' we intend to refer to a body of cognitively significant a priori propositions.

[40] Since Frege thinks that Hilbert's definitions misfire insofar as they do not ensure that the sign for a first-level concept is associated with any property, it is not quite accurate to say that Hilbert has provided enough information entitling us to state which concepts fall within others. For, we can only ascertain whether and how many objects fall under a first-level concept if the concept can be recognized as a property in the first place.

[41] Throughout his exchange with Hilbert and Korselt, Frege tends to treat 'thought-content', 'sense', and 'possessing epistemic value' as synonymous expressions. And I take these expressions to be synonymous with my previously defined 'objectively valid.' In other words, a proper Fregean concept is one whose constituent marks can also be recognized as properties useful in distinguishing what objects, if any, fall under the concept. If a concept satisfies this condition, then it is objectively-valid and possesses epistemic value. To say that a proposition is objectively-valid is to say that it has a non-trivially determinable truth-value and possesses cognitive content. I am well

A second thing to notice is that Sluga (1980) is right when he says that Frege's doctrine concerning the essentially predicative nature of concepts expresses allegiance to Kant's doctrine of the primacy of judgments over concepts. As was the case with its Kantian predecessor, to be a proper Fregean concept is to be a constituent part of a judgeable content with a non-trivial, inter-subjectively assessable and binding truth-value.[42] Recall Kant's characterization of a proper concept as a predicate of a possible judgment. Frege writes, "a concept is for me that which can be predicate of a singular judgment-content" (*FA*, 77). Kant describes a concept as a necessary rule underwriting the truth and falsity of judgments having about particulars, e.g., 'Fido is a dog'. For Frege, properly defined concepts are essentially predicative or unsaturated: "What in the case of a function is called unsaturatedness, we may, in the case of a concept, call its predicative nature" ("Comments on *Sinn* and *Bedeutung*" in *FR*, 174). The "must have" property of these unsaturated entities is yielding "genuine [truth-valued] sentences" whenever conjoined with a complete entity, a.k.a. an object ("Introduction to Logic" in *FR,* 298). We said that proper Kantian concepts are functions enabling us to take up objects as subjects in truth-valued judgments. Proper Fregean concepts are one-argument functions taking objects to truth-values (*CP,* 148, and *BLA*, 36).

Moreover, Frege's allusions to the unsaturated and incomplete nature of concepts are intended to underscore the idea that the process of forming a bona fide concept cannot precede the formation of an entire objectively-valid judgeable content. When contrasting his view of logic with Boole's, Frege writes, "For Aristotle, as in Boole, the logically primitive activity is the formation of concepts through abstraction" (*PW*, 15). For Frege, the logically primitive activity is the formation of a judgeable content:

---

aware that Frege does not always equate the sentences expressing a thought-content or sense with sentences possessing epistemic value or having a determinate truth-value.

[42] By identifying Frege's C&O distinction as a distinction between the constituents of a judgeable content, I am committed to locating the distinction as one drawn primarily within the realm of sense. As previously mentioned, Frege's sense-reference distinction is not a major factor in his use of the C&O distinction against the formalists. Furthermore, in texts where Frege is more careful in distinguishing the sense and reference of conceptual expressions, the predicative or unsaturated nature of concepts tends to be stressed as a feature of the sense of these expressions. See *FR* 295, 363, and especially *PW,* 119 where Frege acknowledges in a footnote: "The words 'unsaturated' and 'predicative' seem more suited to the sense than the meaning [*Bedeutung*]...." Therefore, the fact that my exposition does not firmly place concepts within the realm of reference should not be construed as evidence that it has nothing to offer regarding Frege's mature view on the nature of concepts. We have already noted that Frege's critique of the formalists lasted until the end of his days. I have also argued that we cannot fully appreciate what Frege intends by the C&O distinction or why he is so committed to it unless we first understand the role it plays in that critique. Deciding from the outset that any exposition which cannot adequately recapture all the nuances introduced by the sense/reference distinction can have no bearing on Frege's mature view of concepts presumes either (1) that we already know why Frege cares about the C&O distinction and that the sense/reference distinction is an essential element of that concern or (2) that we ourselves have identified the philosophically significant elements in Frege's remarks on concepts and have entitled those elements 'Frege's mature view.'

> [I]nstead of putting a judgment together out of an individual as subjects and an already previously formed concept as predicate, we do the opposite and arrive at concept by splitting up the content of a possible judgment. (*PW*, 17)

A year later, in the 1882 letter to Anton Marty, Frege first mentions the unsaturatedness of a concept:

> A concept is unsaturated in that it requires something to fall under it; hence it cannot exist on its own. That an individual falls under it is a judgeable content, and here the concept appears as a predicate and is always predicative…Now I do not believe that concept formation can precede judgment, because this would presuppose the independent existence of concepts, but I think of a concept as having arisen by decomposition from a judgeable content. (*FR*, 81)

The unsaturatedness of concepts is a metaphor meant to serve as a reminder that properly defined concepts must function as predicates of singular judgments and that possessing such concepts is best assured by letting them arise via the decomposition of a judgeable whole. Frege goes on to say that he does not believe that a judgeable content can be decomposed in only one way or that any one of these possible ways can "claim objective preeminence" (Ibid.). For example, the judgment expressed by '3 > 2' can be analyzed as an assertion about a relation holding between two individuals or as an assertion that a particular, denoted by the expression '3', bears a certain property, denoted by the expression 'greater than 2'. Decompositions of judgeable contents thus enable us to form perfectly proper expressions for concepts, as well as perfectly proper expressions for an individual falling under that concept, i.e., an object.

In this early account of the unsaturatedness of concepts, Frege is gesturing at his famous context principle: "Never ask for the [objective] meaning of a word in isolation, but only in the context of a proposition" (*FA*, x). There is much debate over whether Frege retained this principle after 1891, once his distinction between sense and reference was fully articulated.[43] Generally, scholars trying to establish an affinity between Frege's views and those of Kant will argue that he did retain the principle, thereby demonstrating Frege's allegiance to Kant's doctrine of the primacy of judgment, while those skeptical of the Kant-Frege connection argue that Frege's mature philosophical insights forced him to abandon the principle.

To see the tensions between Frege's endorsement of the context principle and what he subsequently says concerning sense and reference, we need to examine Frege's use of the principle to license multiply acceptable decompositions of the content of a judgment. Mark Wilson (1992) and Michael Beaney (1997) both argue that Frege primarily used the context principle to justify the introduction of new singular terms that referred to abstract objects, thereby underwriting the existence of these objects and our cognitive access to them. For Wilson, this is what motivates Frege's discussion in the *Foundations* concerning how the term 'the direction of *L*'

---

[43] Dummett (1991b) and Resnik (1980) claim that Frege abandoned the principle and thus any previously held sympathies with Kantian epistemology. Sluga (1980) and Currie (1982) argue that Frege retained the principle and this is indicative of a lifelong commitment to Kantian idealism.

should be introduced. Here Frege maintains that we begin with the sentence 'Line *a* is parallel to line *b*', which asserts a well-established, ordinary fact of Euclidean geometry. He then claims that the original content associated with this sentence can be carved up differently (*FA*, 75). The result is a new sentence 'The direction of line *a* is identical with the direction of line *b*'. Since these two sentences presumably express the same judgeable content, grasping the truth of content associated with the first sentence will enable us to grasp the truth of content associated with the second. Given Frege's context principle, the term 'the direction of line *a*' will have a meaning (referent). This referent is a newly introduced abstract geometric particular: the imaginary point at infinity where all lines parallel to line *a* intersect (Wilson (1992), 129–131, see also Frege's remarks in *FA*, 74–77, and *PMC*, 106). Wilson and Beaney argue that the purpose of Basic Law V is to do for the numbers, as logical objects, essentially what Frege tried to do with his contextual definitions in the *Foundations*.[44]

Now consider what this use of the context principle suggests about Frege's notion of a judgeable content and the later notions of sense and reference. We saw that two sentences can share the same objective content (sense), even though their constituent terms will presumably not have the same content (sense) and certainly not the same referent. For example, in the sentence 'For all *x*, F(x) iff G(x)' there is no constituent term that Frege would consider to be a singular term, whereas in the sentence 'The number belonging to *F* is identical to the number belonging to *G*', there are two singular terms and the term for a two place relation. Given that the content (sense) expressed by a singular term can never be the same as the content (sense) expressed by a general term,[45] it follows that the shared content of the two sentences must be something other than a mere composite of the senses associated with their respective constituent terms. This result is simply incompatible with

---

[44] See Wilson (1992), 134, and Beaney (1997), 15–19. Beaney claims that the purpose of the contextual definition of 'the number belonging to the concept *F*' was to resolve the ontological and epistemological issues related to Frege's abstract number-objects in one fell swoop. According to Beaney, Frege attempts to accomplish this same task with his Basic Law V. Frege's remarks on Basic Law V suggest that the sentences on either side of the equivalence sign express the same content just carved up differently (*FR*, 278–80). In other words, grasping the content expressed by the sentence 'For all *x*, F(x) iff G(x)', which asserts an ordinary logical fact about two concepts, will enable us to simultaneously grasp the content of sentence 'The extension of the concept *F* is identical with the extension of the concept G'. The truth of first sentence guarantees the truth of the second. Thus, the term 'the extension of the concept *F*' must have a referent. This referent is a newly introduced abstract object, the number belonging to the concept *F*. Furthermore, given that grasping the content and the truth of the first sentence can be achieved using logic alone and given that recarving of the content is justified by means of a logical law (Basic Law V), our apprehension of the abstract number-object referenced in the second sentence will also by achieved via logical means alone. Although Beaney argues, pace Dummett and Resnik, that Frege continued to implicitly endorse the principle after 1891, he does not take this argument to lend any strong support to the claim that Frege was a Kantian (Ibid., 21, n. 46).

[45] And everything that Frege says about the C&O distinction suggests that general terms and singular terms cannot express the same content/sense (see especially his 1891 letter to Husserl on *PMC*, 63).

Frege's other remarks suggesting that the sense of an entire sentence is completely exhausted by compounding the senses of its constituent terms (*CP*, 390).[46]

Furthermore, if the point of the context principle is to underwrite existence claims concerning newly introduced abstract objects, then presumably the two sentences expressing the same content must present us with two different states of affairs occurring at the level of reference. For instance, the ordinary Euclidean claim 'Line a is parallel to line b' is supposed to represent or refer to a state of affairs obtaining in the Euclidean domain, that there are two Euclidean straight lines parallel to each other. However, the sentence 'The direction of line a is identical to the direction of line b' is supposed to present us with a state of affairs obtaining in the extended projective domain, that there is an imaginary point where line a and line b intersect. It is difficult to see how Frege could use the context principle and the recarving of the judgeable content "to reveal unexpected objects"[47] once whole sentences are treated as singular terms on a par with the singular terms appearing in a sentence and once the reference of a sentence can be nothing except a truth-value. Due in part to the fact that Frege's later views on sense and reference generate these tensions with the use of context principle in the *Foundations,* Dummett and Resnik maintain that, by 1891, Frege had abandoned any significant use of the principle, along with any serious commitment to Kant's thesis about the priority of an entire judgeable content over its constituent parts.

While acknowledging these tensions between Frege's distinction between sense and reference and the context principle, I maintain that Frege never abandoned his commitment to Kant's view that the notion of an objectively-valid judgment is logically and conceptually prior to the derivative notions of general and particular objective representations (and representeds). Resnik (1980) claims "the appearance in 1891 of the metaphor of unsaturatedness" indicates Frege abandoned his Kantian concern with the original given unity of a judgment (Ibid., 170–1). On Resnik's account, the unity of a judgment is now explained by the fact that a judgeable content is put together using unsaturated and saturating entities. The main problem with Resnik's explanation is that the unsaturatedness metaphor did not first appear in 1891, but in 1882. On my account, the unsaturatedness metaphor does not constitute a rejection of the thesis that judgments possess an originally given unity, but rather a reinforcement of that thesis.

Frege's continuous effort to articulate his distinction between concepts and objects is an implicit endorsement of Kant's judgmental model of cognition. As late

---

[46] See Dummett (1991b) where he attributes this thesis to Frege. Dummett defends this attribution by clarifying what this thesis amounts to: "This doctrine that the sense of whole is compounded of the senses in its parts...means that in grasping the thought expressed by a given sentence, we grasp it only by a sentence having just the same (logical) complexity...[I]n order of explanation the sense of the sentence is primary, but in order of recognition the sense of the word is primary" (Ibid., 94–5). Notice, however, that given role that the context principle plays, according to Wilson and Beaney, we grasp the sense of 'Line a is parallel to Line b' which is the same sense associated with 'The direction of Line a is identical to the direction of Line b' prior to recognizing the sense of 'The direction of Line a'.

[47] Wilson (1992), 129.

as 1891, in his *Function and Concept*, Frege still suggests that the saturated/unsaturated or complete/incomplete distinction in propositional content arises simultaneously with the act of separating an originally given whole. He likens the splitting up of a proposition to the dividing of a line by a point. In order to "make a clean division", we must allocate the dividing-point to only one of the two divisions (*FR*, 134). The segment to which the point is allocated "thus becomes fully complete in itself...; whereas the other is lacking something –viz. the dividing point, which one may call its endpoint, does not belong to it" (*FR*, 134). The unsaturated function or concept can only become complete by replacing the part containing this end-point or another containing "two end-points" (Ibid.).

Besides, Frege's metaphors and elucidations were not entirely lost on his audience. Cassirer, for instance, associated the Fregean notion of a concept as a function with Kant's previous notion of a concept as a necessary rule of unity enabling us to transform undifferentiated phenomena into a stable, judgeable representation of object(s) with properties (Cassirer (1929), 287–288 and 292–293). He also recognized that both notions served to undermine the Aristotelian model of logical analysis, which assumed that judgeable contents must be constructed out of pre-existing concepts, first obtained via abstracting from particulars (Cassirer (1910), 4–26).[48]

Up to now, I have focused on those features of Kant's C&I distinction that carry over in Frege's account of his C&O distinction. One stark contrast, however, is that Frege never connects the distinction between general and singular objective constituents of a judgeable content with any correlative distinction in faculties. On the account presented in Chap. 2, the singularity and givenness criteria demarcating intuitive representations is explained by the fact that these representations are delivered to us via sensibility. In Chap. 3, we saw that the Marburg School rejected this notion of givenness to rescue Kant's epistemology from Helmholtz's naturalized interpretation. Therefore, singular entities are only given insofar as they are the particulars referenced in the propositional facts of science. Frege too explicitly rejects Kant's demand that qualifying as a full-fledged objective particular requires sensible perceptibility or being causally connected to something sensibly perceptible: "I must also protest against the generality of Kant's dictum: without sensibility no objects would be given to us. Nought and one are objects which cannot be given to us in sensation" (*FA*, 101). He also maintains that objecthood is conferred upon entities if they are treated as the particular entities investigated by a science. For

---

[48] In this passage, Cassirer does not mention Kant or Frege by name. However, he is arguing that a theory of logical analysis which attempts to treat the notion of a judgment as prior to the notion of a concept must also involve a different understanding of what concepts are and how they are formed than the one inherited from Aristotle. Cassirer believes that the seeds for this new understanding of a concept are contained in Kant's notion of a concept as a rule and Frege's notion of a concept as a function. What these two notions have in common is that the rule/function is productive in the sense that, without it, there could be no presentation of objects, with their properties and relations, which constitutes a sufficiently stable and objective unity about which we can render cognitively fruitful and inter-subjectively binding judgments (Cassirer (1929), 293–295 and 317). Cassirer also believes that Frege betrayed this essential insight into the nature of a concept whenever he too closely identified a concept with its extension or with the notion of a class.

Frege, the language of a particular discipline thus provides an essential clue for deciding what counts as an object: "When we speak of 'the number one', we indicate by means of the definite article a definite and unique object of scientific study" (*FA*, 49).

Notice, Frege is not saying that the occurrence of the definite article in and of itself indicates that the referent of 'the number one' must be recognized as an object. In other words, he is not claiming here that the ontological categories of the world somehow emboss themselves on the grammatical structure of language in such a way that a correct grammatical analysis can reveal which ontological category the numbers must fall into. Rather, he is saying that arithmetic takes numbers to be the particular entities of its investigation and "we" signal this fact by the use of the definite article. In fact, Frege explains that it was the wide-spread use of the expression 'the extension of a concept *F*' within the scientific community that encouraged him to assume that extensions could be rightfully identified as objects (*PW*, 269). So, for Frege as for the Marburg school, the singularity and givenness criteria of objecthood is substantially determined by what a properly functioning scientific community takes to be the particular entities comprising its subject matter.

We have seen, however, that neither Frege nor Cohen simply allow geometers to investigate whatever entities they like and in whatever manner they choose. Euclidean constructions qualify as perfectly proper geometric objects, whereas Helmholtz' psycho-physiological space-relations for which we are innately predisposed do not. Similarly, geometry cannot move in the direction of studying any sort of purely formal axiomatic systems without threatening its status as a science. The objects of the pure sciences must be given or constructed so as to explain how the concepts and truths articulated by these sciences are applicable within the natural sciences as a whole. In his "On the Law of Inertia" (1891), Frege writes that the semantic assignment of all terms in a technical language of a particular discipline is subject to only one standard: "does it enable the lawfulness of nature to be expressed as simply as possible and at the same time with the perfect precision?" (*CP*, 133). We know that Frege thinks of geometry, Euclidean geometry, as the science encompassing anything spatial. This means that he considers the concepts and truths of Euclidean geometry to be valid concerning the entire domain of objects to which spatial properties and relations are ascribed in the various sciences. For Frege and for Cohen, a necessary condition for granting full-fledged object status to an abstract geometric entity that its construction or presentation account for this scope of validity. This condition is met once geometric objects are identified as the spatial configurations generated by a Euclidean construction procedure (*FG*, 143).

Logic, according to Frege, is the science of pure thought, and pure thought is an independent source of a priori knowledge with the maximally general scope of validity. Thus, pure thought, like pure intuition, contains its own primitive truths, indefinable terms picking out the ultimate building blocks of its discipline, i.e. logic:

> [P]ure thought, regardless of any content given through the senses or even a priori through an intuition, is capable of bringing forth by itself, from the content which arises from its own nature, judgments which at first sight only seem possible on the basis of some intuition. ("Begriffsschrift" in *FR*, 78)

> Just as the concept *point* belongs to geometry, so logic too, has its own concepts and rela-
> tions: and it is only in virtue of this that it can have a content. Toward what is proper to it,
> its relation is not at all formal. ("On Foundations of Geometry" in *FG*, 109)

Genuinely logical concepts and propositions must possess a scope of validity larger
than geometric concepts and propositions. Logical concepts and proposition encom-
pass "everything that can be thought" and anything recognized as an object in all
scientific disciplines (*FG*, 143).[49] If, as Frege wanted to show, arithmetic is a branch
of logic, he must show that the particular entities which it investigates, i.e. the natural numbers, can be given or constructed so as to account for the maximally general
validity of their respective concepts: "if arithmetic is to be independent of all properties, this must hold true of its building blocks: they must be of a purely logical
nature" (*FG*, 143).

Frege's identification of the numbers with the extensions of a concept 'equinu-
merous with the concept *F*' appears to satisfy this demand perfectly. As we have
seen, Frege insists that a proper scientific concept is one that determines, for any
object within the domain of validity accorded to that science, whether it falls under
it or not. So, a properly defined, purely logical concept, e.g. a number theoretic
concept, guarantees the possibility of determining this question for any entity properly cognized as an object:

> A definition of a concept (of a possible predicate) must be complete; it must unambiguously
> determine, as regards any object, whether or not it falls under the concept (whether or not
> the predicate is truly ascribable to it). There must not be any object as regards which the
> definition leaves in doubt whether it falls under the concept; though for us human beings,
> with our defective knowledge, the question may not always be decidable. We may express
> this metaphorically as follows: the concept must have a sharp boundary (*FR*, 259).[50]

Frege repeatedly insists that proper concepts must have sharp boundaries. Indeed,
he claims that definitions of conceptual expressions failing to meet this requirement
are "meaningless" [*bedeutungslos*] (*PW*, 176).

Frege also characterizes a proper object as an entity falling under some properly
defined first-level concept. An object exists, for Frege, if and only if it falls under
some first-level concept. This is implied in his treatment of existence claims as

---

[49] I am indebted to David Sullivan's "The Further Question: Frege, Husserl and the Neo-Kantian
Paradigm" for providing further clarification on what the Marburg school intended by characterizing logic as the science of pure thought and noting is resemblance to Frege's view. Sullivan
describes their view as follows: "Logic promotes itself as the requisite methodological prelude to
the particular sciences, reinforcing its status as the most general of all the scientific investigations…Hence it is the most general of the disciplines: *the sciences of the sciences*" (Sullivan
(2002), 79–80).

[50] It is interesting to note that while Frege complains that Hilbert's definitions of geometric concepts run afoul of this requirement because they do not provide a definitive answer to this question
for empirical objects, e.g. watches and fountain pens. Frege says nothing about their failure to
apply to objects lacking spatial characteristics. Given that Frege assigns a smaller scope of validity
to geometric concepts than to purely logical concepts, it would not be surprising if he only expects
geometric concepts to meet the sharp boundary requirement for objects falling within their domain
of validity, i.e., those having spatial characteristics.

assertions that a first-level concept falls within a second-level concept 'There is n', where n is a natural number greater than zero:

> [A]ffirming the existence of something is to be understood as attributing to the relevant concept the property of *being instantiated* (e.g. to say that God exists is to say that the concept *God* falls under the [second-level] concept *is instantiated*), just as saying that there are n F's is to be understood as attributing to the concept F the property of *being instantiated n-fold*. (Beaney (1997), 103)

It is further implied by a method that Frege endorses for introducing names for objects. If it can be shown that one and only one entity falls under a first-level concept, we are justified in replacing the indefinite article ('a') occurring in the expression for the concept with a definite article ('the') and treating the resultant expression as the proper name of an object (see *FG,* 148 and *FR*, 192). When we add to these considerations, Frege's remarks that "an object" is "that which can be subject of [a singular judgment content]" and that by means of the definite article, we indicate "a definite and unique object of scientific study" *(FA,* 77 and 49), it follows that any entity properly cognized as an existing object will fall under some well-defined scientific concept.

In sum, all proper objects fall under some first-level concept and all proper first-level concepts are sortal concepts. As a result, equinumerosity is a relation that is well-defined over the entire range of first-level concepts, and all entities properly treated as objects within the sciences will qualify as a countable entities. Extensions, thought Frege, were properly cognized as logical objects as evidenced by the long-standing linguistic practices of the relevant scientific community. Identifying the natural numbers as the extensions of a certain concepts demonstrates that the subject matter of arithmetic is logical in character. The truths of arithmetic are truths about these logical objects and so, indirectly, truths about all n-membered extensions of concepts from every scientific discipline. Thus, Frege would have shown that arithmetic has its own, purely logical content or subject matter and that the nature of this subject matter accounted for the maximally general scope of validity of arithmetical concepts and propositions. Unfortunately, Frege's notion of a proper scientific concept, in combination with the idea that extensions are objects, leads to Russell's paradox.[51]

Everything else that Frege says in terms of distinguishing concepts and objects has to do with the logical relations they can and cannot enter into. Concepts can occupy either place of a subordinate relation, and objects can occupy neither. Objects

---

[51] Again, Michael Beaney explains: "The source of Frege's problems was...his conception of them [extensions] as *objects,* when taken together with his assumption that every concept must be defined for all objects. For this latter assumption implies that every concept divides objects into those that do, and those that do not, fall under it (there is no third possibility); with the assumption that extensions of concepts are objects, it implies that extensions can be divided into those that fall under the concept whose extensions they are (e.g. the extension of '() is an extension') and those that do not (e.g. the extension of '() is a horse'). Now consider the concept '() is the extension of a concept under which it does not fall'. Does the extension of *this* concept fall under the concept or not? If it does, then it does not, and if it does not, then it does" *(FR*, 7).

can occupy either place of an identity relation, and concepts can occupy neither. Only objects fall under concepts and thus only objects bear properties.[52] Only concepts can be predicated of objects and thus only concepts can be properties. This sums up what Frege says about the C&O distinction. Now let us see how he uses it.

4.3.2   Frege's C&O Distinction Deployed

Just as my account of the deployment of Kant's C&I distinction required looking at the empiricists and the rationalists through Kantian eyes, so too this account of the deployment of the C&O distinction will portray Frege's opponents, the formalists, through Fregean eyes. For Frege, 'formalist' is a pejorative term and has two connotations. Broadly construed, 'formalist' refers to anyone accused of engaging in practices depriving mathematical terms and sentences of their appropriate sense (*Sinn*) or reference (*Bedeutung*), thus rendering them incapable of expressing general valid, cognitively significant truths. Hilbert is identified as a formalist in the broad sense (*FG,* 18–9, 62–64) Narrowly construed, a formalist is someone who asserts that the objects investigated in number theory are the numerical signs themselves. H. Hankel, E. Heine, J. Thomae, E. Schröder, H. Helmholtz and L. Kronecker are explicitly identified as formalists in the narrow sense.[53] My goal in what follows is to show that Frege's use of the C&O distinction mirrors Kant's use of the C&I distinction. I begin with Frege's critique of his opponent's appeals to creative abstraction and explain how the C&O distinction serves to block such appeals. I then turn to Frege's accusation that almost all of his colleagues, including Dedekind and Hilbert, engaged in the rationalist's fallacy of inferring existence from mere logical consistency.

4.3.3   Fallacy of Creative Abstraction

I have argued that Frege and Cohen rejected Helmholtz's naturalized Kantianism, where the objectivity of mathematical concepts and propositions is grounded on the psychophysiology of the mind-world connection, and affirmed instead that this

---

[52] Frege does maintain that first-level concepts can "fall within" second-level concept and will sometimes liken this relationship to the fact that the first-level concept bears a certain property. However, Frege does not consider the property bearing of first level concepts to be entirely analogous to the property bearing of objects. This is why he insists upon reserving the expression 'falling under' to denote a relation between objects and first-level concepts and coins the expression 'falling within' to denote a relation between first-level and second-level concepts (*PW,* 110–111).

[53] Heine and Thomae are the continual targets of Frege's anti-formalist attacks. Schroder is read as conflating numbers with numerals in *FA* §43. Helmholtz and Kronecker are identified as formalists in "Function and Concept" (TPW, 22n). Frege waffles on whether Hankel is a formalist in the narrow sense (FR, 277n).

objectivity rests on showing how they serve as rules validating the reasoning in the less general sciences. In this section, I argue that, for Frege, the formalists' appeal to abstraction is indicative of this same Helmholtzian confusion as to what scientific objectivity amounts to and how it is secured. I go on to show that Frege's use of the C&O distinction to expose and prevent such appeals recaptures Kant's similar use of the C&I distinction.

In the *Basic Laws of Arithmetic* as in 'Function and Concept,' Frege complains of a "widespread tendency" among logicians and mathematicians to conflate existence and objectivity with actuality. He defines actuality ['wirklichkeit'] as whatever is "capable of acting ['wirken'] directly or indirectly on the senses" (*FR*, 200 and 205). §26 and §27 of the *Foundations* are devoted to Frege's argument that actuality is neither necessary nor sufficient for ascribing objective meaning or content to proper names or concept words.[54] Throughout his work, he repeatedly accuses colleagues of confusing actual mathematical signs – signs for the numbers, signs for arithmetic operations and functions, and the sign for equality – with their objective content:

> [F]or Schröder number is a *symbol* ['Ziechen']...Even by the word 'one' he understands the symbol I, not its meaning ['Bedeuntung']. The symbol + is introduced solely to serve as a visible mark, without any content ['Inhalt'] of its own, for linking up the other symbols; only later does he define addition (*FA*, 55).

> [W]e must go back to the time when higher Analysis was discovered, if we want to know how the word 'function' was originally understood. The answer that we are likely to get... cannot satisfy us, for here no distinction is made between form and content, sign and thing signified ['Bezeichnetes']; a mistake admittedly, that is very often met with in the mathematical works, even of those of celebrated authors (*FR*, 131).[55]

> Opposed to formal arithmetic, there stands nonformal arithmetic. The two differ as follows: In nonformal arithmetic, numerals really are signs: mere tools of research, intended to designate numbers, where the latter are the nonsensible objects of the science. In formal arithmetic, it is the numerals themselves that are the numbers; they are not mere tools but rather the very objects of the investigation (*FG*, 132).

For Frege, the characteristic formalist practice is treating arithmetic signs or notation as the subject matter of arithmetic, which is motivated in turn by adherence to the Kantian dictum that the proper objects of a science must be given by "means of the senses (*FG,*).[56] As we have seen, Frege blames Helmholtz's naturalistic rendering of this dictum for fueling a confused epistemological agenda seeking to ground

---

[54] I am focusing on Frege's account of the C&O distinction before and independently of his subsequent bifurcation of a term's objective content into sense and reference. I am thus presuming that his commitment to the C&O distinction as a precept aimed at warding off the formalists' mistaken notions of abstraction and objectivity can be adequately explained without having recourse to this subsequent Fregean distinction.

[55] In this passage from 'Function and Concept,' Frege refers his reader to the 1884 *Foundations* and 1885 'On Formal Theories of Arithmetic' as previous attempts to articulate and prevent the "defects of the current formal theories in arithmetic."

[56] See too *FA*, p. 101.

the objective validity of arithmetic concepts and propositions on the relatively new science of experimental psychophysiology (*TPW*, 213; *FA*, 38). When reading Frege's *Foundations* protest "against the generality of Kant's dictum" and the correlative precept that a concept's objectivity depends on establishing the psychosensory, intuitional relation that attaches it to a sensibly-given object, I propose reading it as a protest against the prevailing naturalism embraced by Helmholtz and the other so-called formalists (*FA*, 101).

Frege's lengthy quote from Hermann Hankel provides further evidence that he sees naturalized Kantianism driving the methods typifying formal arithmetic:

> Number to-day is no longer a thing, a substance, existing in its own right apart from the thinking subject and the objects which give rise to it…The question whether some number exists can therefore only be understood as referring to the thinking subject or to the objects thought about, relations between which the numbers represent…[I]f the numbers concerned are logically possible, if their concept is clearly and fully defined and therefore free from contradiction, then the question whether they exist can only amount to this: Does there exist in reality or in the actual world given to us in intuition a substratum for these numbers, do there exist objects in which they—relations, that is, for the mind of the type defined—can become phenomenal (*FA*, 104–5).

Frege notes that the first sentence "leaves it doubtful" whether Hankel holds that numbers exist as mental states "in the thinking subject" or "in the objects" giving rise to these states (*FA*, 105). What is not doubtful is that, for Frege, Hankel mistakenly settles questions about the objective existence of the numbers and the legitimate application and scope of arithmetic concepts and operations by appealing to the actual, phenomenal features of arithmetic "symbols ['Zeichen']" (*FA*, 107).

Elsewhere in the *Foundations*, Frege cites Hankel invoking a "pure intuition of magnitude" as the ground for validly applying real number theory throughout "every field" and suggests that he may conceive of arithmetic laws as synthetic a priori (*FA*, 18). This might seem to undermine my claim that Frege sees Hankel and Helmholtz subscribing to a shared epistemology of mathematics. But remember, by the mid-1800s it was generally agreed that a subject's psychophysiology placed constraints on the content of their representational mental states and that this refuted a naïve, pre-Kantian empiricism. In his 1887 'Number and Measuring from an Epistemological Viewpoint,' Helmholtz himself argues that the lawlike sequence characteristic of an ordinal numeric notation can be derived from the fact that memory traces and the content of occurrent mental states stand in an irreversible temporal sequence of succession (*NME*,75–7). I take it that this, along with Schröder's Axiom of Symbolic Stability, is one of the "purely psychological facts" on which he intends to ground the "unlimited extent" and "objective sense" of arithmetic (Ibid., 75). Helmholtz also reminds his readers that his "earlier writings" on the geometric axioms do not entirely "eliminate Kant's view of space as a transcendental form of intuition" (Ibid., 72). Similarly, his account of the arithmetic axioms preserves a role, albeit not an exclusive role, for time as the "inescapable form of our inner intuition" grounding the applicability of these axioms to actual symbolic systems, including the symbols treated in his *Zeichentheorie* (Ibid., 77) For Frege and his contemporaries, an epistemology of mathematics invoking pure intuition or transcendental forms of spatial or temporal magnitudes is entirely compatible with espousing a naturalized and even empiricist Kantianism.

Remember too, what distinguishes Helmholtz's first wave Kantianism is that the "conditions of objectivity" are sought in "an a priori phenomenology or an a posteriori psychology of the mind" (Richardson (2006), 215). This is precisely how Frege portrays Hankel and the other formalists; they all suffer from the misconception that the signs for numbers, arithmetic functions or relations are devoid of objective content unless these signs are identified with, causally related to or mapped onto actual entities or processes bridging a presumed mind-world gap. For Frege, this is why Hankel, Schröder, Stanley Jevons, M. Ballue, J. Thomae, E. Heine, Husserl and H. Schubert end up vacillating between identifying numbers as either actual but empty signs or as bloodless phantoms abstracted from actual entities.[57] He claims that in each of their accounts abstraction plays a crucial role. His use of the C&O distinction to highlight and block these appeals to abstraction is thus directly tied to displacing a Helmholtz's naturalized Kantianism as the prevailing paradigm for epistemology of the science.

Frege accuses the formalists of making three interrelated and illicit appeals to abstraction. First, he goes after those introducing fractional, negative, irrational and complex numbers as symbols and then invoking abstraction to divest these symbols of their unwanted phenomenal features. The formalist is charged with acting "like a god, who can create by his mere word whatever he wants,"[58] and again Frege quotes Hankel as evidence for this charge:

> It is obvious that, for $b > c$, there is no number x in the series 1, 2, 3,...which solves our problem; the subtraction is then impossible. There is nothing, however, to prevent us from regarding the difference $(c - b)$ as a symbol which solves the problem and is to be operated with exactly as if it were a figure number in the series 1, 2, 3...(*FA*, 107).

Dummett (1991a) suggests that 'postulationism' may be a better term than 'formalism' for the target of Frege's critique, since Frege locates Hankel's error—"the error that infects the formal theory of fractions and of negative and complex numbers"—in positing that the laws of arithmetic hold in order to deduce the properties and relations of the newly introduced numerical symbols (*FA*, 108).[59] Dummett is certainly right to point out that Frege accuses Hankel and others of illicitly postulating what ought to be proven. Still, if this is all that we note in Frege's denouncement of Hankel's procedure, we will fail to appreciate why Hankel and his ilk are charged with the falsely assuming that soaring on "the summits of abstraction" entitles them to think that they have guaranteed a sufficiently wide scope of validity for arithmetic, a scope wider than Frege ascribes to geometry or empirical science (*FA*, 119).

Strictly speaking, Hankel does not directly infer from a logically consistent concept to the existence of something falling it. Rather, Hankel introduces a symbol. The existence of the newly introduced numbers is thus grounded in the actuality or

---

[57] See *FA*, 27, 44–6, and 54–6; *CP*, 114–5, 196–8, 222, and 230–2; *TPW*, 162–74.

[58] Hankel as cited in *FA*, 119

[59] See Dummett (1991a), 178 and Beaney (1997), 125.

phenomenal nature of the symbols, and Frege admits this (*FA*, 105). The problem, says Frege, is that the formalists confuse their notion of abstraction with a judgment that a logical relation of subsumption obtains. On the formalists' account, abstraction is a psychological process that licenses disregarding an object's properties because they not are contained in marks of a concept under which it is subsumed. Hankel-like strategies for introducing numbers mistakenly presume that "objects are essentially altered by abstraction, so that objects brought under one concept become more alike" (*TPW*, 85).[60]

Frege's rejection of the formalists' conception of abstraction is even more evident in his response to their use of abstraction to (1) explain the possibility and validity of everyday statements involving the natural numbers; and 2) account for the epistemic fruitfulness of judgments about numerical identity ['Zahlengleichheit'].[61] Both Frege and the formalists feel obliged to resolve a puzzle concerning our ability to count. For entities to be countable entities it seems as if they must be both different and the same. Counting a plurality of utterly distinct entities would only result in the count of one for each distinct entity, never reaching the number belonging to the entire collection. Yet, if we treat the entities as completely identical, we would also never go beyond a count of one (*FA*, 46 and 58). Frege resolves the puzzle by arguing that falling under a concept is necessary and sufficient for presenting objects as countable entities. Falling under the same concept lends objects the unifying element required for counting, while their otherwise distinguishing features confer the requisite diversity (*PMC*, 60–7; *FG*, 141). A statement of number ['Zahlangabe'] "contains an assertion about a concept," answering the question 'how many entities fall under it?' (*FR*, 99; *FA*, 59). The factuality ['Thatsächlichkeit] of statements of number is thus "explained by the objectivity of concepts" (*FA*, ix). The objectivity of concepts is then explained and justified in virtue of the objectivity of assertions about the logical relations holding between concepts (*FA*, 60). In light of this order of explanation and justification—*from* facticity and objectivity of well-recognized assertions in the maximally general science of logic *to* that of everyday number statements concerning empirical concepts and the objects falling under them, Frege shows himself to be a follower of the second way of Kantian interpretation.

In contrast, Frege presents Schröder and Stanley Jevons as followers of the first way. Schröder and Jevons reportedly account for the possibility and objectivity of number statements in virtue of the actuality of the mathematician's notational strokes or the existence ['vorhandensein'] of a plurality of actualities. Each then

---

[60] This quote is taken from Frege's notes on Husserl's *Philosophie der Arithmetik*; see *TPW*, 174–5 for evidence that he considers this to be a view distinctive of formalists generally.

[61] For our purposes, Frege's *Foundations* critique of the three ways in which his colleagues' appeal to abstraction. In subsequent writings, Frege more carefully articulates what he means by insisting that judgments of numerical equivalence should be understood as judgments of strict identity. However, in the *Foundations* and these later writings, Frege is combating J. Thomae's appeal to abstraction to account for the scope, objectivity and epistemic fruitfulness of judgments concerning numerical equivalence. In his 1906 'Reply to Mr. Thomae,' Frege complains that he is still having to "repeat the same arguments" presented in §§34–48 of *Foundations* (*FG*, 126).

invokes abstraction to warrant disregarding any properties not conducive to representing these actualities as denumerable units: .

> For E. Schröder number is modelled on actuality, derived from it by a process of copying the actual units with ones, which he calls the abstraction of number. In this copying, the units are only represented in point of their frequency, all other properties of the things concerned, such as colour or shape, being disregarded (*FA*, 27–8).

> To avoid carrying over into number the distinguishing marks of things numbered, Jevons invokes abstraction: "There will now be little difficulty in forming a clear notion ['klare Vorstellung'] of the nature of numerical abstraction. It consists in abstracting the character of the difference from which the plurality arises, retaining merely the fact ['ihr Vorhandensein']" (*FA*, 55).[62]

Armed with this notion of abstraction, it is easy to see how one might go on to explain and justify the assertions of numerical equivalence about these actual pluralities. This is what Thomae sets out to do when he runs up against Frege's mocking wrath:

> [S]omeone learning how to count abstracts from the differences of the counting blocks which he uses to learn this and equates them (*setzt gleich*) with one another. I believed that I had discerned the fruitfulness of the equality-sign in this possibility or capability of the human mind to abstract from the differences of certain things to equate them with one another (*Elementary Function Theory*, p. 2), but I am thoroughly rebuffed by Frege (*FG*, 117).

Frege's *Foundations* campaign against his colleagues' appeals to abstraction is explicit: "To abstract from the differences between things does not give us the concept of Number nor does it make things identical with one another ['einander gleich']" (*FA*, viii). From 1884 until 1908, he never ceases to chide Thomae and "most mathematicians" for trying to ground the objectivity and scientific fruitfulness of arithmetic on the presumed creative license of a mathematical mind capable of abstracting objects having all and only desirable properties and relations (*TPW*, 174; *CP*, 229, 231 and 252–3; *FG*, 123–5).

So how does Frege deploy the C&O distinction against what he considers fallacious appeals to abstraction? First, he insists that, at best, abstraction is a process yielding a concept, never an individual thing falling under the concept bearing only conceptual marks as properties. This point first appears in the *Foundations* and is directed at Thomae:

> For suppose, that we do, as Thomae demands, "abstract from the peculiarities of the individual members of a set of items", or "disregard, in considering separate things, those characteristics which serve to distinguish them". In that event…what we get is rather a general concept under which the things in question fall. The things themselves do not in the process lose any of their special characteristics (*FA*, 45).

---

[62] See *CP*, 196 and 252, for evidence that J. Thomae, Husserl and H. Schubert are all chastised for invoking abstraction as license for disregarding properties in effort to obtain an actual presentation of a plurality of denumerable units.

It is repeated in his subsequent critiques of Husserl and H. Schubert. Abstraction properly understood, i.e. "the general abstraction of logicians," is a process whereby we "acquire concepts" (*CP*, 254). Throughout this process, the individuals falling under the concepts "remain completely unchanged" (Ibid.). They are neither "transformed into the concept" nor "replaced by it" (Ibid.). This logically proper conception of abstraction is contrasted with the "marvelous" abstraction invoked by Husserl and others that purports to transform "each object…into a more and more bloodless phantom" (Ibid., 254 and 198).

By insisting on the absolute and mutually exclusive distinction of concepts and objects, along with the syntactic criteria marking this distinction, Frege is able to block any move by Schröder and the other formalists to invoke abstraction as a means of ascribing objective content to the "proper name of a thing" ['Eigenname eines Dinges'], specifically, the proper name of a number (*FA*, 62–3). Moreover, in the *Foundations*, Frege explicitly states that an object ['Gegenstand'] is that which can serve as a possible subject of a singular judgeable content ['ein mögliches Subject eines singulären beurtheilbaren Inhalts']. An object is that part of the judgeable content referred to by a proper name in any sentence analyzable so as to express a singular judgment (*FA*, 77). It follows that Frege has effectively blocked the formalists' conception of abstraction from playing a crucial role in accounting for the objective content of such sentences.

Frege also insists, pace the formalists, that we "cannot succeed in making different things identical simply by dint of operations" (*FA*, 46). In other words, abstracting from, or rather disregarding the differences in actual pluralities does not warrant asserting that a numerical equivalence holds between them. So too, abstraction is not a license for mathematicians to treat phenomenally distinct symbols or concatenations of strokes as "definite and unique" objects of scientific study (*FA*, 49). The fact that Frege only allows proper names to occupy the right and left hand side of the sign for mathematical equality is thus aimed at highlighting the formalists' practice of (1) identifying the objects studied in the science of arithmetic with actual symbols or the psychological facts governing their manipulation; (2) vacillating between use of the conceptual word 'unit' and the proper name 'one'; and (3) a misguided empiricist Kantianism that would lead to locating the "the mainspring of arithmetic" in a "faulty notation" and grounding its epistemic fruitfulness in the "miraculous" "creative power" of a "sovereign" human mind (*FA*, 48–9; *FG*, 124). If Kerry and Frege's other colleagues had granted him his requested pinch of salt and accepted his proposal to treat the C&O distinction as a primitive logical fact, abstraction as understood by the formalists would have been barred from playing a significant role in accounting for the objectivity and fruitfulness of not only arithmetic equations, but also identity statements in the sciences generally.

4.3.4   The formalist's Fallacy: Logical Consistency Implies Existence

If formalists called upon abstraction to rid tangible numerical signs of their unwanted properties, what was their rationale for ascribing to these signs their purportedly mathematical properties? How does the chalkboard symbol for the $\sqrt{2}$ obtain the property of yielding 2 whenever multiplied with itself? (*FG*, 147) According to Frege, the formalist responds, "we simply invest them with the properties we wish by so-called 'definitions'" (*BL*, 10). Frege's repeated tirade against "creative" definitions is often presented as evidence for his Platonic realism. This is because it is within this context that the oft-cited passage, "[T]he mathematician cannot create things at will, any more than the geographer can; he too can only discover what is there and give it a name", occurs (cf. *FA*, 108, and *BL*, 11). I will argue, however, that what Frege is actually objecting to is not his colleagues' treatment of numbers as constructions or their having failed to acknowledge that numbers exist independently of our shared cognitive enterprises, but rather that their method for constructing the numbers rests on an illicit slide from logically consistent concept to object instantiating the concept.

Frege's analogy between the mathematician and the geographer is first presented in §96 of the *Foundations*. At the beginning of the prior section, we read:

> Strictly speaking, of course, we can only establish that a concept is free from contradiction by first producing something that falls under. The converse inference is a fallacy and one into which Hankel falls. (*FA*, 106)

How does Hankel commit this fallacious inference? As previously mentioned, Hankel, Thomae and others identified the numbers with symbols. They especially pushed for this identification in the case of any number that was not a positive integer.[63] The problem, of course, is accounting for the mathematical properties of these symbols. Immediately after stating the mathematician-geographer analogy, Frege tells us how the formalists tried to address this problem:

> This is the error that infects the formalist theory of fractions and of negative and complex numbers. It is made a postulate that the familiar rules of calculation shall still hold, where possible, for the newly-introduced numbers, and from this their general properties and relations are deduced. If no contradiction is anywhere encountered, the introduction of the new numbers is held to be justified.... (*FA*, 108)

---

[63] Remember, formalism in the narrow sense was driven by an empiricist's criterion of existence. In the case of the positive integers, the existence question could be settled via abstraction on sensibly-perceptible collections. This explanation cannot, or at least was not, extended to the other numbers. As Frege notes, questions concerning the existence of irrational and complex numbers had been settled by appealing to geometric construction. However, several mathematicians, including Frege, Hankel, and Thomae, disliked this dependence of number theory on geometry. Equating numbers with symbols was thus seen as resolving the existence question in a way that was consonant with empiricist scruples and retained the independence of number theory from geometry (*FG*, 147–8).

There is also textual evidence that Frege is giving a fairly accurate description of what the formalists were proposing. Thomae, for instance, likens the tangible signs for numbers to the pieces in a chess-game. In both cases, their only significant properties and relations are those "assigned to them by the rules of the game" (*FG*, 115).

So, for formalists in the narrow sense, the justification for asserting the existence of a number manifesting the right sort of properties involves three steps: (1) introduce a sign (e.g., $\sqrt{2}$); (2) use this sign in calculations where the ordinary rules of algebra apply and obtain a list of properties that are mutually compatible and do not generate a contradiction with the properties ascribed to the other numerical signs; and (3) ascribe to the sign those properties and only those properties (e.g., the property of yielding 2 when multiplied by itself). While Frege may be slightly overstating their alleged mistake when he says that they assume "freedom of contradiction amounted straight way to existence," given the additional step of producing the tangible sign, it is certainly fair to say that the formalists considered lack of contradiction sufficient for asserting the existence of an object having "properties that it has not already got" (*CP*, 139).[64]

Dedekind, Otto Stolz, and Hilbert are similarly scolded for thinking that defining a logically consistent concept or system of concepts suffices for asserting the existence of objects bearing the marks of those concepts as properties. In his critique, "The Construction of New Objects, according to R. Dedekind, H. Hankel, O.Stolz", Frege is at pains to distinguish the views of Dedekind and Stolz from those of Thomae. Dedekind and Stolz do not equate the numbers with signs and then appeal to creative definition for the purposes of ascribing to those signs the relevant properties to that sign. Instead, Stolz appeals to creative definition to order to construct an entirely "new- at any rate non-sensible—thing", which he then "supplies with a sign" (*FR*, 276).[65] Frege also presents one of Stolz' creative definitions:

1. *Definition* . If in this case (D_1) no magnitude of System (I) satisfies the equation $b \,°x = a$, then it shall be satisfied by *one and only one new thing not found in* (I); this may be symbolized by $a \cup b$, since this symbol has not yet been used. We thus have

$$b\left(a\cup b\right) = \left(a\cup b\right)\, a = a$$

Since the new objects possess no further properties, we can assign them properties arbitrarily, so long as these are not mutually inconsistent. (from Stolz' *Lectures on General Arithmetic* (1885), quoted by Frege in *FR*, 275)

---

[64] The second and last occurrence of Frege's analogy between mathematicians and geographers is in the introduction to the first volume of the *Basic Laws of Arithmetic*. Here too the issue that Frege is concerned with is the formalists' method for ascribing to tangible signs their relevant properties (*BL*, 10–11).

[65] Frege claims that "Dedekind agrees with Stolz in his conception of construction" (*FR*, 277). So, we can assume that whatever Frege takes to be correct or mistaken in Stolz' conception of construction would apply to Dedekind's as well. We will see that while Frege is here applauding the fact that Stolz (and Dedekind) does not conflate a number with its sign and retains the idea that numbers are non-sensible objects, he nevertheless rejects Stolz' construction since it incorporates the formalist fallacy.

He stresses that Stolz never actually states the procedure or materials from which this object is constructed. According to Frege, Stolz' definition first involves the ex nihilo creation of a "stark naked" object, "devoid of the most necessary properties." The second creative step is to enumerate a list of mutually consistent properties and hail the object as "the lucky owner of these properties" (Ibid.). He then raises the same objection that he had previously raised against Hankel in the *Foundations*. On Stolz' method, one does not legitimately infer that the constituent marks are mutually consistent by first producing or constructing the object(s) falling under the concept. Rather, one assumes that the concept is instantiated simply because it has been defined by listing non-contradictory characteristics (*FR*, 275–277).

We can obtain further evidence that what Frege finds most objectionable in the Stolz-Dedekind conception of construction is its implicit endorsement of the formalist's fallacy by returning to his critique of Hilbert. Frege claims that Hilbert's strategy of simultaneously defining geometric concepts and guaranteeing their instantiation by stipulating a set of consistent axioms makes the same mistake as Stolz' creative definitions (*FG*, 19). Initially, this faulty strategy is characterized as attempting to define a concept-term by citing concepts of differing levels in the definiens. The following illustration is intended to get Hilbert to see the error of his ways:

> What would you say to the following?
> Explanation: We conceive of objects we call gods.
> Axiom 1. Every god is omnipotent.
> Axiom 2. Every god is omnipresent.
> Axioms 3. There is at least one god. (*FG*, 18–9)

Frege goes on to explain his distinction between first and second level concepts. 'There is at least one' is the expression for a second-level concept and rightfully employed in sentences intending to assert that a first-level concept, e.g. god, has an object falling under it. Omnipotence and omnipresence are first-level concepts and so these can be legitimately cited as constituent marks of other first-level concepts. On Frege's interpretation, Axioms 1 and 2, taken together, assert that the first-level concept god is subordinate to the other first-level concepts, omnipresence and omnipotence. Axiom 3 states that there exists an object bearing both of these characteristics as properties. The upshot of Frege's illustration is clear. The definitional practices of Hilbert and Stolz pave the way for resurrecting the rationalist's ontological proof for the existence of God (cf. *FG*, 77).

In other words, Frege recognizes the fallacy driving his opponent's creative definitions as the same fallacy that Kant previously saw driving the rationalist's pseudo-proof for God's existence. Furthermore, as I pointed out when discussing Kant's critique of rationalist's argumentative strategy, the criticism here is more general than that the formalist's have simply failed to appreciate that existence is not a proper, first-level, predicate. The more general mistake is trying to extract from a purely arbitrary, logically consistent idea the existence of an object corresponding to it.[66] After having informed Hilbert of the distinction between first and second level concepts, Frege cuts to the chase: "Our views are probably most sharply opposed with respect to your criterion of existence and truth" (*FG*, 19). Once again he shows how Hilbert's view allows us to deduce the existence of an object with god-like attributes so long as these attributes appear mutually compatible. Finally, he presents the bogus rule of inference that seems to capture Hilbert's criteria for asserting an existence claim:

> If (generally, whatever A may be) the propositions:
>
> *A* has the property Φ
> *A* has the property Ψ
> *A* has the property X
>
>   Together with all their consequences do not contradict one another, then there exists an object that has all of the properties Φ, Ψ, X. (*FG*, 20)

On Frege's account, Hilbert's grounds for asserting the existence of an object having the right sort of properties are identical to that of Hankel, Thomae, Dedekind, Stolz, and the rationalists. They all assume that stipulating a collection of mutually consistent marks establishes the existence of an object manifesting those marks as properties.[67]

So far I have tried to show that when Frege complains about the creative definitional practices of his contemporaries, the issue that he is primarily concerned with is the illicit inference from logical consistency to existence. I have also tried to show that Frege sides with Kant on this issue and recognizes it as the error underlying the rationalist's ontological argument. I will now argue that Frege's rejection of creative definitions is independent of any worry about whether the numbers and other mathematical objects should be recognized as completely independent platonic entities, as opposed to constructions. First, I have already presented my case that Frege believed that geometric objects were correctly identified as Euclidean constructions. Second, I have argued for aligning Frege with Cohen's Kantianism and that this is incompatible with reading him as a platonic realist or even as a realist about empirical objects. Third, Frege dismisses the question, whether his own method of introducing the numbers as extensions can rightfully be called a constructive procedure, as a trivial one and entirely beside the point:

---

[66] See Sect. 2.2.2, as a reminder that this is what Kant identified as the "unnatural procedure" and "mere innovation of scholastic subtlety" behind the rationalist's proof (A603/B631).

[67] Hilbert responds to Frege's accusation that he is inferring existence from logical consistency by declaring that he is guilty as charged: "Of course I must also be able to do as I please in the matter of positing characteristics; for as soon as I have posited an axiom, it will exist and be 'true.' And herewith I come to another point in your letter. You write, 'Axioms I call propositions…From the fact that axioms are true it follows that they do not contradict one another.' I was extremely interested to read just this proposition in your letter, because for as long as I have been thinking, writing, and lecturing about such things, I have always been saying the opposite: If the arbitrarily posited axioms together with all of their consequences do not contradict one another, then they are true and the things defined by these axioms exist. For me, this is the criterion of truth and existence" (*FG*, 12).

> Somebody might indicate that we ourselves nevertheless constructed new objects, viz.
> value-ranges (Vol. I, §§ 3, 9, 10). What then did we do there? Or rather, in the first place,
> what did we not do? We did not enumerate properties and then say: we construct a thing that
> is to have these properties.[68]....Can our procedure be termed construction? Discussion of
> this question may easily degenerate to a quarrel over words. In any case our construction (if
> you like to call it that) is not unrestricted and arbitrary; the mode for performing it, and its
> legitimacy, are established once for all. (*FR,* 277 and 279)

What Frege considers significant about his construction of the numbers is that it
proceeds in accordance with well-established laws of logic, just as proper geometric
constructions proceed in accordance with the well established axioms and postulates of Euclidean geometry, and that it does not rest upon the assumption made by
his colleagues that enumerating a list of mutually consistent properties proves the
existence of a corresponding object.[69]

It took some effort to demonstrate that the inference from logical consistency to
existence was, for Frege, the common denominator infecting the work of both narrow and broad formalists. Showing that he intends for the C&O distinction to highlight and block this purported fallacy requires no effort at all. In the *Foundations,* he
states his intent outright. If his readers remember "never to lose sight of the distinction between concept and object", they will see "that a widely-held formalist theory
of fractional, negative, etc. numbers is untenable" (*FA,* x). Immediately after
describing Hankel's strategy for introducing fractions, etc. and "the error that
infects" this strategy, Frege accounts for the prevalence of this error:

> That this mistake is so easily made is due, of course, to the failure to distinguish between
> concepts and objects. Nothing prevents us from using the concept "square-root of one"; but
> we are not entitled to put the definite article in front of it without more ado and take the
> expression "the square root of one" as having a sense. (*FA,* 108)

Frege use of 'sense' here is somewhat misleading. He had not yet clearly distinguished sense and reference. His point is that the formalist's strategy of listing
mutually consistent characteristics may entitle them to say they have defined a concept, but this "is still no guarantee that anything falls under the concept" (*FA,* 114).

He repeats this same point at least three times in the *Foundations* and throughout
his other writings.[70] He mentions it in response to Dedekind's and Schroder's creative definitions (*FR,* 209–10), and it is always appended to Frege's rants against
creative definitions:

> [T]he mathematician cannot really create anything by his act of definition. Nor can we by
> mere definition conjure into a thing a property it does not have in the first place...Confusion

---

[68] From here Frege goes on to describe how Basic Law V countenances the transformation of a
proposition concerning a relationship holding between two first-level concepts into an identity
claim concerning value-ranges. He justifies transformations of this type by citing their long-standing use in logic and mathematics: "What we are doing by means of our transformation is thus not
really anything novel" (*FR,* 278).

[69] One cannot ignore the irony of Frege's confidence, since he would receive Russell's letter
informing him of the problem with Basic Law V during the printing of this passage.

[70] See *FA,* §37–8, §53 and §109.

easily arises here by failing to distinguish between concept and object. If we say: 'A square is a rectangle in which adjacent sides are equal', we define the concept *square* by specifying what properties something must have to fall under this concept. These properties I call marks of the concept. But, it should be noted, these marks of the concept are not its properties...Whether there are such objects is not immediately known from the definition. (*BL*, 11; see also *FG*, 148)

Frege consistently deployed his C&O distinction to focus attention on his contemporaries' illicit slide from logical consistency to existence, just as Kant had similarly deployed his C&I distinction against the rationalists.

Let me summarize just how analogous the two deployments are. Both Kant and Frege grant to their opponents that stipulating a list of mutually consistent marks warrants the assertion of trivial tautologies about conceptual subordination. In Sect. 2.2.2, I explained that Kant concedes to the rationalists that their notion of a supreme being qualifies as a logically possible idea or empty conceptual form so long as it is defined via a list of mutually compatible marks. He also acknowledges that, depending upon which constituent marks they chose to pack into the definition of this idea, a proposition such as 'God is omnipotent' turns out true on pain of contradiction. However, propositions of this kind only assert that a subordination relationship holds between two, cognitively empty conceptual forms, since it has yet to be proved whether 'god' or 'omnipotence' refers to any properties useful in detecting which objects fall under them. Therefore, the rationalists have only succeeded in arbitrarily defining the term 'god', with result that certain propositions turn out to be necessarily true in virtue of that definition.

Frege also concedes to his opponents that listing component characteristics may be an appropriate strategy for defining conceptual terms: "Dedekind then cites the properties that a thing must have to be part of a system, i.e., he defines a concept by its marks" (*FR*, 210). He also admits that "we can form self-evident propositions" based on "arbitrary stipulations" (*CP*, 274). What kind of propositions are these? Frege often complains that his colleagues efface the distinction between property and characteristic mark, thereby confusing the relationship that a mark bears to its concept with the relationship that a property bears to its object (*FR*, 81; *PW*, 104–5). The former is subordination and the latter is subsumption (*PW*, 103; *FA*, 64–5). Now, let us consider Hilbert's definition of 'point on a straight line'. By stipulating several non-contradictory axioms, Hilbert presents the component characteristics of this concept, for instance, the component characteristic 'stands in a betweenness relation with other points on the line.' Frege thus agrees that this entitles us to form the self-evident proposition 'All points on a straight line are points standing in the betweenness relation.'[71] However, what this proposition actually asserts is that the Hilbertian concept of 'point in a straight line' is subordinate to the Hilbertian concept 'stands in a betweenness relation with...' and the truth of this proposition depends entirely on Hilbert's choice in defining the terms. This much Frege willingly

---

[71] Recall that, for Frege, propositions of the form 'All *F's* are *G's*' only assert that the concept *F* is subordinate to the concept *G*. Such propositions do not necessarily assert anything about the particulars falling under the concepts (*FA*, 60–1).

grants to his adversaries. What he denies is that their strategy results in propositions saying anything about individuals, their properties, or their relations. As a matter of fact, Frege denies that this strategy results in mathematical concepts and propositions with any "epistemic value" at all (*CP*, 274).

Both Kant and Frege challenge their colleagues' assumption that logical consistency is enough to show that concepts possess cognitive significance. According to Kant, the rationalists were confused on this point because they neglected the role of construction in defining mathematical concepts. He argued that the objective validity of mathematical concepts depended on constructing a corresponding formal intuition, a singular entity whose construction could account for the universal applicability of the concept to the inter-subjectively accessible and judgeable world of empirical objects. Euclidean constructions counted as the archetypal formal intuitions.[72]

A century later, Frege accuses his fellow mathematicians of conducting themselves like metaphysical rationalists and forgetting that the cognitive value of mathematical concepts depends on establishing their applicability to a suitably wide range of inter-subjectively accessible and judgeable objects, the objects treated in the various sciences. Because Hilbert did not ensure that his geometric concepts were applicable to the entire range of spatial entities, they did not qualify as concepts in the proper sense of the word. Frege also faults the type of formalism endorsed by Heine and Thomae because it cannot account for the application of numerical concepts throughout the sciences as whole: "Now it is applicability alone which elevates arithmetic from a game to the rank of a science. So applicability necessarily belongs to it" (*TPW*, 167). And like Kant, Frege maintains that showing a priori mathematical concepts qualify as full-fledged, cognitively fruitful concepts necessarily involves presenting the corresponding singular entities, whose construction somehow lends to these entities the generality to account for the concepts' scope of validity.[73] Frege and Cohen agreed with Kant that Euclidean constructions satisfy this condition for geometric concepts. Frege intends for his introduction of the natural numbers as extensions (and later value-ranges) to satisfy this condition for arithmetic concepts. So, for both Kant and Frege, the notion of a properly valid concept serves to highlight the fact that their opponents are operating at best with arbitrarily stipulated, logically possible, general ideas, which are devoid of cognitive content.

---

[72] See Sect. 2.12 and Sect. 2.3.

[73] I believe this also explains why Frege sometimes says that proving the logical consistency of mathematical concepts requires producing the object(s) falling under them (cf. *FA*, 106 and *FR*, 275). On Frege's account, although it may be possible to prove the logical consistency of Hilbert's and Stolz' "senseless" concepts by some other means, showing that we have logically consistent, *cognitively significant* mathematical concepts will require producing the objects that fall under them. We must produce the mathematical objects mediating the application of the corresponding mathematical concepts to all of objects treated in the less general sciences (cf. *FG*, 63–64, and *CP*, 264).

Kant then called upon his distinction between a full-fledged concept and full-fledged intuition to serve as warning to those who would argue directly from the general representation of a conceived something with certain properties to the actual given-ness of a particular something having those properties. For Frege too, the distinction between a proper concept and proper object serves as reminder that even if his colleagues are operating with an objectively valid general idea, they should not confuse this concept with the singular entity bearing conceptual marks as properties. As we saw, Frege complained that his colleagues blurred the distinction between mark and property and thus the difference between subordination and subsumption. In his 1882 letter, Frege agrees with Marty on the importance of distinguishing between the act of judging and the judgeable content, but explains why "the distinction between individual and concept seems to me more more important":

> In language the two merge into each other. The proper name 'sun' becomes a concept name when one speaks of suns, and a concept name with a demonstrative serves to designate an individual. In logic, too, this distinction is not always observed (for Boole only concepts really exist). The relation of subordination is quite different from that of an individual falling under a concept. (*FR*, 80–1)

Here we see that Frege expects C&O distinction to mark off the difference between obtaining a first-level concept capable of having one, several or no objects falling under it and obtaining a particular entity with the relevant properties. How were the formalists guilty of making this mistake? They would define or abstract a numerical concept,[74] then claim this concept possessed the sought after arithmetical properties. Frege reminds them that the concept of a right-angled triangle is neither a triangle nor right-angled (*FG*, 148). The C&O distinction thus blocks the formalists' illicit move from obtaining a perfectly proper concept to asserting the existence of an individual falling under it.

The interpretative strategy that I have pursued concerning Frege's C&O distinction is similar to that pursued in Burge (1986) and Ricketts (2010). For Burge, we cannot adequately assess the validity of Frege's more counter-intuitive doctrines until we recognize their crucial *practical* importance in achieving the ends that Frege has set for himself. It is because Burge cannot discover any such significance for the absolute nature of the C&O distinction and the rigidity of inferential links between surface syntax, logico-grammatical function and referent that he ends up dismissing Frege's claims on this subject as unnecessary metaphysical appendages.

Ricketts claims, by contrast, to have uncovered the practical import of Frege's dogmatic adherence to the distinction, thus providing an additional rationale for Frege's refusal to be moved by Kerry's counter-example. For Ricketts, Frege's conception of logic is revolutionary because it treats the inference from generality to particular—from 'Everything is mortal' to 'Socrates is mortal'—as a distinct, fundamental mode of inference. Convincing Kerry and his colleagues to revise their everyday linguistic practices so as to respect the C&O distinction as a fundamental logical fact is crucial in completing Frege's revolution, given that the distinction

---

[74] Frege would deny that they were successful even at this, but let's disregard that for moment.

encodes or "embodies" the primacy of this distinctive form of inference and the multi-level "quantificational understanding of generality that Frege sets against [Aristotelian and Boolean] conceptions of logic" (Ricketts (2010), 149).[75] On Ricketts' interpretation, §46–§54 of *Foundations* and Frege's requested pinch of salt are aimed at reshaping "his audience's logical gestalt to initiate them into this viewpoint without having to instruct them in Begriffsschrift" (Ibid.).[76]

The pay-off for reading Frege in the context of the Helmholtz-Cohen debate is that we have now uncovered an additional reason for Frege's insistence on the absolute distinction between concepts and objects. Besides launching a revolution, Frege found himself in the midst of one, a contest over who would serve as chief interpreter for the Kantian vocabulary from which the primitive terms for the nascent science of the epistemology of science would be derived. On my interpretation, Frege's requested pinch of salt is aimed at reforming his audience's linguistic practices so as to facilitate the shift from Helmholtz's first way to Cohen's second way of Kantianism for the sciences. The distinction encodes, among other things, a reworking of Kant's C&I distinction particularly designed to supplant formalist conceptions of abstraction and objectivity and the naturalized Kantianism that Frege sees harboring them.

References

Beaney, M. (Ed. and Trans.). (1997). *The Frege Reader.* Oxford: Blackwell Publisher's

Brandom, R. (2009). *Reason in philosophy: Animating ideas.* Harvard: Harvard University Press.

Burge, Tyler (1986), 'Frege on Truth' in Haarparanta and Hintikka, 97–154.

Cassirer, E. (1910). Substanzbegriff und Funktionsbegriff, Trans. and Repr. In W. C. Swabey & M. C. Swabey (Eds.), *Substance and function and Einstein's theory of relativity* (pp. 1–346). New York: Dover Publications, Inc, 1923.

Cassirer, Ernst (1929) *The philosophy of symbolic forms* (Vol. 3, R. Manheim, Trans.). New Haven: Yale University Press, 1957.

Coffa, J. A. (1991). *The semantic tradition from Kant to Carnap* (Linda Wessels, Ed.). Cambridge: Cambridge University Press.

Cohen, H. (1871). *Kant's Theorie der Erfahrung* (1st ed) selected portions (Prasse, J., Gallagher, K., Merrick, T., Trans.). Berlin: F. Dummler.

Cohen, H. (1883). *Das Princip der Infinitesimal-Methode und seine Geschichte,* selected portions (Hildebrand, K., Merrick, T., Trans.). Berlin: F. Dummler.

Cohen, R. S., & Elkana, Y. (1977). *Hermann von Helmholtz: Epistemological writings.* Dordrecht: D. Reidel Publishing.

Currie, G. (1982). *Frege: An introduction to his philosophy.* Brighton: Harvester Press.

Daston, L., & Galison, P. (2007). *Objectivity.* Brooklyn: Zone Books.

Demopoulos, W. (Ed.). (1997). *Frege's philosophy of mathematics.* Cambridge, MA: Harvard University Press.

Dummett, M. (1981). *Frege philosophy of language* (2nd ed.). Cambridge: Harvard University Press.

---

[75] Ricketts (2010), 149. Specifically, the distinction operating in conjunction the Context Principle and Leibniz Law embodies this primacy.

[76] Ricketts (2010), 149.

Dummett, M. (1991a). *Frege: Philosophy of mathematics*. Cambridge: Harvard University Press.

Dummett, M. (1991b). *Frege and other philosophers*. Oxford: Clarendon Press.

Ewald, W. (Ed.). (1996). *From Kant to Hilbert: A source book in the foundations of mathematics*. Oxford: Clarendon Press.

Frege, G. (1952). *Translations from the philosophical writings of Gottlob Frege (TPW)* (P. Geach, and M. Black, Eds.). Oxford: Basil Blackwell.

Frege, G. (1964). *Basic laws of arithmetic: Exposition of a the system (BLA)* (Ed., and Trans. with introduction by Montgomery Furth). Berkeley: University of California Press.

Frege, G. (1953), *Die Grundlagen der Arithmetic* (J. L. Austin, Trans., as *The Foundations of Arithmetic (FA)*. Evanston: Northwestern University Press.

Frege, G. (1979). *Posthumous writings (PW)* (H. Hermes, Kambartel, F., & Kaulbach, F. Eds., P. Long, & White, R. Trans.). Oxford: Basil Blackwell.

Friedman, M. (2000). *A parting of the ways*. Chicago/La Salle: Open Court.

Haaparanta, L., & Hintikka, J. (Eds.). (1986). *Frege synthesized*. Dordrecht: D. Reidel Publishing Company.

Heis, J. (2010). Critical philosophy begins at the very point where logistic leaves off: Cassirer's response to Frege and Russell. *Perspectives on Science, 18*(4), 382–408.

Helmholtz, H. (1887). Numbering and measuring from an epistemological point of view (*NME*) in Ewald (1996), pp. 727–752.

Hilbert, D. (1899). *Foundations of geometry* (E. J. Townsend, Trans.). Chicago: Open Court Publishing, 1921.

Kluge, E. -H. (Ed. and Trans.). (1971). *Gottlob Frege: On the foundations of geometry and formal theories of arithmetic*. New Haven: Yale University Press.

Macbeth, D. (2005). *Frege's logic*. Harvard: Harvard University Press.

Macfarlane, J. (2002). Frege, Kant, and the logic in Logicism. *The Philosophical Review, 111*, 25–65.

Natorp, P. (1887). On the objective and subjective grounding of knowledge (D. Kolb, Ed. and Trans.). *Journal of the British Society for Phenomenology, 12*(3, October 1981), 246–266 and Reprinted in Luft (2015), pp. 164–179.

Patton, L. (2005). The critical philosophy renewed: The bridge between Herman Cohen's early work on Kant and later philosophy of science. *Angelaki, 10*(1), 109–118.

Picardi, E. (1996). Frege's anti-psychologism, in Schirn (1996), pp. 307–328.

Poma, A. (1997). *The critical philosophy of Hermann Cohen* (J. Denton, Trans.). Albany: State University of New York Press, 1997.

Reck, E. (Ed.). (2002). *From Frege to Wittgenstein*. Oxford: Oxford University Press.

Resnik, M. D. (1980). *Frege and the philosophy of mathematics*. Ithaca/London: Cornell University Press.

Richardson, A. W. (2006). *"The fact of science" and critique of knowledge: Exact science as problem and resource in Marburg neo-Kantianism*, in Friedmann and Nordmann (2006), pp. 211–226.

Ricketts, T. G. (2010). Concepts, objects, and the context principle. In Ricketts & Potter (Eds.), *The Cambridge companion to Frege* (pp. 149–219). Cambridge: Cambridge University Press.

Schirn, M. (Ed.). (1996a). *Frege: Importance and legacy*. Berlin: Walter deGruyter.

Schirn, M. (1996b). *On Frege's introduction of cardinal objects as logical objects*, in Schirn (1996a), pp. 114–173.

Sluga, H. D. (1980). *Gottlob Frege*. London: Routledge/Kegan Paul.

Sullivan, D. (2002). The further question: Frege, Husserl and the Neo-Kantian Paradigm. *Philosophiegeschichte und Logiscle Analyse, 5*, 77–95.

Tappenden, J. (1995). Geometry and generality in Frege's philosophy of arithmetic. *Synthese, 102*, 319–336.

Weiner, J. (2002). Section 31 Revisited: Frege's Elucidations, in Reck (2002), pp. 149–182.

Wilson, M. (1992). *Frege: Royal road from geometry*. Reprinted in Demopoulos, 1997.

Chapter 5
Some Sanctifying Precepts for Science and Religion

Abstract This chapter presents the lessons learned from the Helmholtz-Cohen-Frege debate and from Cohen's corpus about wisely navigating revolutionary changes in science, religion, and other cultural practices. Section 5.1 draws on the work of Bas van Fraassen and Michael Friedman to explain the problems that Kuhn's *The Structure of Scientific Revolutions* poses for the notion of scientific progress. Because Helmholtz, Cohen, and Frege initiated and reacted to radical modifications to the mathematical sciences and the Kantian epistemological framework for housing them, the material covered in Chaps. 1, 2, 3, and 4 delivers tactical rules for addressing these Kuhnian challenges. The section ends with summarizing these rules. Section 5.2 turns to examining the political and epistemic crises threatening Cohen's German Jewish community. Drawing on the work of Kristi Dotson and Miranda Fricker, I argue that these crises include instances of epistemic injustice. I show how Cohen's works in ethics, history of philosophy, and Jewish philosophy serve to launch a conceptual revolution aimed at redressing this injustice. The section ends by summarizing what Cohen teaches us about reasonably responding to epistemic crises within Abrahamic religious traditions. Section 5.3 then applies these rules and lessons to an epistemic crisis currently confronting my own religious community, namely, the debate over the redefinition of Christian marriage.

Keywords Bas van Fraassen · Michael Friedman · Kristi Dotson · Miranda Fricker · Scientific progress · Epistemic injustice · Science and religion

At the turn of the twentieth century, Frege warned that irresponsibly deviating from the traditional sense of a word is tantamount to sinning against the scientific practice of which one is a part. We saw that both Cohen and Frege anticipate Thomas Kuhn's claim that radically modifying terminology in the wake of revolutionary developments threatens a science's epistemic cache. How can scientists presume they are communicating with one another and accumulating knowledge about a domain of entities, if the senses of the terms delimiting that domain are allowed to vacillate? We also know, however, that had Frege prevailed in convincing his colleagues to

© Springer Nature Switzerland AG 2020   161
T. Merrick, *Helmholtz, Cohen, and Frege on Progress and Fidelity*,
Philosophical Studies in Contemporary Culture 27,
https://doi.org/10.1007/978-3-030-57299-0_5

adhere to the traditional Euclidean sense of primitive geometric terms and axioms, Einstein would not have had the mathematical and conceptual resources to articulate the theory of relativity.[1]

We thus confront a dilemma and not just one for the sciences but for any practice or tradition presuming to pass down a body of knowledge from one generation to the next. If, on the one hand, practitioners refuse to modify the sense of a term to better accommodate novel insights and well-confirmed empirical data, the tradition loses traction with experience and is in danger of becoming a mere exercise in hierarchically enforced dogmatism. In the words of Kant and Frege, the tradition's basic laws, concepts and propositions are being deprived of genuine cognitive content. If, on the other hand, practitioners follow Helmholtz's lead and quickly modify the sense of terms to best accord with such novelties and data, they risk undermining the normativity and scope of validity historically accorded to the tradition's claims. In the words of Michael Friedman, the consensus on the "rules of the game" grounding the tradition's identity, the objectivity of its claims, and norms of rationality is starting to erode.[2]

According to Cohen and Bas van Fraassen, religious traditions confront this same situation. In his 1908 *Ethics of Maimonides,* Cohen argues that classical Jewish thinkers understood Judaism as tasked with communicating universalizable ethico-religious truths.[3] In typical Kantian fashion, he asks how such a collective cognitive task is possible in light of advancing scientific knowledge and philosophical theorizing. This question is especially pressing for religions with a high view of scriptural authority:

---

[1] According to Friedman (2001), Einstein's "great innovation" was primarily conceptual rather than mathematical and made possible by realizing that geometry was "not forced upon us by either reason or experience, but rather as resting on a free choice, a convention of our own" (Ibid., 23). This realization was due to reading Henri Poincaré's conventionalist account of geometric axioms, an account violating the Fregean stricture that geometric axioms must retain their Euclidean sense (Shapiro (1997), 165).

[2] Friedman describes Kuhnian normal science as "a consensus or agreement on a single set of rules of the game, as it were, which set the parameters of inquiry for all practitioners of the discipline" (Friedman (2001), 19). A paradigm shift constitutes a disruption of this consensus.

[3] Cohen does not think that this particular task exhausts the significance or existential import of Judaism. He does, however, maintain that it is an essential part of the tradition and an authentic Jewish vocation, one to which he is called (Poma (1997), 159). Peter Harrison (2015) presents a strong argument that conceiving of religious fidelity as primarily adhering to and promulgating "a system of beliefs and practices" is a distinctly modern conception (35). Harrison argues that the terms 'science' and 'religion,' understood as referring to distinct bodies of knowledge obtained via differing methods and modes of justification, did not emerge until the sixteenth century and that contemporary use of the English term 'science' to refer strictly to the natural sciences only dates back to the nineteenth century: "Over the course of the sixteenth and seventeenth centuries we will witness the beginning of a process in which the idea of religion and science as virtues or habits of mind begins to be overshadowed by the modern, systematic entities 'science' and 'religion.'" (*Harrison* 2015, 14). If Harrison is right, Cohen's belief that resolving this neo-Kantian epistemological conundrum is crucial to Judaism's flourishing marks him as characteristically *modern* Jewish philosopher.

[I]s philosophical ethics—and is there any other kind?—compatible with ethical norms that seem to be inherently religious, as they are an integral part of religious doctrine? This problem is aggravated by the nature of Jewish tradition, which, much more than Christianity or Islam, is dominated and controlled by laws of Scripture, and even more by laws of oral transmission, regulating in detail the ethical conduct of all individual and social activity (*EM*, 326).

Bracketing Cohen's relative comparison, the problem he describes is a serious one for all three Abrahamic religions. How can ethico-religious propositions, propositions deriving from a long-standing practice of traditionally authorized biblical hermeneutics, be reconciled with advancing moral knowledge? And for Cohen—contra Kant and most twentieth century logical empiricists and Wittgensteinian fideists—there is such a thing as advancing propositional moral knowledge.[4]

On Cohen's account, viable religious traditions are in large measure ethical traditions,[5] and so the problem of ensuring genuine cognitive content for religious traditions generalizes to ethics:

[I]f ethics is to become and remain seriously and genuinely a problem of cognitive pursuit—then ethics may not be separated from any other conscientious intellectual endeavor and may not be exposed disparately. Only within the epistemological can ethics flourish and advance as a science (*EM*, 3).

For classical ethical and religious traditions to flourish, they must be responsive to developments in any sound cognitive enterprise, namely, the natural sciences. Such vulnerability is at odds, however, with the robust universality, strict normativity and fundamental nature accorded to ethical concepts and propositions:

Here we arrive at a crossroad which we feel prompted to compare to the prophetic metaphor: Heaven and Earth, nature and science, may pass away, if only God's word, if only ethics shall remain. On the other hand, if ethics must become a science, and nothing but a science—will it not thus become subject to the destiny of all natural sciences, by being submitted to scientific methodology? (*EM,* 9).

Cohen is clearly well-positioned, given the Kuhnian crisis he witnessed, to appreciate what is at stake in tethering the epistemic status of a religious tradition to synthetic a priori mathematical propositions, let alone the a posteriori propositions of natural science. Yet, he maintains as do I that a religion's ongoing vitality and cultural relevance demands that adherents learn to wisely navigate their way through

---

[4] Since the reaction against logical empiricism and the increased critiques of Enlightenment assumptions, it is not uncommon for philosophers to deny the so-called sharp distinction between propositional statements of fact and expressions of value. For some Christian philosophers, Dallas Willard (2018) and R. Scott Smith (2014), denying the fact-value distinction is part and parcel of recognizing the possibility and actuality of moral knowledge. To my mind, however, there has not been sufficient attention paid to the problem that Cohen introduces: reconciling advances in our moral understanding with the value judgments expressed in classical Christianity.

[5] Insisting on the essential relationship between monotheism of Judaism and philosophical ethics is a constant theme of Cohen's later work. He maintains that "Judaism…makes no distinction between religion and ethics" and that any "scientifically reasoned ethics…must be grounded in the idea of the One God" (*RH*, 45–6).

the "Charybdis" of cognitively empty dogmatism and the "Scylla" of unduly hetero-dox scientism (*EM*, 9).

Like Cohen, van Fraassen sees religious traditions as cognitive pursuits in that they too undergo epistemic crises and paradigms shifts analogous to those punctuat-ing the history of modern western science,[6] and there is ample evidence he is right. Philosophers and historians of science have successfully refuted the idea that the relationship between science and Christianity is one of inherent and perpetual con-flict (Plantinga (2011); Harrison (2015)). That said, Christian traditions have not and cannot easily accommodate revolutionary scientific developments within their doctrinal, liturgical and hermeneutical practices. We are all familiar with the so-called Galileo affair, but we may be less aware that Catholic philosophers are still expressing disappointment with its handling by more contemporary church authori-ties. In response to a 1992 report and speech delivered by John Paul II to the Pontifical Academy of Sciences, philosopher of science Ernan McMullin writes:

> They did acknowledge fault on the part of the Church's theologians in dealing with the Copernican challenged; they did praise Galileo's contributions to science and to exegesis. But they also perpetuated some of the defensive stratagems employed by apologists of an early time, stratagems easily discredited on historical grounds. And they passed over in silence the trial of Galileo... (McMullin (2005), 2).

McMullin's concern is that without a frank and complete analysis of the Galileo affair, Christian scholars, clergy, and laity will fail to learn the theological and ecclesiological lessons it has to teach. Protestant traditions also tend to adopt a strong defensive posture when confronting perceived threats from the sciences. Consider the episodic but recurring dismissal of faculty from North American Evangelical colleges and universities when they question their tradition's reading of the Book of Genesis in light of Darwinian evolution and the implications of popula-tion genetics.

Providing a complete analysis of these ecclesial reactions would take us too far afield, but two points are worth noting. First, it would be a mistake to view such reactions as motivated by nothing other than an interest in solidifying papal or insti-tutional power. In fact, they are motivated at least in part by a fidelity to inherited readings of Scripture and, in turn, the tradition's consensus on proper biblical hermeneutics. Consider, for example, the Grand Duchess Cristina of Lorraine's reaction upon hearing of Galileo's astronomical discoveries and their implied sup-port of Copernicanism. She objected on grounds that it contradicted a prima facie reading of Joshua 10:13, where it is the sun, not the earth, which miraculously stops moving (Sobel (2000), 62–3). Notice too how Galileo responds. He does not cite

---

[6] Van Fraassen maintains that the account of St. Paul's conversion to Christianity reads like Kuhn's account of scientists undergoing a paradigm shift. For a scientist, the "transition to the new view of nature, the conversion to that view does not admit of justification within prior understanding of standards of acceptability" (van Fraassen (2002), 151). Prospectively speaking, the shift appears irrational and absurd. Similarly, for St. Paul, "the gospel made literally no sense" from his prior perspective. It is only from his post-conversion perspective that elements within this new gospel can be seen as continuous with "what he had accepted beforehand" (Ibid.,250 n36).

additional empirical data or evidence on the reliability of his telescope. Instead, he proposes an alternative hermeneutical framework and a method for deciding the relative weight of epistemic authority that should be accorded to the Scriptures on questions of concern to natural science (Galileo (2012), 61–94). In short, the Grand Duchess poses one horn of Cohen's dilemma: what happens to the scriptural authority if we subject it to Galilean scientific methodology? Galileo responds by proposing a paradigm shift, a modification of the rules of the game governing normal religion for seventeenth century Italian Catholics.[7]

Showing proper deference to the tradition's understanding of Scripture, liturgies and sacraments also accounts for the difficulties Protestant institutions are experiencing due to rapid developments in the life sciences. Mounting evidence that the human species does not trace back to a single copulating pair and that embryonic sex development is more complex and variable than previously understood strike at the heart of standard Protestant readings of the Bible, readings that support core doctrines on human uniqueness, marriage, sex and gender. Physicist Freeman Dyson reports that biology has now surpassed physics in terms of allocated resources and the number of major discoveries (Dyson (2007), 1). He predicts that biology will "remain the biggest part of science through the twenty-first century" and notes the profound implications of biological discoveries for conceptions of human welfare (Ibid.). I contend that many Christian institutions are undergoing an epistemic crisis spawned by the life sciences that is proportional to the historical crisis spawned by the advent of modern physics. According to van Fraassen, a necessary mark of a "radical or revolutionary" change within a tradition is that the proposed change appears "absurd" to practitioners prior to its acceptance (van Fraassen (2002), 111). What does this mean when talking about religious traditions? When it comes to proposals aimed at enabling western Christian traditions to accommodate progressive scientific developments, 'absurdity' usually means incompatibility with extant interpretations of Scripture and sacramental praxis.

The second point worth noting is that while the historical record does not support the perennial conflict thesis on the relationship of science and Christianity, it also does not support the non-overlapping magisterial thesis advanced by evolutionary biologist Stephen Jay Gould. According to Gould, science and religion have equivalent sovereign epistemic authority over disjoint domains of inquiry:

> [T]he net, or magisterium, of science covers the empirical realm: what is the universe made of (fact) and why does it work this way (theory). The magisterium of religion extends over questions of ultimate meaning and moral value. These two magisteria do not overlap, nor do they encompass all inquiry… (Gould as cited in Stenmark (2010), 278).

In contrast to Gould's general picture of non-overlapping fields of inquiry, the Helmoltz-Cohen-Frege debate shows that differing conscientious intellectual endeavors, however well-delineated, often intersect and overlap one another. This is why nineteenth century developments in geometry generated a hermeneutical crisis

---

[7] Given subsequent events and the relatively recent praise for Galileo's views on biblical exegesis, it is safe to say his proposal was not readily adopted.

for Kantian philosophy. A paradigm shift in one domain of inquiry can easily initiate an epistemic crisis in another.

Another problem with Gould's thesis is it rests on the assumption that science trades only in facts not values, and this positivist view of science is no longer tenable. Feminist philosophers of science and critical race theorists have demonstrated how androcentric and Eurocentric value judgments played a crucial role in framing many of the questions and answers provided by western modern science (see Lederman and Bartsch (2001); Zack (2002); Harding (2008); and Anderson (2020)). Philosophers of all stripes are looking for ways to intentionally and responsibly incorporate value judgments into scientific research (see Plantinga (1996); Kitcher (2001); Anderson (2004); and Friedman and Nordmann (2012)). Finally, it is difficult to conceive how biomedical sciences can proceed without at least implicitly subscribing to claims about the meanings of 'human health' or 'normal functioning,' and as disability theorists have been quick to point out, these meanings are saturated with insufficiently scrutinized value judgments.

So the issue is not whether the domains of scientific and religious traditions overlap and intersect; they do. Rather, the issue is how best to navigate this shared territory in the wake of an epistemic crisis. I believe the debate between Helmholtz, Cohen, and Frege has something to teach us in this regard. My intent is not to provide a blueprint or recipe for resolving epistemic crises, since I agree with van Fraassen that this is impossible. Wisely responding to an epistemic crisis requires exercising intellectual courage, humility and a host of other intellectual and moral virtues, and we should accept the consensus of virtue theorists that exercising the virtues is not reducible to behavioral rule-following. Nevertheless, guiding principles and proverbs, along with historical exemplars of virtue and vice, have long been employed as a means of cultivating virtue. My goal is to extract a few of these principles from the nineteenth century crisis in geometry and Kantian philosophy. My hope is that these can aid us twenty-first century practitioners in exercising a reasonable and virtuous response when confronting similar crises in our respective scientific or religious traditions.

This chapter is organized as follows. Section 5.1 presents a detailed exposition of van Fraassen's and Friedman's descriptions of and proposed solutions to the Kuhnian problem. Their descriptions and solutions of the problem are evaluated vis-à-vis the Helmholtz, Cohen, and Frege debate. Here the focus is on scientific traditions and identifying precepts useful in avoiding the Fregean sin of communicative confusion when undergoing revolutionary changes involving substantial semantic modifications. Section 5.2 turns to Cohen's *Ethics of Maimonides* and related works on the philosophy of religion. The goal here is identifying additional principles particularly relevant to Abrahamic religious traditions in the midst of an epistemic crisis, namely, North American Protestant traditions at the turn of the twenty-first century.[8] Section 5.3 introduces the debate over redefining marriage

---

[8] I focus on North American Protestant religious institutions simply because these are the ones with which I am most familiar. I leave it to others to decide how much of what is said is applicable to other religious traditions and ecclesial institutions or communities.

within the United Methodist Church as a test case for applying these principles. I argue that this debate satisfies van Fraassen's criteria for an epistemic crisis. I remind my reader of the burden of proof on someone proposing a paradigm shift or conceptual revolution within a Christian tradition: she must show how the Bible and other authoritative literary sources can be interpreted so as to view the proposed modification as a reasonable continuation of the tradition. I end by sketching how this burden could be met.

5.1 Redeeming the Concept of Scientific Progress

How exactly does Kuhn's account of scientific revolutions challenge the idea of scientific progress and how is this challenge related to the Fregean sin against science? As we have seen, Frege's charge against the nineteenth century formalists is that they are diverging from the methods and meanings characterizing a particular mathematical tradition without ensuring recognition of its past, present and future scientific status. Geometry, on his account, is rooted in the synthetic constructive procedures countenanced by the traditional Euclidean axioms and postulates. These procedures secure both the sense [*Sinne*] of geometric terms and the range of empirical phenomena over with geometric propositions are valid. Therefore, to the extent that Helmholtz and Hilbert fail to employ traditional synthetic methods, they sever geometric terms from their historically recognized objective content. The use of analytic methods and applying geometric terms to non-Euclidean spatial domains introduces a semantic discontinuity that threatens the presumed and requisite communicability among practitioners: "If someone does not accept [the parallels axiom], I can only assume that he understands these words ['straight line' and 'intersect'] differently" (*PW*, 247). Moreover, unless and until the fruitfulness of non-Euclidean geometry for the empirical sciences is demonstrated and the past success of Euclidean geometry explained, the revolutionary practices of his colleagues not only undercut the epistemic authority accorded to geometry in the future, but also its past successes: "Do we dare to treat Euclid's elements, which have exercised unquestioned say for 2000 years, as we have treated astrology?" (*PW*, 169). In other words, Frege is demanding that his formalist colleagues tackle two interrelated problems that van Fraassen and Friedman see emerging from Kuhn's analysis of scientific revolutions: royal succession and trans-paradigm communicative rationality. As we will see, salvaging the notion of scientific progress depends on adequately responding to these problems.

5.1.1   Van Fraassen and Friedman on the Kuhnian Problem

Van Fraassen and Friedman both view western modern science as an exemplary collective intellectual endeavor. For van Fraassen, this is a crucial aspect of his proffered empirical stance. What makes science exemplary is its ability to tolerate doubt and dissent among practitioners:

> In science disagreement is not impiety, and doubt is not treason, no matter the content. That feature is very salient in the empiricist case for taking empirical science as our paradigm of rational inquiry (van Fraassen (2002), 43).

This admiration for scientific inquiry is retained even if its propositional deliverances have been, are, and may yet be false:

> How do we live in a world in which, to our best knowledge and belief, all our best most fundamental scientific theories are false? We live in it by the lights of science as practice, as search, as rational form of inquiry par excellence. For the materialist, science is what teaches us what to believe. For the empiricist, science is more nearly what teaches us how to give up our beliefs (Ibid., 63).[9]

For van Fraassen, materialists or naturalists prize science because it presents us with the best means for "finding out what the world is like" (van Fraassen (2015), 66, citing Maddy 2001).[10] While both the empiricist and naturalist stance embrace fallibilism with respect to our current scientific theories and reject a robust scientific realism as unduly metaphysical, van Fraassen maintains that the naturalist's admiration for science focuses more on its content, a content assumed to be transparently clear, true enough, and complete enough to prescribe beliefs. He thus likens the stance of the scientific naturalist to that of a scriptural fundamentalist; both "eschew exploration of interpretative alternatives, or even to recognize the need or relevance thereof" (Ibid., 81). In contrast, van Fraassen's empiricist admires science not for its ability to dictate our beliefs about the world, but because it exemplifies a kind of communal intellectual humility, a collective willingness to tolerate dissenting views and acknowledge that previously asserted beliefs were mistaken.[11]

The problem for both van Fraassen and Friedman is that this tolerated dissent and acculturated humility seems possible only within the context of a more general consensus, a consensus forged by a shared Kuhnian paradigm. Indeed, for Friedman, it

---

[9] Van Fraassen derives his claim that our current best scientific theories are false from the fact that quantum mechanics and general relativity are, as presently articulated, mutually inconsistent.

[10] Van Fraassen takes Maddy (2001, 2007) as the exemplar of twenty-first century naturalism. Once we have thoroughly unpacked van Fraasen's critique of naturalized epistemologies and stances, it should be clear to my reader that this critique would similarly apply to Helmholtz's naturalized Kantianism as described in Sect. 3.1.

[11] There are numerous debates over defining 'intellectual humility' in contemporary literature on the intellectual virtues. Here I speak of "a kind of intellectual humility" because I assume that tolerating dissent and abandoning unsupported beliefs are indicators of intellectual humility under at least one of these definitions.

is this consensus and the kind of communication it affords that best explains why science is a model of rational inquiry:

> In the scientific enterprise, unlike other areas of intellectual and cultural life, there is such a thing as normal science—periods of "firm research consensus" in which a given paradigm, conceptual framework, or set of rules of the game, as it were, is "universally received" by all practitioners or a given field or discipline...In the sciences, then, we are actually able to achieve a situation of communicative rationality far exceeding that possible in other areas of intellectual and cultural life; and this is undoubtedly the reason that the scientific enterprise has been taken to be a particularly good model or exemplar of human rationality from the Enlightenment on... (Friedman (2001), 59).[12]

Borrowing from Jurgen Habermas, Friedman distinguishes between instrumental and communicative rationality. Where the former aims at "successful self-maintenance" and "intelligent adaptation" to the contingencies of one's environment, the latter aims at dialogically testing subjectively-held perceptions and ends to ensure they are adequately tied to "the unity of the objective world" and the "intersubjectivity of [one's] context of life" (Habermas cited in Friedman (2001), 54). Scientific communicative rationality is thus inherently dialogical and encourages dissent as a means of forging a consensus among practitioners about which perceptions and ends are sufficiently world-directed and intersubjectively valid. Under the conditions of normal science, a paradigm supplies the "universally received" principles of reasoning and argumentation that allows for this dissent and the confidence of arriving at a consensus (Friedman (2001), 55). However, it is precisely the erosion of this universally received or higher-order consensus that characterizes revolutionary scientific change. So, for both van Fraassen and Friedman, unless we can describe the conditions under which a successor paradigm might be viewed as a rationally-motivated, consensual decision by practitioners, scientific inquiry loses its exemplary status as an admirable model of communal human inquiry.

At a minimum, then, progress within a scientific discipline implies that the prevailing paradigm is seen as a "rational endorsable continuation" of its past (van Fraassen (2002), 112). The history of modern western science seems to indicate that this is indeed the case. Drawing on the rich resources of Reimannian geometry, one can show that as spatial regions approach the infinitely small, they conform to specifications of Euclidean geometry. The historic validity and utility of Euclidean geometry is thus explained and justified relative to its successor. Similarly, Newtonian mechanics is formally derivable as a special case of Einsteinian mechanics, the former becoming approximately valid for phenomena with velocities that are exceedingly small in comparison to the speed of light (van Fraassen (2002), 115; Friedman (2001), 58–9). The new paradigm thus establishes itself as the "rightful" successor by demonstrating it can recapture the instrumental pay-offs of its

---

[12] Here and elsewhere in his *Dynamics of Reason,* one could legitimately accuse Friedman of being too caught up in the Enlightenment metanarrative on the humane and liberating effects of modern western science. Given my purposes here, I do not challenge this aspect of Friedman's account. For those interested in my critique of the Enlightenment metanarrative, see Merrick (2014).

predecessor and explain its validity over what is now seen as a more well-defined and narrowly circumscribed domain of inquiry as compared with the subject matter as conceived under the old framework (van Fraassen (2002), 112). On Friedman's account, the fact that this sort of legitimacy is demanded of the sciences is another reason for according them special status:

> In the sciences, unlike other areas of intellectual and cultural life, we are never in a position simply to throw out all that has gone before (if only rhetorically) and start again anew with an entirely clean slate. On the contrary, even, and indeed, especially, in periods of deep conceptual revolution, we still strive to preserve what has gone before (the preceding paradigm or framework) as far as possible (Friedman (2001), 58).

The problem is that this demand is almost always met *retrospectively* and from the successor's perspective.

For both van Fraassen and Friedman claim, the Kuhnian challenge to scientific progress and the problem of royal succession retains its force unless and until one provides a satisfactory account of *prospective* rationality and continuity:

> From the posterior point of view, the prior can be made intelligible and the change ratified. From the prior position, however, the posterior view was absurd and the transition to it possible but incapable of justification (van Fraassen (2002), 65).

The move to an 'absurdly' new way of conceiving of one's discipline or tradition must be somehow simultaneously viewed by practitioners as a reasonable transition within the tradition. Otherwise, purported progress within the field is a matter of epistemic luck or the artifice of a history rewritten from the victor's perspective.

Van Fraassen and Friedman thus concede to Kuhn's and Feyerabend's critics that the incommensurability thesis is mistaken, if taken to mean that pre- and post-revolutionary practitioners occupy mutually isolated discursive communities, which makes communication impossible. The fact of retrospective translatability testifies otherwise.[13] Still, they insist, the thesis holds in the weaker sense that it is exceedingly difficult to explain how trans-paradigm communication occurs such that practitioners might be sufficiently and rationally motivated to move off the island in the first place (van Fraassen (2002), 115–6; Friedman (2001), 59–60). For Friedman, accounting for trans-paradigm, rationally-motivating communication is "the most fundamental problem raised the Kuhnian account of scientific revolutions" (Ibid., 99).

Before evaluating van Fraassen's and Friedman's proposed solutions to the problems of royal succession and trans-paradigm communicative rationality, it will help to see why they reject Kuhn's own solution and the notion of scientific progress attending it.[14] According to Kuhn, the incommensurability between paradigms is

---

[13] Friedman notes that Kuhn rejected the claim that Newtonian mechanics was even retrospectively translatable within the Einsteinian framework (Friedman and Nordmann (2008), 284). The extent to which Friedman agrees and disagrees with Kuhn on this point becomes clear once we turn to Friedman's proposed solution for prospective trans-paradigm communication.

[14] For Friedman, this assessment of Kuhn's proposal is explicit and frames Friedman's own formulation of what kind of trans-paradigm communication is required to account for prospective ratio-

substantial enough to show that the idea of scientific progress as an accumulation of knowledge must be jettisoned. Similarly, one must abandon the belief that science is progressing in the sense of converging upon a single truth about the really real:

> There is, I think, no theory-independent way to reconstruct phrases like 'really there'.....I do not doubt, for example, that Newton's mechanics improves on Aristotle's and that Einstein's improves on Newton's as instruments for puzzle-solving. But I can see in their succession no coherent direction of ontological development (Kuhn (1996 [1962]), 206).

Kuhn's solution is to equate scientific progress with the continual development of sophisticated tools useful for solving the puzzles that characterize the discipline (Friedman and Nordmann (2008), 239–40)

Friedman maintains that Kuhn's solution is inadequate because, at best, it shows that scientific enterprises can only lay claim to trans-paradigm *instrumental* rationality, when what is required is establishing the possibility of *communicative* rationality (Friedman (2001), 54). Instrumental rationality involves a successful means-end deliberation in cases where the ends are previously set. Establishing the superior puzzle-solving virtues of a proposed successor theory will only be rationally compelling for those who take puzzle-solving to be the end or aim of scientific inquiry. But puzzle-solving in Kuhn's sense is not a long-standing, trans-paradigm goal of science:

> Quantitative accuracy and precision were not widespread ends of inquiry within Aristotelian-Scholastic natural philosophy...[and] did not become meaningful widespread ends of inquiry—that is, practically achievable such ends—until the "second scientific revolution" of the late eighteenth and early nineteenth centuries...(Friedman (2001), 55).

Friedman is right to reject Kuhn's proposed solution. Scientific progress as conceived by Kuhn does not go far enough in capturing the world-directedness and broad scope of intersubjective validity upon which a strong admiration for science depends. On his account, scientific theorizing is not as answerable to the world as it is to the puzzles that happen to be of current interest to practitioners. Kuhn takes what he sees as an aim of nineteenth and twentieth century scientists and makes it the aim of science per se. As a consequence, he effectively forecloses the possibility and responsibility of those in the midst of an epistemic crisis to make a good faith effort at dialogically engaging those who came before and those who might come after about the proper and practically achievable ends in a longstanding scientific tradition. In other words, Kuhn does not go far enough in preventing scientists from committing the Fregean sin.

Given van Fraassen's and Friedman's conception of proper scientific traditions as forms of rational inquiry par excellence, their solution to the Kuhnian problem must deliver a more robust notion of scientific progress. Therefore, it is not surprising that Van Fraassen (2006) maintains, "there *is* an accumulation of knowledge about nature in science" (290). What is surprising is that he thinks this can be said while still agreeing with Kuhn that accepting a new paradigm involves

---

nality and continuity. For van Fraassen, this assessment is implicit but can be ascertained from the solution he presents.

countenancing seismic semantic change: "the new theory…is allowed to re-describe nature entirely in its own terms" (299). For Friedman, one must show that the decision to adopt the new theory can satisfy the criteria of a prospective communicative rationality (Friedman (2001), 101). Having explained van Fraassen's and Friedman's articulation of the Kuhnian problem and the parameters of an adequate solution, let's see what they propose.

5.1.2 Van Fraassen's Solution

For van Fraassen, the first step is realizing that any objectifying epistemology will prove inadequate. By 'objectifying epistemology,' he means any epistemology eschewing the use of value judgements and striving for a strictly factual description of cognitive functioning. Quine's "Naturalized Epistemology" is cited as the obvious example here, but it should be clear to my reader that Helmholtz's naturalized Kantianism would count as well.[15] According to van Fraassen, Alvin Plantinga's account of warranted beliefs as beliefs produced by proper cognitive functioning might also count as an objectifying epistemology, depending on how 'proper' is parsed:

> We may see [Plantinga's epistemology] as a factual theory (metaphysical or theological rather than empirical) about the cognizing subject and cognitive functioning…[W]e can speak of humans functioning properly in the area of cognition only if we think of them as either created, trained or educated, in accordance with a certain design—whether the design of a Creator or of a society that trains and educates its members…[T]his can still be a factual assertion, either about creation or society (Ibid., 76).

An epistemology ceases to be objectifying if any evaluative terms are reasonably viewed as an insertion of the epistemologist's own value judgment. The problem with objectifying epistemologies is that normative expressions like 'functional cognition' or 'warranted belief' derive their meaning from extant empirical, metaphysical, or theological theoretical paradigms. They thus lack the resources for explaining the possibility of rational deliberation or choice in a crisis situation; any proposed radical modification to the factual theory to which the objectifying epistemology is indebted will automatically be viewed as cognitively dysfunctional or unwarranted (Ibid., 79). On van Fraassen's account, a Kuhnian epistemic crisis and its resolution is best described as a "radical conversion" rather than a disquieting assimilation of new facts. It follows that explaining the rationality of these conversions "is an unsolvable problem for objectifying epistemologies" (Ibid., 81)

The second step involves looking to the work of Continental existentialists and American Pragmatists for a more accurate description of our epistemic pursuits and for richer resources to understand how crises are reasonably resolved. Historic

---

[15] Compare van Fraassen's description of objectifying epistemologies to Helmholtz's attempt to derive epistemic theses from facts about perception and psychophysiology. See Sect. 3.1.2

paradigm shifts tell us something about the human condition, namely, that we can easily find ourselves in a situation where continuing to pursue our epistemic aims requires something other than the rule-guided practices and inferences licensed under a normal science situation. It may require a "leap of faith" (Ibid., 92). Invoking both Heidegger and Otto Neurath, van Fraassen describes human knowers as fallible finite beings thrown into collective knowledge-gathering enterprises that have often already set sail. In other words, we must learn to live without the kind of epistemic security that either a foundationalist or an objectifying epistemology may have promised (Ibid., 82–3; van Fraassen (2015), 69). We must also admit that our epistemic goals and means of pursuing them, even within the same discipline or tradition, are subject to reassessment and negotiation. Given these lessons from our intellectual history, van Fraassen concludes we should accept the truth expressed in the voluntarist epistemologies of William James and Wilfred Sellars: a proper description of our epistemic efforts must reference agent-centered aims, volitions, and values. Any epistemology or philosophical stance trying to eliminate such references can only address a "severely abstracted aspect of our epistemic life" (van Fraassen (2002), 88).

So how does the move to a voluntarist epistemology get us closer to a solution? Agent-centered epistemic concerns and value judgments can be invariant across paradigms in a way that paradigm-dependent norms of rationality are not (Ibid., 92). This means that practitioners may appeal to these shared concerns and values when deciding how to respond to the revolutionary changes impacting their field of inquiry: "there is a continuing inner and outer dialogue throughout the conversion experience" (Ibid., 94). Van Fraassen cites Einstein's dialogue with his fellow physicists, but we can also cite the shared values expressed in Frege's exchange with Hilbert, Thomae, and other mathematicians. All parties were concerned about ensuring the broadest possible scope of validity and applicability for mathematical concepts and propositions. All parties were concerned with vouchsafing the objectivity of these concepts and propositions and their fruitfulness for natural science. The debate then centered on alternative means of achieving these goals.

Van Fraassen's voluntarism also provides us with a conception of rationality broad enough to include the leaps of faith made during trans-paradigm periods. For van Fraassen, efforts to identify general laws and principles capable of determining a uniquely rational belief or choice are based on and perpetuate faulty depictions of our epistemic life. Philosophers should focus instead on identifying tactics and strategies for achieving particular epistemic ends in particular contexts and striking some sort of a balance when these ends compete with one another (Ibid., 82).[16]

---

[16] James, for example, takes believing truths and avoiding error to be two epistemic ends. As van Fraassen points out, however, pursuing these ends "pulls us in opposite direction" (Ibid., 87). Cultivating excessive credulity maximizes the former whereas encouraging wholesale skepticism maximizes the latter. What is required are strategies useful in hitting the golden mean or for figuring out when hope and the need for new information outweighs the risk of error. Such strategies must be sensitive to the context and value-judgments about the worth of certain information. Van Fraassen concludes that once we sufficiently attend to the concrete particularities of epistemic life

Although tactical rules and strategies of this sort are too tailor-made and indeterminate to function as decidable rules in a comprehensive rational choice theory, they are perfectly suited to function as rules of thumb for reasonably navigating one's way through an epistemic crisis. Accepting the voluntarist's picture of our epistemic life and solving the problem of prospective rationality means recognizing that the changes individuals or communities undergo in the midst of an epistemic conversion are not rational in the sense of being "rationally compelled." They can, however, be understood as "rationally permitted" (Ibid., 92).

Once again, Frege's exchange with his colleagues during the revolutionary developments in math and logic provides support for van Fraassen's account. Consider Frege's request that his colleagues grant him a pinch of salt and affirm his C&O distinction. Frege scholars are surely right that nothing in Frege's arguments for the distinction would make its acceptance rationally obligatory. However, as I tried to show in Chap. 4, Frege goes a long way in proving that acceptance of the distinction is a rationally permissible move for late nineteenth century logicians, mathematicians, and epistemologists to make, particularly for those operating within a broadly Kantian philosophical framework.

Still, it is one thing to see a radical modification in the discursive praxis of one's scientific or religious tradition as rationally permissible and quite another to take the leap of conversion. To illustrate the difference, I will piggyback off of van Fraassen's example of the conversion of St. Paul. Imagine an alternative road to Dasmascus story where Saul of Tarsus becomes convinced that the Way is a permissible Jewish sect and so quits persecuting its members. Yet, he does not see it as an intelligible way *for him* to pursue the epistemic aims of Judaism, namely, to know and serve the God of Abraham. He thus neither converts nor becomes an evangelist for Christianity. To solve the problem of prospective rationality, van Fraassen needs to account for a rationally motivated conversion. Recognizing this gap in the account sketched thus far, he looks to Jean-Paul Sartre's work on the emotions to bridge it.[17]

According to Sartre, a change in one's emotional state produces gestalt-like changes in how one perceives the salient features of a situation or experience. Van Fraassen has already argued for recognizing the ineliminable role of value judgments in our epistemic pursuits. He goes on to argue that 'experience' is often used as a success word and so we should distinguish the following uses of the word: (1) experience as a mere happening, e.g. stepping on a garden hose; (2) experience as

---

and the ineliminable role of value judgments "the very conception of a methodological cookbooks with precise recipes...loses all plausibility" (Ibid., 88–9).

[17] This appeal to emotions is the sketchiest part of van Fraassen's proposed solution, as commentators and van Fraassen (2011) himself have subsequently noted. That said, the basic idea is easy to understand and has been more fully developed in contemporary cognitivist theories of emotions. The brief exposition I provide here implicitly draws upon this literature and explicitly from Jones (2011). Jones fleshes out this part of the van Fraassen's solution in a manner that the latter both appreciated and found compelling. While I agree that the emotions can and do play the cognitive role described here, I do not think that this appeal to the emotions is sufficient to solve the problem of prospective rationality. I take up this issue when critiquing van Fraassen rejection of Friedman's solution.

the immediate subjective awareness of that happening, e.g. 'I am stepping on a snake'; and (3) experience as an inter-subjectively validated awareness of that happening, e.g. 'I stepped on a hose and thought it was a snake' (Ibid., 134–5).[18] For van Fraassen, to say 'experience' is a success word is to say that only experience taken the third sense counts as experience proper and that these are the kinds of experiences to which scientific theories are rightly beholden. I would add that it also means deciding which experiences, taken in all three senses, warrant further inquiry and this is where the emotions can play a vital role. An altered emotional state can account for the rationality of a leap toward seeing the world as depicted in an 'absurd' theory because emotions are capable of shifting our attention to new experiential saliences and alerting us to the importance of rendering them intelligible.[19] Van Fraassen concludes that converting to the successor's worldview is reasonable insofar as it is motivated by a value-laden, emotion-filled, rational hope of attaining this intelligibility: "[T]he posterior position is one that we now value as a more insightful way of seeing the world we live in" (Ibid., 109).

Having addressed the question of prospective rationality, van Fraassen turns to the problem of royal succession: how is it possible to see the enthroned paradigm as continuing the epistemic goals and accomplishments of its predecessor, particularly since the latter is now recognized as riddled with error? Van Fraassen responds by expanding on Paul Feyerabend's use of the Jesuit critique of Protestant fundamentalism.

Feyerabend argued that empiricist foundationalists viewed experience much like Protestants viewed scripture; both are viewed as self-evident givens, the meaning of which is transparently clear. Following Feyerabend, van Fraassen identifies a certain fundamentalist disposition at work in both traditions, a disposition towards utterly rejecting the idea that understanding experience or biblical content requires interpretive work on our part.[20] The empiricist fundamentalist thus subscribes to *Sola Experientia* in a manner entirely analogous to the religious fundamentalist's *Sola Scriptura*: "any claim to knowledge, any support for opinion, must come from experience; experience trumps all" (Ibid., 120). However, as the Jesuits pointed out, there is a problem when it comes to applying *Sola* type rules, for their application presumes answers to the following: (1) what is and is not genuine experience [Scripture]; and (2) given that differing interpretations do emerge, what or who decides between them? The rule itself cannot answer these questions except on pain of circularity. Feyerabend interprets the Jesuit challenge as highlighting the inherent limitation and ambiguity of rule-following. For it is not the rule itself that answers

---

[18] The distinction between (2) and (3) is significantly similar to the Kantian distinction between subjectively-valid and objectively-valid judgments discussed in Sect. 2.1.

[19] The emotion of wonder has been long recognized by philosophers and historians of science as playing precisely this role (see Descartes (1985), AT XI 373, p. 350; Bynum (1977), 26)

[20] Van Fraassen similarly treats Maddy's naturalistic stance as a version of scientistic fundamentalism. Maddy's naturalizing Second Philosopher rejects the idea that the content of our best currently scientific theories stand in need of interpretation just as scriptural fundamentalists reject the idea that biblical content needs interpreting (Van Fraassen (2015), 81–2)

these questions, but a tacit understanding of what it means to embrace the tradition under consideration. Empiricist foundationalists and Protestant fundamentalists know how to answer these questions because this is simply part of what it means to be a member in good standing of their respective communities.

On van Fraassen's construal, Feyerabend's critique of empiricist foundationalism supports the claim that what counts as a rational move cannot be entirely decided via appeal to a definitive rulebook, which in turn indicates what tactical rule is at work in successfully resolving a Kuhnian crisis. Practitioners employ *Sola* type rules to achieve a certain epistemic end during a period of normal science and another during periods of epistemic crisis. In periods of normalcy, there is no need to debate what qualifies as a genuine experience or authentic biblical content, since the shared understanding of the inherited tradition has already settled such matters. In this context, *Sola Experientia* and *Sola Scriptura* function to "enforce orthodoxy" and prohibit running after any "alternative interpretations ingenious minds can concoct" (Ibid., 142). Here the rule serves as "a counsel of epistemic conservatism" (Ibid., 141). However, as the anomalies begin to pile up and a rationally permissible rival is being primed for inauguration, the rule functions as a tool of critique, encouraging practitioners to identify those parts of the predecessor which can be discarded as mere interpretative mistakes. In a crisis situation, the rule serves as a counsel against epistemic dogmatism and paves the way for "reasoned and proportionate change" with an eye toward "consensual revision" (Ibid., 142).

It follows from van Fraassen's response to the problem of royal succession that the burden on someone proposing a successor is convincing fellow practitioners that nothing essential is lost in translation. Of course, this will require foregrounding what one takes those essentials to be. In the religious case, van Fraassen cites the Christian scriptures as attempt to satisfy this burden with respect to Judaism:

> With reference to our earlier religious parallel we can recall here the early Christians' portrayal of themselves as fulfilling the Jewish sacred scriptures. The New Testament depicts a sustained act of (re)interpretation of those scriptures so as to allow this portrayal (Ibid., 246, note 1).[21]

In the case of science, he holds up the derivation of Newtonian mechanics from Special Relativity as the "loveliest" and "neatest" example of meeting this burden (van Fraassen (2006), 299). As previously mentioned, the proof consists of showing that Einstein's equation reduces to Newton's equations as the speed of light is allowed to approach infinity. So while this proof does not preserve the truth of Newton's equations—"for no finite speed is less than a fraction of the speed of light"—it does explain, by the successor's lights, why Newtonian mechanics performed so well in describing observable velocities falling well below the speed of light (Ibid.).

---

[21] There are several objections to van Fraassen's use of this example, not the least of which is that it seems to advance a supercessionist Christian theology whereby Christianity is the rightful continuation of all that is good in Jewish thought and practice. I believe that these objections can be satisfactorily addressed and attempt to do so in Sect. 5.2.

Van Fraassen's treatment of this proof as an exemplary response to the problem of royal succession helps clarify his conception of scientific progress. For van Fraassen, scientific progress is the accumulation of knowledge about two things: (1) "concrete, observable things, events, and processes in nature" and (2) "the abstract structures studied in mathematics" through which the structure of these observables is characterized (Ibid., 297). He is careful to distinguish his position on scientific progress from the position advanced by a structural realist and the position advanced by Friedman. In general, scientific realism holds that the "aim of science is to provide us with a literally true story about what there is in the world, and this aim is actually achieved to a great extent, because it is pursued by effective means to serve that end" (Ibid., 288). A structural realist holds a slightly weaker position, maintaining that the aim of science is providing a true story about the world's *structural* features. Given this view on the aim of science, establishing trans-paradigm progress would require showing that a successor theory can preserve a predecessor's truths about these structural features. Van Fraassen rightly notes that this criterion for progress cannot be satisfied because the revolutionary conceptual modifications contained in the successor often entail a radical reconfiguring of the predecessor's presumed ontology or physical referents. Furthermore, as we just saw, there are examples of elegant solutions to the problem of royal succession that do not preserve any supposed literal truths of its predecessor. The realist's criterion for a successful solution to the Kuhnian problem is hence too demanding and unnecessary. On van Fraassen's account, the aim of science is not delivering truths about any structure *underlying* phenomena, but delivering representations or models *of* phenomena, models credentialed within a scientific community in virtue of exhibiting certain "empirical successes" (Ibid., 299).[22] Establishing trans-paradigm progress thus requires showing a successor can preserve and explain its predecessor's successes, and this is precisely what the derivation of the Newtonian equations accomplishes. Any reference to the truth of their respective ontological commitments is beside the point.

Van Fraassen further maintains that accounting for these successes does not depend on demonstrating any trans-paradigm continuity between the "theoretical principles" employed by the two theories. This is where he sees himself differing from Friedman (Ibid., 300). I contend, however, that van Fraassen has either overstated this difference or failed to ensure the possibility of trans-paradigm communicative rationality. I also agree with Friedman that unless it can be shown that communicative rationality is possible and indeed sometimes occurs, we are not entitled to parade scientific inquiry as an exemplar of rational inquiry. Since van Fraassen's empirical stance involves taking this attitude toward science, I will argue that he is too quick in dismissing or distancing himself from Friedman's proposal.

---

[22] By 'empirical successes,' van Fraassen means those results that "*credential*" scientific theories and credits Suppe 1993 for use of this term. Suppe (1993) describes this credentialing process as "the standardized ways that observational, experimental and theoretical results are written up for journal publication" (161). These then are the results that van Fraassen believes a successor is obliged to retain or recover.

Besides, evaluating van Fraassen's critique of Friedman is a good point of entry for unpacking the latter's solution and summarizing how both of their solutions stand up in light of the Helmholtz-Cohen-Frege dialectic.

5.1.3 Friedman's Solution

Van Fraassen raises two objections to Friedman's insistence on the need for a priori principles governing scientific praxis through crisis situations. First, he argues that deciding what parts of a predecessor must be recovered under a successor's reign is "highly selective." What guides this selection is not any concern about preserving the theoretical apparatus supporting empirical successes but merely the successes themselves (Ibid., 300). Second, given that solving the problem of royal succession does not require establishing continuity at the theoretical level, there is no "*clear rationale*" for insisting that a new theory explain or duplicate use of the principles supporting past successes. Van Fraassen speculates, correctly I believe, that Friedman's concern with a priori principles is partly motivated by the conviction that such principles are necessary to achieve a rational resolution during times of paradigm transition. He then dismisses this conviction as an instance of the mistaken view that "rationality must consist in rule-following" (Ibid., 303).

Let's take these objections in turn. Van Fraassen never explicitly states what he means by 'theoretical principles' but it is safe to assume they include: (1) mathematical laws or structures presupposed in articulating the theory and supporting its inferences, e.g. the Euclidean axioms and structure presupposed in Newtonian mechanics; and (2) coordinating principles necessary for relating mathematical abstracta to concrete empirical phenomena, e.g. the Newtonian Laws of Motion. Why make this assumption? Because on Friedman's account these just are the a priori intratheoretical principles the continuity of which must be explained to satisfactorily resolve the Kuhnian problem, and it is this more stringent criterion that van Fraassen rejects.

The fact that Friedman insists on establishing trans-paradigm mathematical or theoretical continuity demonstrates his allegiance to the conception of scientific objectivity defended by Cohen and Frege.[23] Recall Frege's opposition to Helmholtz-inspired attempts to derive the objectivity of arithmetic from facts about the psychophysiology of perception. For Frege, as for Cohen, the objectivity of empirically

---

[23] Friedman self-consciously articulates his solution to the Kuhnian problem in reference to the Marburg Neo-Kantians, namely, Cassirer. Friedman agrees with Cassirer that demonstrating scientific progress entails establishing "the continuity of purely mathematical structures" (Friedman and Nordmann (2008), 246). However, my claim that Friedman is indebted to Cohen's and Frege's notion of objectivity does not mean that there are no appreciable differences in what Cohen, Frege, Cassirer and Friedman take to be the necessary and sufficient conditions for establishing the objectivity and progressive nature of a properly scientific inquiry. These differences will soon become clear.

confirmable facts is derived from a priori logical and mathematical principles. Helmholtz' naturalized Kantian epistemology thus rests on a mistaken understanding of scientific objectivity. On Frege and Cohen's account, empirical phenomena cannot be the subject of inter-subjectively valid judgments unless the phenomena is already constituted in accordance with a priori laws or principles. Friedman agrees. According to Friedman, scientific theories contain a priori "constitutive" or "coordinating" principles that function as necessary conditions for genuinely empirical laws. To say these principles are necessary is to say that, without them, empirical laws would lack cognitive content or a decidable truth-value (Friedman (2001), 74). Friedman thus rejects a thoroughgoing anti-apriorist, naturalized epistemology like Quine's and would also reject Helmholtz's naturalized Kantianism insofar as it anticipates Quine (Ibid., 6–7; 28–29).[24]

However, in contrast to Frege, Friedman believes that mathematics and logic alone cannot supply the resources necessary for showing how pure logico-mathematical or geometric structures are applicable to the phenomena treated in natural science. In other words, Friedman does not share Frege's view that geometry, arithmetic and logic are full-fledged sciences in their own right.[25] For Frege, mathematicians and logicians themselves are responsible for guaranteeing the real possibility of mathematical and logical propositions. They must construct or introduce mathematical or logical abstracta in such way that their applicability to the subject matter of the natural sciences is readily apparent. For Friedman, as for many contemporary historians and philosophers of science, a lesson learned from the paradoxes derailing Frege's logicist agenda is that mathematicians and logicians should focus on formal (non-contentual) systems. Looking back on the debate between Frege and Hilbert over fixing a definitive sense and reference for geometric terms and propositions, many now declare Hilbert the winner. Accounting for the real possibility of abstract mathematical structures may be of concern to philosophers, but need not concern nor constrain mathematical practice per se.[26] So while Friedman agrees with Frege and Cohen that a priori mathematical laws and struc-

---

[24] This is not to say that Friedman rejects Helmholtz's epistemological views wholesale. As we saw in Sect. 3.1, Helmholtz recognizes that an empirical confirmation of Euclidean vs. non-Euclidean axioms must presuppose the free mobility of measuring rods. Friedman reads this as an acknowledgment on Helmholtz's part that scientific reasoning must include a priori Kantian principles, albeit not the ones identified by Kant himself. He thus sees Helmholtz offering two epistemologies: one that subsumes philosophy entirely within cognitive psychology and hence prefigures Quine's naturalism and another that is closer to Friedman's own (Friedman (2001), 7; 29–30). Cohen and Frege, on the other hand, read Helmholtz as a thoroughgoing anti-apriorist naturalizing epistemologist and reject his views accordingly.

[25] See Sects. 4.1 and 4.3.1 for Frege's arguments that geometry and logic are sciences in the full sense of the word. Compare this with Friedman's account of the "modern axiomatic conception of mathematics" inherited from Hilbert (Friedman (2001), 78).

[26] I would like to note, based on anecdotal evidence, that twenty-first century physicists and mathematicians tend to deny that mathematics *is* a science in the proper sense of the word. So perhaps Frege was correct that moving to a more formalistic conception would threaten the scientific status of pure mathematical research.

tures are necessary conditions for ensuring the objectivity of a posteriori scientific propositions, they are not sufficient.

It is on this point too that Friedman contrasts his solution to the Kuhnian problem with the solution advanced by Cohen's student Cassirer. After the 1919 confirmation of Special Relativity and demonstrated fruitfulness of non-Euclidean Riemannian geometry, Cassirer acknowledged that one could no longer defend Kant's claim that the Euclidean axioms were a priori in the sense of a necessary and unrevisable part of scientific theorizing. Still, he continued to insist on the more general Marburgian claim that a priori mathematical laws and structures provide the objectifying conditions for science. He also maintained that scientific progress in a post-Newtonian age may be adequately accounted for simply by demonstrating the "continuity of purely mathematical structures" (Friedman (2008), 246). Friedman follows Cassirer in reconceiving the Kantian a priori as referring to principles and concepts that function as constitutive, objectifying conditions and in recognizing that these principles and concepts will change as science progresses. This is the upshot of Friedman's claim that one must now adopt a relativized version of the Kantian a priori. So, for example, Euclidean principles and concepts are a priori relative to pre-twentieth century Newtonian science, but not within an Einsteinian framework.

Friedman parts company with Cassirer on whether establishing the continuity of a priori mathematical structures suffices for securing the progress of science:

> Like the Marburg school [and unlike Kuhn] … I want to confine the discussion to the conceptual realm and avoid ontology; unlike the Marburg school, however, I agree with Kuhn that purely mathematical continuity and convergence is not sufficient. I set up the problem, accordingly, by appealing to the relationship between (purely abstract) mathematical concepts and sensible experience (Ibid., 249 note 25).

For Friedman, solving the Kuhnian problem entails solving the problem of trans-paradigm communicative rationality. A tradition or practice manifests communicative rationality only if it strives to ensure that subjective (individually-held) perceptions or experiences are somehow held accountable to the unity of the objective world. In a post-Fregean, Hilbertian world, demonstrating that scientific practice exhibits this kind of accountability requires more than what Cassirer offers. Cassirer's proposal would only establish the continuity of *logically* possible frameworks; what is needed is establishing trans-paradigm *real* possibility. To establish the latter, one must account for the continuity of a priori coordinating principles, since these are the principles constituting the unity of the objective world insofar as they bridge the gaps between logically possible mathematical concepts, the mathematized phenomena investigated within science and ordinary sensible experience.[27]

---

[27] At least this was the case in 2001. Most recently, Friedman and Nordmann (2012) has argued that the coordinating principles that he had in mind are not sufficient in and of themselves to relate the abstract mathematical structures presumed in the sciences to ordinary sensible perceptions of the world (48). Friedman's dissatisfaction with his previous proposal is that he wants to retain what he takes to be an important Kantian principle that the faculty of sensibility has an a priori structure which is not derived from the faculty of the understanding or from any productive activity of

Regardless of whether van Fraassen sides with Frege that mathematics can and should account for its real possibility in and of itself, or with Cassirrer that establishing mathematical continuity is sufficient, or with Friedman that establishing the continuity of a priori coordinating principles is required, he needs to side with one of them and Friedman seems the reasonable choice. Remember that on van Fraassen's account science progresses in the sense of accumulating knowledge about concrete observables and abstract mathematical structures (van Fraassen (2006), 297). When critiquing Friedman's position, he writes as if our scientific knowledge about concrete observables doesn't depend on embedding them within these structures. But this is not an accurate description of western modern scientific practice and van Fraassen recognizes this: "By the end of the 19th century the mathematization of the physical world picture was nearly complete. Is Hertz' mechanics a theory of physical systems or of mathematical structures? Hard to say" (Ibid., 287). Given the mathematization of nature in science, it is practically if not theoretically impossible to disentangle concrete phenomena from the mathematical properties and relations ascribed to them in virtue of their being embedded within an abstract structure. If distinguishing between the observably concrete and mathematically abstract is problematic in regards to the ontology of scientific phenomena, it is even more so in regards to their epistemic access. How can we account for the continuous accumulation of scientific knowledge of phenomena without saying something about the continuity of the mathematical structures or coordinating principles making this knowledge possible? So far as I know, van Fraassen does not tackle this question. Moreover, when describing the empiricist (nonrealist) structuralism that he endorses, van Fraassen insists that "the crucial relation to focus on is that between the 'Forms' (mathematical representations) and the phenomena..." (Ibid., 293 note 18). Yet, this is the relation that Friedman's a priori coordinating principles are intended to address. I conclude that Friedman's stronger requirement for an adequate explanation of royal succession *does* have a clear rationale and one that van Fraassen himself should accept. Establishing the trans-paradigm continuity of a priori coordinating principles is a necessary prerequisite in accounting for the accumulation of knowledge about the phenomena studied in science and its empirical successes because the phenomena and successes in question are so thoroughly mathematized.

Van Fraassen's second objection to Friedman's interest in trans-paradigm a priori principles is that it derives from a faulty "metaphysical instinct...that rationality must consist in rule-following" (Ibid., 303). Here again I believe van Fraassen too

---

thought. Given that coordinating principles can only relate the mathematics to the subject matter of natural science, the independence of sensibility is not preserved or not accounted for. As Friedman notes, the Marburg school, beginning with Cohen, interpreted Kant so as to focus on explaining the facts of science rather than the facts of ordinary sense perception. In sum, they reject the idea of an independent, richly structured faculty of sensibility that Friedman is at pains to retain. For my part, I don't see the force of Friedman's hesitancy to follow Cassirer or Cohen in rejecting an independent faculty of sensibility, especially since they do see the need to relate our experience of the world through science with our experience of it through other cultural endeavors, which Friedman himself acknowledges (Ibid., 50 note 10).

quickly dismisses Friedman's proposed solution and fails to see how close it comes to his own. As we have seen, van Fraassen and Friedman uphold scientific inquiry as an exemplary form of communal rational inquiry. On van Fraassen's view, it is exemplary because it eschews dogmatism and cultivates intellectual humility:

> All our factual beliefs are to be given over as hostages to fortune, to the fortunes of future empirical evidence, and given up when they fail, without succumbing to despair, cynicism, or debilitating relativism (van Fraassen (2002), 63).

Science teaches us to hold our factual beliefs loosely and to test them against experience and the views of those whose experience or interpretations of experience differ from our own (Ibid., 48–9). What enables practitioners to give up their beliefs without succumbing to despair or debilitating relativism? Van Fraassen does not explicitly say. But given his description of the emotional angst attending epistemic crises, it is reasonable to infer that he would echo Kuhn's and Friedman's assessment that, under normal conditions, the prevailing theoretical framework holds out the hope of renewed fortunes for beliefs enjoying more empirical-support and for consensus among one's scientific colleagues. The question of course is what bears the weight of this hope when the framework is in crisis, and we saw that van Fraassen looks to Sartre's theory of emotions for the answer.

The problem with van Fraassen's appeal to the Gestalt-like effects of the emotions is that it cannot fully explain why a group of people or the majority of those within a scientific community take the leap of seeing the world as depicted in a potential successor. One would need to posit a collective transformational and epistemically significant emotional experience, and van Fraassen adduces little to no evidence that such experiences have occurred within the history of science.[28] Furthermore, though he references a "continuing inner and outer dialogue throughout the conversion experience," he does not elaborate on its form or content (van Fraassen (2002), 94). In short, van Fraassen has not said enough about how a *communal* conversion might be viewed as prospectively and *dialogically* rational. This is where Friedman's treatment of scientific inquiry as a paradigm of communicative rationality is especially helpful.

Scientific inquiry exhibits or aims at "communicative rationality" only if the consensus towards which practitioners strive is two-fold. First, there must be consensus about the world-directed objectifying conditions. This is secured by a priori coordinating principles connecting abstract mathematical structures to possible empirical observations. Second, there be a consensus about the rules employed to resolve disputes:

> Communicative rationality…is essentially public or intersubjective. It aims, by its very nature, at an agreement or consensus based on mutually acceptable principles of argument or reasoning shared by all parties in a dispute (Friedman (2001), 55).

---

[28] I believe that van Fraassen's appeal to the power of emotions is more plausible in explaining epistemic crises and their resolution in the history of religious traditions. I will elaborate on this point in Sect. 5.2.

During periods of normal science, disagreements between practitioners do not lead to despair because there is a consensus on the principles or parameters for reasonably resolving disagreements. The prevailing framework or "firm research consensus" thus provides not only intradisciplinary standards of objectivity, but also standards and conditions for intradisciplinary communication (Ibid.). When the framework starts to erode, the communicative rationality of a scientific practice in both aspects is threatened. Historical support for this point is found in Frege's two-pronged attack on colleagues deviating from the method and meanings characteristic of Euclidean geometry. He complained that stripping geometric terms of their traditional Euclidean sense and reference deprived geometric propositions of their status as objectifying conditions for the empirical sciences (*PW*, 247). He also complained that this threatened the trans-historical, trans-cultural communication hitherto enjoyed and presumed by mathematicians (Ibid., 268).

Van Fraassen's and Friedman's joint admiration for scientific inquiry commits them to a fairly robust conception of scientific progress. In addition to van Fraassen's conception of progress as the accumulation of knowledge about phenomena, we now need to include Friedman's conception of progress as better approximating a Kantian-Habermasian regulative ideal. This ideal is patterned on Kant's description of a kingdom of end: an idealized community of rational agents and inquirers that resolves disputes by relying on the "non-coercively uniting consensus creating power of argumentative speech" (Habermas as cited in Friedman (2001), 54). It follows, or so I am arguing, that scientific inquiry is entitled to the admiration van Fraassen expects it to evoke only if the dialogical proceedings resulting in a collective decision to adopt a new framework are shown to approximate this communicative ideal. Citing individualistic emotional experiences, however, transformational and epistemically salient, does not suffice.

According to Friedman, in order for trans-paradigm dialogical exchanges to approximate the Kantian-Habermasian ideal, they need to be guided by a priori principles of reasoning, and I see no reason why we or van Fraassen should disagree. We saw that van Fraassen rejects Friedman's insistence on the use of a priori principles of reasoning on grounds that it derives from a mistaken assumption that rationality consists in rule-following. Yet, van Fraassen's own solution to the problem of royal succession involved a strategic or tactical deployment of rules. The goal in deploying these rules is convincing one's peers that a potential successor is a reasonable and responsible way of continuing the epistemic goals and interests of the practice or tradition in crisis. The rules *Sola Scriptura* or *Sola Experientia* are thus deployed in trans-paradigm contexts to show "the way to reasoned and proportionate change, providing a rational form for consensual revision" (van Fraassen (2002), 142). Since van Fraassen himself resorts to rule-guided discourse as a means of forging a non-coercive rational consensus, it is best to interpret his complaint against Friedman as a complaint against the idea that rationality consists in arriving at a unique, rationally compelling consensus or decision.[29] But as we will see, Friedman never suggests that use of trans-paradigm a priori principles is intended to or capable of bringing about such a result.

Just as Friedman modifies the Kantian a priori in the case of principles securing the cognitive content and a decidable truth-value for empirical propositions, so too in the case of principles of reason. To grant rules of reasoning a priori status does not imply, as it did for Kant, that they are "absolutely universal principles of reasoning common to all human beings as such" (Friedman (2001), 55).[30] Rather, it is to claim, along with Wilfred Sellars (1997 [1956]), that human beings operate within a space of reasons and these are the principles determining the structure of that space. Here again, Friedman reminds us that a priori principles are relative to a particular discursive context. Sets of governing a priori constitutive principles and principles of reason can also overlap. For instance, the constitutive principles of a prevailing mathematical-physical theory determine "an empirical space of reasons: a network of inferential evidential relationships…that defines what can count as an empirical reason or justification for any given real possibility" (Ibid., 85).[31] In times of crisis, the status of these principles and the space of reasons they determine is precisely what is at issue. When this occurs, the best the proponent of a new framework can do is present it as a "reasonable and responsible live option" for continuing the scientific tradition in question (Ibid., 103).[32] None of the considerations that a proponent might bring to bear will amount to "a rational *compulsion* to embrace the new paradigm" (Ibid.). Regardless of the reasons that a proponent may adduce in favor of a successor, he will still need to rely on "a little good will" from his

---

[29] For additional textual support that this is crux of van Fraassen's complaint with the idea that rationality necessarily involves rule-following, see his analysis of Pascal's Wager (van Fraassen (2002), 94–101).

[30] As Friedman explains, Kant's belief that there were such universal principles of reasoning about phenomena was based on the idea that are "shared cognitive faculties necessarily common to all human beings at all times and places" (Friedman and Nordmann (2012), 50). Given that this idea was intimately related to Kant's claim regarding the universal necessity and unrevisable truth status of the Eucldean axioms, we must introduce more historical contingency and spatio-temporal specificity into our account of the a priori norms of reasoning governing a discursive context (Ibid.).

[31] Similarly, van Fraassen or Feyerabend might say that the prevailing biblical hermeneutics for a Protestant community determines a religious space of reasons, defining what can count as a Sola-scriptura supported reason or justification for any given disputed matter within the community.

[32] Friedman's focus throughout *Dynamics of Reason* is on paradigm shifts as they occur within scientific traditions. However, I agree with van Fraassen and Cohen that there are non-scientific traditions or long-standing practices that can achieve the kind of consensus that Friedman associates with Kuhn's normal science and that similarly undergo epistemic crises and paradigm shifts. Therefore, I believe that what Friedman says here is applicable to any historically identifiable tradition that assumes the goal of either transmitting and accumulating knowledge or striving to attain knowledge. I can find no textual evidence that Friedman himself subscribes to the accumulation model of scientific progress. But given how closely his views parallel those of the Marburg school and given his depiction of scientific inquiry as an exemplar of communicative rationality, it is reasonable to conclude that he takes it to be an inquiry that at least strives toward knowledge of one sort or another (see Friedman (2001), 68).

colleagues (Ibid.). Therefore, while Friedman argues that solving the Kuhnian problem means showing that epistemic crises within science have and can be resolved in manner exhibiting communicative rationality, this should not be taken to mean that they are rationally compelled resolutions.

What then is involved in arguing for a successor as reasonable and responsible live option and how do these arguments make use of relativized a priori principles of reasoning? According to Friedman, proponents must map "some kind of (communicatively) rational route" from the old framework to the new (Friedman (2001), 101). It is not enough to show that old constitutive concepts and principles can be recovered as a special or limiting case under the successor. One must also show that the new concepts and principles can be prospectively viewed as "a natural continuation of" the old ones (Ibid.). The rational route towards seeing the new framework as a natural continuation of the old need not be paved with material supplied strictly from the latter. For reasons that we now understand, it cannot be: "practitioners of the new framework indeed speak a language incommensurable or non-translatable with the old" (Ibid.). That said, actual scientific practitioners are not restricted to the constitutive principles of reasoning and spaces of reasons delimited by their respective scientific fields; they operate within a broader cultural and intellectual milieu. The dispute over whether to adopt a new paradigm can and often does take place within this broader discursive context. By locating the dispute within or alongside of discursive contexts or spaces of reasons structured in accordance with their own principles of reasoning, proponents of a new framework can invoke these principles as shared rational norms with an eye toward forging a consensual resolution to the dispute. The form and content of a space of reasons unconstrained by the constitutive principles and norms of reasoning comprising an anomaly-ridden framework can also supply the material necessary for depicting a new one as its reasonable and natural continuation.

In *Dynamics of Reason,* Friedman argued that philosophy was especially well-positioned to function as this extra-scientific discursive context. The idea is that philosophical inquiry contains a relatively weak but stable consensus "on what [discursive] moves and arguments must be taken seriously" (Ibid., 107). Moreover, the philosophical space of reasons typically neighbors and encroaches on scientific space:

> [P]hilosophical reflection interacts with properly scientific reflection in such a way that controversial and conceptually problematic philosophical themes become productively intertwined with relatively uncontroversial and unproblematic scientific accomplishments (Ibid.).

Because of this close and productive contact, philosophy can play a crucial role when the sciences encounter controversies and conceptual problems of their own. For historical support, Friedman cites Einstein's engagement with longstanding philosophical debates over absolute versus relative motion and the nineteenth century debate over the status of Euclidean geometry. By relating the core concepts of relativity theory to these debates, Einstein was able to present it as a rational and natural continuation of classical Newtonian mechanics (Ibid., 107–108).

Additionally, we can cite Frege's engagement with the Helmholtz-Cohen debate over Kantian exegesis, an engagement enabling Frege to map a rational route from Kant's C&I distinction to his own C&O distinction.

We are also in a position to more fully understand how Frege's distinction can be viewed as a legitimate constitutive principle of scientific reasoning, even though Benno Kerry and subsequent logicians adduced good reasons for not accepting it. Given the notion of rationality that Friedman and van Fraassen are operating with, the fact that a proposed seismic change to scientific discourse is rejected and progress still ensues does not count against the legitimacy of the proposal. Consider Friedman's account of the contingent tree-like structure of scientific progress:

> [T]he present conception of scientific rationality need not imply the elimination of all genuine contingency from scientific progress, in the sense that there is a single pre-ordained route through the set of all possible constitutive principles, as it were, which the evolution of science necessarily follows at each stage. On the contrary, we can, if we like, imagine a branching tree structure at every point, so that alternative future evolutions of our fundamental constitutive principles are always possible (Friedman (2001), 68).

Based on this account, one is free to say that Frege's colleagues simply declined to grant him the requisite pinch of salt or good will necessary for making the leap to seeing the C&O distinction as a fundamental logico-epistemological principle. The distinction can still be viewed as a natural descendant of an actual constitutive principle of scientific reasoning and discourse, namely, Kant's C&I distinction, a view I argued for in Chap. 4. The contingent fact that the history of logic, science and philosophy of science did not trace the route running through a branch containing the C&O distinction does not count against its status as a constitutive principle with real possibility. For critics to deny it this status, they would need to show that had the scientific community taken this route, it would not have progressed towards approximating the Kantian-Habermasian ideal or in accumulating knowledge about observable scientific phenomena.[33]

Though Friedman initially thought the problem of prospective communicative rationality could be satisfactorily addressed simply by locating trans-paradigm scientific discussions within concurrent and overlapping philosophical discussions, he has since decided otherwise. Recall that redeeming a post-Kuhnian concept of scientific progress depends on presenting the history of modern western science as a history of practitioners holding themselves collectively accountable to progressing norms of objectivity and consensus-oriented dialogical exchange. To show that the history of science is one of progressive scientific rationality, one only need show that the scientific community is approximating the limit of "ideal standards of universal, trans-historical communicative rationality" (Friedman and Nordmann (2006), 68). Friedman now believes that a sufficiently complete account of this history requires looking not only at how scientists engaged the problems, goals, and norms animating philosophical discussions, but also how they engaged with the

---

[33] To my knowledge, no one has marshalled an argument against the C&O distinction that involves establishing this counter-factual, and it is not within the scope of the purpose of this book to present or critique such an argument.

problems, goals and norms animating concurrent religious, political, and technological discussions (Friedman and Nordmann (2012), 51).

Friedman's assessment that a complete response to the Kuhnian problem should
include describing how the epistemic concerns and norms of science productively
interact with the epistemic concerns and norms of religious traditions is especially
relevant to completing what I set out to do in this chapter. As stated in the introduction, Cohen maintains that flourishing religious traditions are conscientious cognitive enterprises and that this implies they must be open to incorporating developments
in other such endeavors, namely, mathematical natural science and philosophy. He
further maintains that scientific inquiry and education cannot be properly pursued
independently of moral teachings rooted in a monotheistic tradition:

> In face of this idolization of universal education (and, unfortunately, also of specialized,
> scientific education) we assert: there can be no universal education, nor can there be any
> European culture or any ethics without the idea of the One God as the God of morality
> (Cohen (1993 [1907]), 46).

For Cohen, it is the idea of the One God, the God of all humanity, which grounds a
rational hope and confidence that our cultural practices can approximate the ideal
human community described in the Hebrew prophets' messianic vision and Kant's
Enlightenment vision of a kingdom of ends (Cohen (1993 [1910]), 86–8). Friedman
also characterizes scientific progress as asymptotically approaching Kant's
Enlightenment ideal and now maintains that the story of scientific progress cannot
be told apart from the history of religious concerns. We also saw van Fraassen claim
that epistemic crises occur in science and religion, and I argued that rapid, revolutionary developments in the one can cause an epistemic crisis in the other. In Sect.
5.3, I argue that the debate over the definition of marriage occurring in many North
American Protestant churches constitutes an epistemic crisis and that it is due in in
part to advances in developmental biology and healthcare. All this suggests there
will be considerable overlap in the strategic rules for wisely navigating epistemic
crises in science and religion, and the primary goal of this chapter is identifying a
few of these rules.

In Sect. 5.2 we turn our attention to Cohen's philosophy of religion and his own
efforts to underwrite Judaism's cultural relevance by showing that it can incorporate
eighteenth and nineteenth century advances in science and philosophical ethics. I do
not intend to provide a comprehensive examination of Cohen's philosophy of
Judaism. Rather, I offer a survey of his work in this area aimed at uncovering additional strategies for resolving epistemic crises in religious traditions, specifically
those with a high view of biblical and traditional authority. Cohen serves as an
example, if not an exemplar, of a would-be faithful practitioner of an Abrahamic
tradition attempting to constructively respond to a crisis situation within which he
and his religious community find themselves. First, however, let me summarize the
rules for reasonably responding to epistemic crises that have emerged thus far.

5.1.4 Tactical Rules for Reasonably Resolving Epistemic Crises

1. Concepts and principles expressing the norms or conditions for objectivity and communicative rationality within a tradition are answerable to sensible experience and revisable in light of those experiences. At the turn of the twentieth century, Frege posed the rhetorical question "Do we dare to treat Euclid's elements, which have exercised unquestioned say for 2000 years, as we have treated astrology?" We can now appreciate the illegitimacy of this question. First, despite the long and productive reign of traditional Euclidean geometry, the undeniable scientific fruitfulness of non-Euclidean geometry was just around the corner. Here we must follow Cassirer, van Fraassen, and Friedman in stressing the fallibility and hence potential revisability of all aspects of human theorizing and epistemic endeavors. Second, deciding that non-Euclidean geometric concepts and principles should assume the role previously held by Euclidean ones does not imply, as Frege suggests, that the latter are relegated to the status of mere "historical curiosities" (*PW*, 169). For as van Fraassen says, successfully resolving an epistemic crisis depends on recognizing a successor as a rationally endorsable continuation of a tradition's past. As applied to the shift from classical to relativistic mechanics, Friedman persuasively argues that Einstein met this burden by showing that Riemannian geometry is a natural continuation or development of the Euclidean geometry holding sway for two millennia. Euclidean geometry thus retains its status as an ancestral bearer of scientific knowledge. The same would hold for other relative a priori predecessors within science and other knowledge-transmitting practices. Pace Frege, substantially revising the sense, reference, or scope of validity once ascribed to basic concepts and principles does not utterly negate their epistemic status or worth.

2. Though norms of objectivity and rational communicability are answerable to experience, they cannot simply be read off of experience. Both Cohen and Frege rightly pushed back against Helmholtz's naturalized Kantianism because it could not recapture Kant's emphasis on the normativity and inter-subjective validity accruing to proper scientific propositions. As explained in Chap. 3, Cohen's 1871 *Kant's Theory of Experience* aimed at displacing psychologized readings of Kant. Contra Helmholtz, Cohen insisted that Kant's 'a priori' referred to a source of normativity and not an innate feature of human cognitive functioning. Based on the content of Helmholtz-Cohen exchange, it is easy to see Cohen reiterating Kant's response to Humean skepticism about necessary causation and similar concepts: a concept may function as a necessary norm and constitutive element of scientific theorizing *about* experience even if it is not derivable from experience. In fact, Cohen's book is now seen as a turning point in Kant scholarship and the vanguard of orthodox neo-Kantianism precisely because it insisted on the "distinctively Kantian claim that any norm must have an a priori basis" (Anderson (2005), 291).

Frege and Friedman are similarly committed to this distinctly Kantian claim. Like Cohen, Frege rejects Helmholtz's attempt to locate the ground or norms of objectivity among the stuff of sense-impressions: "Now objectivity cannot, of course, be based on any sense-impression, ...but only, so far as I can see on the reason" (*FA*, 38). For Frege, as for Cohen, any attempt to derive the objectivity and normative necessity characteristic of scientific reasoning and propositions from sense-impressions or the observable actualities of empirical science is simply wrong-headed.[34] For Friedman, appending 'a priori' to concepts and principles signals that they function as constitutive elements of a prevailing scientific paradigm. These elements furnish the rules governing the activities of normal science and the conditions enabling empirical concepts and principles to "squarely and precisely face the 'tribunal of experience'" (Friedman (2001), 45). This description of how science works is intended to refute Quine's naturalized epistemological holism. Quine objected to drawing any distinction between the a priori and empirical parts of a theory, maintaining that all parts directly confront the tribunal of experience. Friedman counters that Quine's holism cannot make sense of the conceptual revolutions punctuating the history of science nor the experiences of those living through them. The phenomenology of scientific epistemic crises is not accurately captured by Quine's idea of a well-entrenched empirically-supported part of the framework undergoing modification, but rather by Rudolf Carnap's (and Friedman's) idea that practitioners are confronting a choice about which norms of objectivity and communicative rationality should govern their collective inquiry. This choice, a choice about which communal epistemic norms should be adopted, cannot be rationally decided merely be appealing to the tribunal of experience (Ibid. 41–2).[35]

Even van Fraassen, who bristles at Friedman's invoking of a priori principles, is committed to the essential use of normative judgments in directing our knowledge-gathering efforts and to the claim that these judgments cannot be derived from sensible experience in and of itself. First, he agrees with Friedman that Quine's naturalism and other epistemologies restricting themselves to descriptive factual assertions and eschewing reference to agent-centered value-judgements cannot make sense of the epistemic trauma experienced when a paradigm begins to erode. Second, recall his discussion on the need to distinguish 'experience' as a success word from 'experience' as referencing nothing other than the immediate subjective awareness of a sensibly-given happening. We can conclude that van Fraassen would share Cohen's neo-Kantian perspective that experience, insofar as it is referenced in scientific theories and respected professional journals, contains a normative component that cannot be attributed to mere sensory input or to the psychophysiological processing of that input.

---

[34] See Sects. 4.1 and 4.2 for the textual support and argument that Frege's epistemology of science bears significant similarities to Cohen's Erkenntniskritik.

[35] We can add this to the list of arguments already canvassed for Friedman's insistence on retaining a relativized notion of the Kantian a priori.

3. <u>Fallible human knowers and the knowledge-transmitting practices of which they are a part bear responsibility for the normative 'a priori' concepts and principles governing those practices.</u> The point behind this maxim is that practitioners of a science and other epistemic enterprises cannot rid themselves of this responsibility by trying to ground communal norms in some sort of 'given' presumably untainted by human hands. We have already discussed the problems with attempts to base epistemic and hermeneutical norms on the purported self-evident givenness of experience or the Bible, the *Sola Experientia* and *Sola Scriptura* of empiricist and Protestant fundamentalism. We must also reject similar appeals to the purported self-evidence of basic laws or axioms currently functioning as prescriptive norms for a practice.

To be sure, Cohen reads Kant to say that "all cognition must be based on principles which are to be regarded as basic truths" and affirms this as the sine qua non for objective and inter-subjectively valid human reasoning (Cohen (1993 [1910]), 78).[36] We also saw that Frege and the Marburg school insist that the basic truths of logic and mathematics express the preconditions of objectivity and communicability within their own fields and all of natural science.[37] The question then is what justifies identifying certain truths as basic and as functioning in this prescriptive capacity. As discussed in Chap. 2, Kant addressed this question by deriving these truths from a schematization involving universally shared features of human cognition, the forms of sensibility and categories of the understanding. By Cohen's and Frege's lights, this particular Kantian answer needed to be rejected because it encouraged attempts to provide a psychophysiological justification of basic truths like the one Helmholtz proposed. Furthermore, as Friedman points outs, Kant's answer is no longer viable because of its entanglement with the supposed necessity of the Euclidean axioms

So how do Frege and the Marburg school address the justification question? According to Tyler Burge, Frege's answer is standard for "a representative of the Euclidean, rationalist tradition in the epistemology of logic and mathematics": basic truths and axioms are self-evident (Burge (2005), 317). He admits, however, that Frege "never appeals to [self-evidence] in justifying his own logical theory or logical axioms" and that his arguments for the necessity of the Euclidean axioms tend to be based on pragmatic and methodological considerations (Ibid., 338; 345). In contrast to Burge, I have argued for aligning Frege with Kant's idealist tradition, specifically the branch stemming from Cohen. I

---

[36] For Cohen, this Kantian precept holds not only for reasoning within mathematics and natural science but also for those aiming at a rational, objective investigation of the Bible: "To accept sense perception as the ultimate ground of knowledge is to relinquish any claim to an objective basis of knowledge as represented, for instance, by the axioms of mathematics. For believers in miracles and spiritualists invoke sense experience, and literalists accept the sense datum of the written word as their criteria of truth. But those who would make reason their criterion of truth must apply this principle to the study of Holy Scriptures as well. And our devout philosophers of reason have actually done so" (Cohen (1993 [1910]), 78).

[37] See Sect. 4.1 for my argument that Frege and Cohen agree on this point.

take Frege's strongest arguments for the necessity of the Euclidean axioms and for recognizing the C&O distinction as a basic logico-epistemological truth to be entirely pragmatic and methodological. The justification for identifying the Euclidean axioms, traditionally understood, as basic truths governing all that is spatially intuitable is that these are the prescriptive norms that have been recognized and authorized by geometers for the past 2000 years. Functioning in this capacity for the natural sciences, Euclidean concepts and principles have been indispensable for yielding empirical successes. In other words, Frege appeals to the authority of historic geometric practice and evidence of the epistemic goods that the Euclidean axioms, as conditions for the possibility of empirical knowledge, have already delivered. His resistance to non-Euclidean concepts and principles is also pragmatic and methodological. Up until 1919, they had not marshalled the empirical successes that could constitute a serious threat to the Euclidean reign. Besides, none of Frege's colleagues had provided an account whereby non-Euclidean concepts and principles might be seen as reasonable, natural descendants of their Euclidean predecessors, at least not an account to his liking. When it comes to arguing for his C&O distinction, Frege shows that it can be seen as reasonable extension of Kant's C&I distinction and, if adopted, would block the methodological errors that he attributes to the formalists. In sum, to the extent that Frege offers any justification for the necessary norms governing scientific reasoning and practice, these justifications almost always involve ineliminable references to human practices.[38]

The idea that fallible human knowers are nevertheless responsible for the a priori principles governing their respective communal epistemic endeavors is more explicit in the works of the Marburg school. First of all, any references Kant himself may have made to a fixed, static given as a source of normativity or as justifying the a priori status of certain principles drops out or is reinterpreted. When describing what is distinctive about Cohen's understanding of Kant's critical project, Cassirer writes that the 'given' towards which philosophy must "orient itself" is not the "material determinateness of things" or "the unity of consciousness" of an existing noumenal subject (Cassirer (2005), 98). Nor is it the logico-mathematical structure that determines the space of reasons within which theorizing about "natural-scientific 'realities'" currently occurs. According to Cassirer, Cohen understood, prior to confirmation of general relativity, that the

---

[38] I stress this point for two reasons: first, to support my claim that human practitioners are responsible for choosing the norms guiding their practice by citing Frege's endorsement of this claim; second, to further challenge Burge's treatment of Frege as a metaphysical Platonic realist. Burge argues that Frege's discussion of the 'third realm' commits him to Platonic realism, specifically the view that there exists an abstract structure of interrelated of logical, mathematical and empirical truths and that the existence and nature of this structure can be explained without referencing "human language, human inference, human practices (including the *activity* of judgment) or other patterns of human activity in time" (Burge (2005), 305; italics in the original). However, as Burge himself seems to suggest, the arguments that Frege mounts for recognizing certain propositions as a priori basic truths contained within this structure are not the kind of arguments that one would attribute to a confirmed Platonic realist.

set of a priori principles comprising the system of scientific validity will be revised as science progresses:

[F]or Cohen, the orientation to science does not imply any commitment to its temporal, contingent form. The 'givenness' that the philosopher recognizes in the mathematical science of nature ultimately means the givenness of the *problem* (Ibid., 100 italics in the original).[39]

The given towards which philosophically cogent scientific reasoning orients itself ends up being not the prevailing fact of science, but the unending task of bringing the 'is' of our current conception of nature ever closer to the 'ought' of how it should be conceived: "In its actual form the philosopher seeks and recognizes an ideal form, which he singles out, to confront it with the changing historical configuration as a standard for measurement" (Ibid.).[40] By characterizing the given of scientific reasoning as a task of this sort, the task of continually striving to bring the 'isness' of nature into better conformity with the rational ought legislated by an ideal community of human inquirers, Cohen bridges the notorious gap in Kant's philosophy between the interests of theoretical and practical reason: "[A]s the theoretical *a priori* promotes reference back to 'experience' and its possibility, so the idea of a 'kingdom of ends' is the maxim to which the phenomenal order of nature conforms…" (Ibid., 104).[41]

Secondly, according to the Marburg school, properly pursuing the given tasks of science and ethics will result in becoming increasingly aware of and enacting our autonomy. We saw that, for Natorp, mathematicians and physicists enact this autonomy by acknowledging that the ultimate justification for the basic concepts and principles expressing the norms of objectivity within their respective fields and any subfields resides solely in the "autonomous legislation" of the practice itself; there is no other basis of authority (Natorp (1887), 252).[42] According to Cassirer, Cohen improves on Kant's concept of freedom by ridding it of any

---

[39] Almut Sh. Bruckstein affirms Cassirer's depiction of Cohen's philosophy of science. In her commentary on Cohen's intention of treating ethics as a science and his voicing of the concern that this would subject ethical norms to untoward revision, Bruckstein explains that, for Cohen, "scientific inquiry constitutes a self-corrective process, in which relative truths are ever-changing" (*Ethics of Maimonides (EM)*, 9). The challenge that Cohen sets for himself is explaining how ethical propositions and the moral teachings of Abrahamic traditions can be open to revision in light of progress within natural science without making their truth status entirely dependent on the methods characterizing natural science.

[40] Consider too Friedman's claim that "the real object of empirical cognition" for the Marburg school is "a never completed 'X' towards which the methodological progress of science is converging" (Friedman (2000), 79).

[41] Although Cohen argues for a deep similarity in the form of scientific and ethical inquiry and the interconnectedness of these cognitive tasks insofar as they are pursued conscientiously, he insists on not conflating them. This is reportedly Hegel's mistake and the mistake underlying pantheistic or panentheistic philosophies generally. As previously mentioned, Cohen argues for treating scientific, ethical and religious inquiry as scientific inquiries in the broad sense of cognition based on a priori principles or basic truths. Throughout this section, I have focused on the similarities of method, aims and histories of these distinct epistemic endeavors. In the next section, I explain how Cohen distinguishes them and to what purpose.

[42] To recall how this emphasis on the legislative autonomy exercised by mathematicians and mathematical physicists is intended to block Helmholtz's epistemological agenda, see Sect. 3.2.2.

connotation as the kind of causation originating in a noumenally existent will, treating it instead as a strictly teleological and regulative concept: "As ethical subjects, we act not from freedom but toward freedom" (Cassirer (2005), 104). Autonomy or freedom is still conceived along broadly Kantian lines as the right and responsibility to rationally set and pursue one's ends or purposes: "[T]he idea of 'autonomy' becomes 'autotely'..." (Ibid.). The given for ethical inquiry is the unending task of becoming a free self by lawfully transforming nature and society such that "the other" receives due honor and the self-legislation, self-determination, self-preservation and self-responsibility characteristic of human-kind is realized (Zank (2006), 14; Poma (1997), 117–119). Thus, for Cohen, any progress in tackling the ongoing problems of science and ethics is marked by an increased understanding of our responsibility for setting and evaluating the norms governing those epistemic practices and pursuits.

Finally, for all the subtle differences between the Kantianism that Friedman espouses and that of the Marburg School, he agrees that this increased self-understanding is implied by the concept of scientific progress. A scientific community is progressing only if it is approximating "ideal standards of universal, trans-historical communicative rationality," and to say science is approximating this ideal is to say "our reason grows increasingly self-conscious and thereby takes responsibility for itself" (Friedman (2001), 68).

4. Anyone proposing to modify the meaning of a basic concept or principle functioning as a norm of objectivity or rule of communicative rationality incurs the burden of explaining how this modification might be viewed as a natural, rationally endorsable continuation of its predecessor. As we have seen, Cohen, Frege, van Fraassen, and Friedman mount strong arguments that satisfying this responsibility is the best if not the only way to reasonably resolve an epistemic crisis. All agree that radical shifts in the meaning of a tradition's core concepts threatens its status as a knowledge accumulating and transmitting practice. Moreover, there is sufficient historical evidence that epistemic crises occur within the history of science and religion and that practitioners often do try to resolve them in the manner that van Fraassen and Friedman describe. First century Jewish Christians tried to show that emulating Jesus' life and teaching was a reasonable and faithful continuation of their Jewish heritage by offering an alternative reading of the Hebrew scriptures to establish this was the case. Einstein engaged with concurrent philosophical discussions in an effort to present general relativity as a reasonable continuation of Newtonian physics. And the Helmholtz-Cohen-Frege debate is a prime example of a trans-paradigm dialogical exchange aimed at reaching a consensus about the rules of the game governing mathematical practice by drawing upon the shared principles of sound scientific and philosophical reasoning inherited from Kant. Based on the historical evidence, my exposition of the Helmholtz-Cohen-Frege exchange and my analysis of van Fraassen's and Friedman's proposed solutions to the Kuhnian problem, I conclude that attempting to meet this burden qualifies as a good faith effort to address the problems of prospective rationality and royal succession. Therefore, anyone following this maxim is neither willfully nor negligently introducing

communicative confusion into her tradition and so is not guilty of committing the Fregean sin.

5.2   Abrahamic Traditions and Accumulating Knowledge

The last section introduced several significant similarities between scientific and Abrahamic religious traditions. Van Fraassen and I claimed that history shows both can undergo epistemic crises. If this is true, both must be conceived as traditions charged with the task of transmitting some sort of knowledge. Moreover, for any tradition to undergo an epistemic crisis, it must employ concepts or principles considered fundamental to or constitutive of its particular epistemic pursuit and communal identity. For example, on van Fraassen's and Feyerabend's account, Sola Scriptura functions as a principle for deciding which experiences and testimonies[43] may be incorporated into the body of inherited knowledge claims for fundamentalist Protestantism (van Fraassen (2002), 119–120). Understanding how to apply this rule and a commitment to doing so is a mark of membership within the tradition (Ibid., 126). Van Fraassen also claims that, throughout Jesus' lifetime up through the conversion of St. Paul, the idea that becoming a Christian meant becoming a Jew functioned as unquestioned doctrine and a central part of the self-understanding of the community of Jesus followers. For this reason, he treats the subsequent rejection or modification of this doctrine as an "other-than-science" conceptual revolution (Ibid., 70).[44] Based on these examples and the historical evidence already adduced, we should agree with van Fraassen that a typical feature of epistemic crises is that the tradition or community in question is bound to an "old or 'classical' framework" in danger of being replaced (Ibid., 71). Therefore, if religious traditions or communities experience or are susceptible to a Kuhnian shift, it follows that members see themselves obliged to certain concepts and principles constituting an inherited or classical framework. It also follows that these concepts and principles can encounter anomalies or counter-instances calling for a substantive change.

While Cohen does not discuss the similarities of science and Judaism in terms of their susceptibility to a paradigm shift, he does say that ethical and religious traditions are like the natural sciences in that they too are tasked with accumulating and

---

[43] 'Testimony' is a philosophical term of art with a corresponding body of literature on how it should be defined. Here I use 'testimony' as does Zagzebski (2012), to refer to "all cases in which a person A says that p to another person, B, who then believes p at least partly on the say-so of A" (121). As Zagzebski notes, much of the religious propositions and directives that a member of an Abrahamic tradition believes is based on the trust she places in the forebearers of her tradition (Ibid., 199–203). It also true that this body of authorized testimonial knowledge, which includes canonical readings of scripture, is used to assess the credibility and epistemic weight of additional testimony.

[44] The New Testament scholars whom I have consulted confirm van Fraassen's description of the events surrounding the decision for full inclusion of Gentiles into the community of Jesus followers as fraught with epistemic trauma and the decision itself as tantamount to a paradigm shift.

transmitting knowledge. Citing no less an authority than medieval Jewish philosopher Moses Maimonides, he writes: "cognition ['Erkenntnis'] is the task and telos of religion, and consequently of ethics" (*EM*, 66).[45] He further maintains that mathematical natural science, philosophical ethics, and Judaism should all be conceived as sciences:

> What holds true for every science holds no less true for religion. Insofar as religion, too, consists of concepts and is based on concepts, its ultimate source can only be reason. This connection with reason determines and conditions its connection with philosophy, understood as the universal reason of human knowledge (*RoR*, 5–6).

Here 'science' refers to any conscientious cognitive inquiry proceeding in accordance with concepts. These concepts are subject to philosophical rational scrutiny and its demand that the body of human knowledge, at the idealized limit, exhibit a lawlike unity and universal inter-subjective validity.

According to Michael Zank (2006), the Marburg school generally held that natural science, ethics, art, and religion were distinct but interrelated epistemic endeavors. Natorp referred to these endeavors as foundational 'directions of culture.' The challenge was demonstrating that each cultural direction made a unique and yet unified contribution to the advancement of human knowledge. This challenge was especially pressing and difficult because of the overriding epistemic authority accorded to the content and methods of the mathematical and experimental sciences (Ibid., 3–8). Zank argues that Cohen took up this challenge in earnest in his 1904 *Ethics of Pure Will*. This argument is intended to rebut those interpreting Cohen's *Ethics of Maimonides* and other writings on Judaism as a serious departure from his previous interest in systematic ethics and advancing critical idealism. Because my aim is mining Cohen's *Ethics of Maimonides* for what it can teach us about reasonably managing epistemic crises in religious traditions and communities, there is no need for me to take a definitive stance on this interpretive debate. It will become clear, however, that I tend to side with those stressing the continuity and coherence in Cohen's corpus. Having gleaned what we can from Cohen's debate with

---

[45] In contemporary philosophical literature, it is not uncommon to find 'cognition' and its cognates distinguished from 'knowledge' and its cognates. The former typically refers to the processes yielding percepts, beliefs, and other mental states or behavior related to knowledge acquisition, and where *cognitive* studies aim at accurately describing these processes, states or behaviors. In contrast, the latter introduces a normative component and typically refers to the processes yielding 'justified', 'warranted', or 'valid' processes, states or behaviors, and where the study of *knowledge*, a.k.a. epistemology, aims at identifying these normative processes, states, or behaviors. This terminological distinction was not clearly and systematically adhered to in the works of the nineteenth century philosophers referenced throughout the book, and certainly not in their English translations. Therefore, I have used the terms 'cognitive' and 'epistemic' somewhat interchangeably, allowing the context to signify that normative questions were at issue. In her translation of *Ethics of Maimonides,* Bruckstein sometimes renders Cohen's 'Erkenntnis' and 'Wissen' as cognition. It is clear, however, that Cohen's overriding concern is with knowledge in the normative sense and Bruckstein's excellent translation and commentary makes that abundantly clear. Therefore, I include her translations as is and only amend them for the purposes of emphasis or when further clarification is required.

.Helmholtz on resolving the nineteenth century crisis in geometry, let us turn to his work in the philosophy of religion.

5.2.1   The Crisis Confronting Cohen's Religious Community

So what crisis is Cohen's Jewish community facing and in what sense is it an epistemic crisis? Remember that a scientific tradition or any other longstanding epistemic community[46] is in crisis when the consensus about the authoritative meaning of constitutive concepts or principles starts to erode or practitioners begin to doubt their genuine cognitive worth. For Cohen, the constitutive concept of any religion and especially Judaism is its concept of God: "A religion's right to exist is derived from its concept of God" (*RH,* 45). He presents a threefold explanation for the lack of consensus and doubts concerning this concept within his community. First, as the natural sciences registered multiple empirical successes, broadened their explanatory scope, and increased their epistemic authority, all references to God became scientifically suspect. Cohen describes the impact this had on educated European circles generally and Jewish circles in particular:

> Modern academicians view philosophy and religion with equal coolness, particularly with regard to the problem of God. This goes so far that a philosopher's scientific credentials are already suspect if he does not observe an official silence concerning theological questions. This attitude, by now almost predominant in our circles, also constitutes the greatest threat and danger to our existence, whose whole foundation is the idea of the One God (*RH,* 44).

According to Cohen, his Jewish colleagues were embracing a methodological and epistemological naturalism inconsistent with faithful Jewish thought and practice in an effort to be respected members of the European intelligentsia. Embracing this

---

[46] I use the term 'epistemic community' to refer to any social collective that has as one of its aims the accumulation, production, or transmission of knowledge. Given this broad designation, the term appears within philosophical literature in reference to scientific disciplines, religious traditions, courts, and a host of other panels or institutional bodies. Precisely how to specify the identity conditions for such communities is a matter of dispute. For now, it suffices to stipulate these conditions as is implied in Cohen's, Frege's, van Fraassen's and Friedman's treatment of an impending paradigm shift as a perceived threat to the ongoing existence or identity of the community. In other words, under normal non-crisis conditions, an epistemic community is roughly identified and individuated in virtue of (1) a well-defined subject matter concerning which it aims to accumulate, produce, or transmit knowledge; (2) consensus on the epistemic goals and means of progressing towards them; and (3) shared norms of validity and principles for settling intra-communal disputes. I take it that the practices that Natorp classifies as directions of culture would all qualify as epistemic communities in my sense. More importantly, based on my rough criteria and Cohen's description of Judaism, it qualifies as well. The same holds for various Christian traditions as described by Zagzebski (2012) and Swinburne (2007). The following discussion about religious traditions as epistemic communities is restricted to Jewish and Christian traditions because of Cohen's focus and because Christian traditions, i.e. particular Protestant communities, are those with which I am most familiar. I leave it to others to decide how much of what is said pertains to other religious communities.

naturalism was tantamount to accepting that the classical Jewish idea of God lacked any cognitive significance.[47] The Hebrew Bible and rabbinical tradition speak of God as creating all natural or becoming things and determining their telos (*RoR*, 64–5). More specifically, the tradition speaks of knowing and loving God as the telos of humanity (*EM*, 66; 109).[48] Throughout the eighteenth and much of the nineteenth century, it remained an open question whether biological research could proceed without invoking the concept of final causation and the implication of an intelligent designer of natural organisms.[49] However, as Darwin's account of biological change as the effect of natural selection gained widespread acceptance, references to divine creation and divinely ordained teloi appeared increasingly explanatorily idle and rightly subject to Occam's razor. From 1868 to 1899, Ernst Haeckel, a professor of zoology at the University of Jena, argued that Darwinian evolution was the final vindication of a scientific naturalism falsifying traditional religious concepts and rendering them culturally obsolete. Cohen worries that because many within his circle fail to see how their traditional idea of God relates to progress in natural science, they too consider it cognitively meaningless and rightly lain aside.[50]

Second, just as the Jewish idea of God seemed irrelevant or even a hindrance to garnering scientific knowledge, so too in regards to ethics. While Cohen endorsed the widely held view that autonomy served as a constitutive concept for any philosophically sound ethical theory, he realized this view challenged the epistemic authority of biblically and theologically grounded moral claims: "The consensus of present-day Western culture see in the autonomy of the human mind the main argument against divine provenance of ethical judgment" (*EM*, 27). The crisis is heightened for Cohen's community because Judaism "makes no distinction between religion and ethics" (*RH*, 45). Jewish orthodoxy and orthopraxy takes God to be

---

[47] Cohen fluctuates between usage of the '*concept* of the One God' and the '*idea* of the One God' to refer to God as understood within Judaism. For reasons that will soon be clear, the '*idea* of God' ['der Gottesidee'] is the preferred term for understanding the absolutely unique function that Cohen subscribes to this idea in grounding and directing scientific, ethical, and religious thought and practice.

[48] For more on the self-understanding of Abrahamic traditions as bequeathed with the responsibility of declaring the purpose of humanity divinely revealed to them, see Erlewine (2010), 12–16.

[49] In the *Critique of the Power of Judgment,* for example, Kant argues for the necessary use of teleological judgments to account for the purposiveness of nature. While this use does not warrant the proposition 'God exists,' it does warrant the idea of God as a necessary postulate for the purposes of developing a comprehensive science of nature: "Objectively...we cannot establish the proposition there is an intelligent original being; we can establish it only subjectively for the use of our power of judgment in its reflection on the ends of nature, which cannot be conceived in accordance with any other principle than that of intentional causality of a highest cause" (*CPJ* 5:399). The idea of God thus functions as a necessary methodological presupposition within the life sciences. For more on Kant's philosophy of biology in relation to design arguments, see *Goy* (2014).

[50] In "Religiöse Postulate," Cohen laments that many in his circle of Jewish scholars maintain that the problem concerning the foundation of philosophy and science is resolved and so no longer need to speak of their ancient God: "und es ist sehr traurig, wenn gerade wir in breiten Scharen dieses tiefste Problem für erledigt, und unseren alten Gott für abgetan halten" (Cohen (1907a), 5).

"the Author and Guarantor of the moral universe" (Ibid.). Can a faithful practitioner of Judaism accept the position that autonomy grounds the validity of moral claims or judgments? If so, is there any warrant for treating the body of Jewish theological doctrines and ethico-religious directives as a genuine body of knowledge? For those, like Cohen, who insist on the validity and significance of this traditional body of religious knowledge, the question thus remains: "what place is occupied by the autonomy of human reason, if we admit that its most precious achievement the validity of ethics, is predicated upon God? (*EM*, 29). These are a few of the epistemic and practical conundrums confronting Cohen's community at the turn of the twentieth century.

A third reason for the crisis is a reason Cohen gestures at but lacks the conceptual resources to explicitly spell out, namely, that members of his community were subject to epistemic oppression. Kristi Dotson (2014) defines 'epistemic oppression' as "the persistent epistemic exclusion that hinders one's contribution to knowledge production" (Ibid., 115). Epistemic exclusion is an unwarranted infringement on someone's participation in an epistemic community, where full participation implies the ability to persuasively utilize and contribute to shared epistemic resources and to revise them when necessary.[51] One type of epistemic oppression that Dotson describes is testimonial smothering. In these cases, speakers self-censor or smother their testimony because they recognize their audience is not in a position to understand the intended meaning of what they have to say. To qualify as an instance of epistemic oppression, this lack of understanding or "testimonial incompetence" on the audience's part must be due to "pernicious ignorance" (Dotson (2011), 250). Pernicious ignorance is an ignorance produced or sustained by the sociohistorical marginalization of certain people groups.[52] In sum, testimonial smothering occurs when speakers perceive that efforts at frank communication will most likely fail or misfire due to pernicious ignorance and that any miscommunication will result in further exclusion or marginalization. To avoid these predictable results, speakers silence or censor themselves (Ibid., 249). Testimonial smothering and other forms of epistemic oppression can lead and have led to the suppression of entire intellectual traditions (Ibid., 243).[53]

---

[51] Epistemic resources are resources that a community draws upon in order to engage in effective communication, to make collective and communicative sense of one's experiences, and to supply shared standards of fairness and accuracy (Dotson (2014), 116). Miranda Fricker (2007) refers to such resources as 'the collective hermeneutical resource.' Given these characterizations and Friedman's characterization of a conceptual framework enabling those within a scientific or political community to achieve or approximate communicative rationality, I take 'epistemic resource,' 'collective hermeneutical resource,' and 'conceptual framework' to be sufficiently synonymous for my purposes and treat them accordingly.

[52] For example, Charles Mills defines 'white ignorance' as "a non-knowing, that is not contingent, but in which race—white racism and/or white racial dominance and their ramifications—plays a crucial causal role" (Mills (2007), 20). White ignorance is thus a species of pernicious ignorance.

[53] See Dotson (2011, 2012, 2014) and Fricker (2007, 2012, 2013) for well-articulated descriptions of distinct types of epistemic oppression and injustice. My intent is to show that one aspect of the crisis that Cohen's Jewish community is undergoing fits within this general category. Arguing for

We can now interpret Cohen's description of the self-censoring occurring within his religious community as a description of epistemic oppression. Recall his account of the official silence observed among Jewish intellectuals

> [A] philosopher's scientific credentials are already suspect if he does not observe an official silence concerning theological questions. This attitude, by now almost predominant in our circles, also constitutes the greatest threat and danger to our existence, whose whole foundation is the idea of the One God (*RH,* 44).

The silence described here is due to a prevailing scientific naturalism that would seem to pose an equal threat to German religious communities tout court. The fact that the danger to Cohen's community is reported as much more severe than that experienced by contemporaneous Christian communities begs an explanation. Part of the explanation is that Kant, Hegel, and others had constructed a metanarrative of scientific and cultural progress that highlighted the contributions of ancient Greek and Roman culture and Protestant Christianity. In contrast, Judaism is repeatedly represented as a thoroughly dogmatic, ritualistic, and heteronomous enterprise, contributing virtually nothing to an enlightened European society. Throughout the nineteenth century, German scholars in a wide variety of fields operated within a broadly Kantian framework and, as Robert Erlewine notes, this framework denigrated the God of Judaism:

> According to Kant, the Jewish God "desires merely obedience to commands, "attaches "prime importance to mechanical worship," and has no genuine concerns with the moral disposition or conscience. The God of Judaism, at least in Kant's rendering, seems virtually indistinguishable to the account of God in superstitious and statutory forms of faith (Erlewine (2010), 110).

Cohen himself reports that because of the anti-Semitic bias expressing itself in the prevailing scientific, ethical, and political epistemic resources, German Jews avoid mentioning the particularities of their faith tradition, especially its idea of God:

> [F]or decades, we have neglected this fundamental idea [the idea of the One God] in our religious discussions. We prefer to talk exclusively of our moral teachings because they seem to provide legitimate proof that we are decent people (*RH,* 45).

He insists, however, that the cost of trying to establish one's "scientific credentials" and "moral decency" by means of this coerced silence[54] is the increased ghettoization of Judaism's rich intellectual legacy and its possible extinction: "[D]ecent morals do not, by any means, constitute sufficient grounds on which to base a religion. A religion's right to exist is derived from its concept of God. And this concept must be constantly reaffirmed and perfected" (Ibid.). Because of testimonial smothering and most likely other forms of epistemic and political oppression, the Jewish idea of

---

the various types of oppression being instantiated would take us too far afield. I will be content if I convince my reader that epistemic oppression is one factor threatening the identity and recognition of Judaism as a productive epistemic community in its own right.

[54] For Dotson, a speaker's decision to remain silent in response to epistemic oppression is a coerced silence, and I see no good reason to disagree.

God is being neither perfected nor reaffirmed, and so Judaism, understood as a once flourishing epistemic community, is in crisis.

According to Zank, Cohen realized sometime around 1880 that "anti-Semitism was a force to be reckoned with" and decided that demonstrating the unique contribution of the Hebrew prophets to contemporary directions of culture and German cultural consciousness was the best way of combatting it (Zank (2000), 11–12). In a similar vein, I have argued that Cohen realized anti-Semitic epistemic oppression was a force to be reckoned with and decided to dispel the pernicious ignorance fostering it. This oppression, in conjunction with developments in science and philosophy and an uptick in European anti-Semitism, created a perfect storm in the latter half of the nineteenth century. In 1904, Cohen responded by co-founding the Society for the Advancement of the Science of Judaism and, in 1908, his "Ethics of Maimonides" appeared in a two volume collection on Maimonides published by the Society. As we now know, Cohen's and the Society's efforts did little to prevent the atrocities soon committed against Jewish communities. I believe, however, that Cohen's essay still has much to teach practitioners of Abrahamic traditions about proposing a reasonable way forward when they find their community in the midst of an epistemic crisis.

5.2.2   Cohen's Response to the Crisis

Recall the fourth tactical rule for resolving a crisis: when proposing a substantive change to the meaning of a tradition's constitutive concept or the received interpretation of its authoritative texts, show this change is a natural, rationally endorsable continuation of tradition's previous aims and successes. In *Ethics of Maimonides,* Cohen takes up this challenge. He presents a reading of Maimonides' philosophy that wrests it from its original medieval Aristotelian framework and relocates it within a critical idealist framework based on Cohen's reading of Plato and Kant. This relocation effort enables Cohen to present a sustained critique against the naturalism gaining popularity within educated European circles. He argues that the critical idealist tradition of Plato and Kant better accounts for the production of knowledge in natural science and the normativity of ethics than Aristotelian naturalism. Reading Maimonides through the lens of critical idealism also illuminates how an authoritative Jewish conception of God and its ethical corollaries might be rendered compatible with a properly Kantian ethics and epistemology. Cohen is not satisfied, however, with merely establishing this compatibility; he wants to show that the Jewish tradition is an epistemic endeavor in its own right, making a crucial contribution to an ever-increasing body of human knowledge.[55] Toward this end, he

---

[55] Remember that, for Cohen and the Marburg School, this body of knowledge is the idealized limit or regulative idea of a systematic whole consisting of interrelated concepts, ideas, principles and propositions from a variety of conscientiously pursued, methodologically sound cultural practices (see *EM*, 2–3; Cassirer (2005); Friedman (2000); Zank (2006)).

extracts an idea of Jewish messianism from the Hebrew Bible and maintains this idea is necessary to compensate for the formal character of philosophical, i.e., Kantian, ethics. With this reading of the Hebrew prophets and Maimonides' theological ethics, Cohen offers members of his religious community and members of the European intelligentsia a rational route for recognizing Jewish thought and practice as having a legitimate share in the cultural advances of the nineteenth century.

(a) *Eliminating the competition: naturalism is not the way forward*

In *Ethics of Maimonides,* Cohen's refutation of Aristotelian naturalism focuses primarily on its inadequacies as an ethical theory. He cites three main objections against Aristotelian ethics: internal inconsistency, positing an unwarranted dualistic conception of eudaemonia, and operating with a strictly immanent teleology that neither accounts for nor promotes genuine moral progress. Cohen writes: "Aristotle's ethics is nothing but his eudaemonia" (*EM*, 124). He then offers a textually supported analysis of Aristotle's treatment of eudaemonia that saddles him with the paradoxical claim that "the purpose of ethics consists in refraining from ethics; ethics aims at abstention from all virtuous human conduct" (*EM*, 151).[56] Cohen's reductio on Aristotelian ethics runs as follows. For Aristotle, eudaemonia is the highest human good and hence the purpose of ethics. Eudaemonia consists in the life of philosophical thought, and for Aristotle, "thinking and acting are diametrically opposed (*EM*, 150). This opposition manifests itself in Aristotle's distinction between intellectual and practical virtues and the fact that he treats ethics, insofar as it concerns cultivating virtuous human *actions*, as unrelated to the pursuit of scientific knowledge (*EM*, 149). Eudaemonia thus consists in the leisurely (inactive) contemplation characteristic of Aristotle's Prime Mover:

> This leisure is identical with contemplation for which it provides the *condition sine qua non*. A life of contemplation is thus…a life of reason. Hence, this eudaemonia of thinking and of reason constitutes divine excellence, or, if you wish, the characteristic of God (*EM*, 149–150).

Moreover, a perfectly self-sufficient unmoved mover need only contemplate his own thoughts: "He thinks thinking (that is, He thinks Himself): for he is thinking, the very principle of thinking" (*EM*, 150). For Aristotle, the idea that a god would trouble himself with virtuous human or divine activity, with the practical virtues, is "ridiculous" (*EM*, 150). It follows that the eudaemonia characteristic of the divine

---

[56] To say that it is a textually supported analysis is not to say that it is the most charitable analysis. As we have seen, Cohen, like Frege, has numerous objections to the naturalistic philosophical frameworks implicitly or explicitly endorsed by their contemporaries. We have also seen that one strategy for challenging the use of these frameworks is presenting a devastating critique of a well-respected figure within the "empiricist" tradition. For example, Frege goes after the naturalism espoused by Helmholtz and other formalists by likening it to Mill's view and then dismissing Mill's view as 'pebble and gingerbread arithmetic.' The effectiveness and validity of the strategy does not so much depend on providing the most accurate and fair interpretation of Mill's or, in this case, Aristotle's actual position as in providing an interpretation sufficiently close to the positions held by one's contemporaries that objections lodged at the former will strike at the latter.

life and a life that the leisure class of mortals should strive to acquire is a life where one neither practices nor contemplates virtuous human conduct. Based on this analysis, Cohen draws his own conclusion about Aristotelian ethics: "Both for man and for God, this eudaemonia actually raises a wall of separation blocking the way to ethics" (*EM,* 151–152).

As textually grounded and devastating as the first objection may be, it is not obvious how it strikes at the heart of a naturalism currently generating a crisis in Cohen's religious community. To see the connection, we need to consider Cohen's second objection, namely, that Aristotle is operating with two notions of eudaemonia: the eudaemonia of leisure described above and the "diametrically opposed" eudaemonia of action (*EM,* 151). Cohen argues that eudaemonia in the second sense reduces to mere proper biological functioning and the psychosocial and behavioral expectations deriving from a particular political order. It is eudaemonia in this second sense, in conjunction with the naturalistic ethics and metaphysics related to it, that he sees as a clear and present danger to the progress of a genuinely enlightened European culture,[57] as well as the ethico-religious ideal presented in the prophetic literature of the Hebrew Bible.

According to Cohen, the reason why adherents to the Aristotelian tradition end up biologizing or politicizing eudaemonia is because Aristotle himself failed to appreciate the methodological significance of Plato's conception of the Good. Cohen credits Plato for introducing ethics as a scientific inquiry and recognizing the essential role of a priori ideas and principles in grounding such inquiries. Cohen's description of Platonic ideas and hypotheses[58] makes it clear he sees them as precursors, if not identical, to the Kantian a priori:

> [T]he idea constitutes the necessary premise, or grounding, for every scientific investigation. It contains the rationale, the basis and the foundation—the account that cognition renders of itself ['die Rechenschaft der Erkenntnis von sich selber']. There is no cognition without accountability, and no rendering of account without proper grounding. Science is nothing but the science of grounding (*EM,* 8).

---

[57] As we will see, Cohen should be credited, along with Martin Buber and Emmanuel Levinas, for introducing and developing the claim that compassionately suffering with the Other is an essential part of Jewish ethics and ethics generally. That said, there is no getting away from the fact that Cohen's framing of the pressing philosophical issues and proposed resolutions is decidedly Eurocentric. A question this book raises but does not attempt to answer is to whether the epistemological insights or wisdom that one hopes to glean from Cohen's handling of these issues should thus be discounted. To place my reader in the best position for considering this question, I have not censored or whitewashed Cohen's remarks so that they appear less Eurocentric or patriarchal than they are in fact.

[58] Zank (2006) contends that Cohen treats 'idea qua hypothesis,' 'thing-in-self,' and 'task' as synonyms. Cohen's discussion of Platonic ideas as hypotheses grounding various scientific enterprises is evidence for this contention. For the purposes on my exposition, I will treat ideas and hypotheses as equivalent notions. As we will soon see, the the significant contrast is between the ideas or hypotheses grounding the mathematic natural sciences and those grounding ethics and the science of Judaism.

The validity ['die Richtigkeit'] and fecundity ['die Fruchtbarkeit'] of any such investigation is predicated upon congruence with the premise, which in turn constitutes its effective, continuous foundation (*EM*, 7).

Just as mathematical ideas constitute the necessary premises for a scientific study of natural and observable phenomena, so too the Good contains the grounding and prerequisite for a valid and fruitful science of ethics (*EM*, 7–8; 13).

There is a critical difference, however, between the idea of the Good and the ideas grounding mathematized natural and social sciences. The content and necessity of the Good is significantly less accountable to sensible experience and less revisable in light of this experience. We saw in Cohen's response to Helmholtz that he was open to replacing Euclidean concepts and axioms with non-Euclidean ones so long as their fruitfulness for mathematical physics was established and the continuity with Newtonian mechanics retained and explained. We also saw that soon after 1919 Cassirer acknowledged that the Kantian a priori must be relativized, meaning that the mathematical concepts and principles functioning as validating conditions for science can change as science progresses. Cohen's description of the distinction between Platonic ideas generally and the idea of the Good indicates that he accepted the relativized status of a priori mathematical concepts and principles at least a decade prior to 1919: "Whereas the idea in general merits the predicate 'reliability', the idea of the Good may be defined as that hypothesis which constitutes the telos or end of reason, and thus plainly terminates the report of its accountability" (*EM*, 11).

What does it mean to say the mathematical ideas grounding a scientific study can only attain the status of reliability but the idea of the Good constitutes the telos of reason itself? Cohen insists that the is-ought distinction marks the difference between the theoretical use of reason as employed in the empirical natural and social sciences and its practical use in ethics.[59] The given epistemic problem or task for theoretical reason is delivering increasingly exact and valid descriptions of how things are, that is, descriptions of the state of affairs constituting our natural and social environments. Platonic ideas functioning as necessary premises or hypotheses for the descriptive sciences of nature and culture are predicated 'reliable' insofar as they bear fruit and account for continual progress with respect to this task. In contrast, the given problem for practical reason is delivering an increased knowledge of how things ought to be. Moral knowledge also bears an essential relationship to performing the actions required for modifying our natural and social environments and, most importantly, ourselves to better align with the ideals

---

[59] According to Poma, Cohen's view on the correct neo-Kantian formulation of the relationship between logic—the formal and transcendental principles grounding the natural sciences—and ethics was finally worked out and presented in his *Ethics of Pure Will*, the second edition of which was published a year before the essay on Maimonides' ethics. In *Ethics of Pure Will*, Cohen writes that Kant's eternal worth lies in recognizing that the being or becoming of things is the content or subject matter of the natural sciences and that this content must be kept separate from the content of the ethics, namely, the subject of what ought to be (Poma (1997), 106–107). Cohen further maintains that this is where "Kant is in agreement with Plato" (Cohen as cited by Poma).

expressed in the Good. *Contra* Aristotelian eudaemonism, Cohen's Platonism maintains that genuine ethical and virtuous behavior is inseparable from knowledge of the Good.

Mathematical ideas set the initial conditions or parameters for a scientific investigation generally but do not in and of themselves determine the end or completed content of that inquiry: "Hypothesis...lends itself to the experiment of conversion in content" (*EM*. 10). Plato's designation of the idea of the Good as a 'sufficient' hypothesis or 'non-hypothesis' is intended to mark the supreme axiological status properly ascribed to this idea and its content. The content of the Good is not under construction as it were, but rather serves as a "prophecy of wisdom" and "grounding of ethical cognition" to which the empirical sciences and empirical selves are ultimately held accountable (*EM*, 11–12). The Good not only furnishes the conditions for a valid and fruitful science of ethics, but also the objective vision of the ideal outcome of that science. It thus expresses a substantive teleological content missing in other ideas and hypotheses. To say that the Good constitutes the end of reason itself is to say that it posits a telos for human thought and action that is neither described by nor derived from any collection of empirically confirmable facts, which is to say that it posits a "Good beyond the being of nature" (*EM*, 17). In sum, Cohen interprets Plato's idea of the Good as a transcendent, regulative ideal for the science of ethics and any other correctly oriented exercise of human reason and praxis: "Plato's Helios Parable [establishes] the idea of the Good as the foundation of all Being, and hence of cognition" (*EM*, 16–17)

Cohen takes care to ensure that Plato's identification of the Good as "an unalterable guiding precept" beyond the being of nature is not dismissed as the sayings of a metaphysical or religious dogmatist unmindful of our actual situation. Indeed, this is the interpretive error giving rise to the Aristotelian tradition. To redress and prevent this mistake, Cohen insists on interpreting Plato methodologically and teleologically, not ontologically. That is, Cohen's Plato does not conceive of the Good as any sort of metaphysical substance.[60] Cohen also maintains that while the science of what ought to be (ethics) is distinct from the science of what is (mathematicized empirical science) the content of the latter must be harmonized with the former. This harmonizing effort gives rise to the "sublime and profound problems" comprising the subject matter of ethics as a science. Addressing these problems necessarily involves discovery and the accumulation of new knowledge, for "every new solution" yields "new challenges" resulting in "new questions posed unceasingly" and demanding "ever-innovative rejuvenation" (*EM*, 17). The practitioner of ethics, as understood within the critical idealist tradition, should anticipate ever-increasing moral knowledge. This body of knowledge will inform and be informed by individual and collective actions aimed at progressively transforming the circumstances

---

[60] Poma argues that Cohen's interpretation of Plato's theory of ideas helps account for changes in the first and second editions of *Kants Theorie der Erfahrung*, specifically with respect to treatment of Kant's notion of the thing-in-itself. Although Cohen had resisted any metaphysically realist interpretation of this notion, it was reflecting on the transcendental meaning of Platonic ideas that led to interpreting the thing-in-itself as "a task" (Poma (1997, 45–46).

surrounding us to better approximate the Good.[61] Recognizing ethics as a scientific enterprise in this sense constitutes a, if not the, legitimate reading of the legacy expressed in Plato's claim that the Good constitutes the foundation of the world and the Kantian claim regarding the primacy of practical reason.

On Cohen's account, Aristotle ends up being the unmindful one. Because Aristotle fails to recognize the methodological import of Platonic ideas generally, namely, their role as necessary premises grounding a properly scientific inquiry, he fails to recognize the idea of the Good as a regulative ideal for human thought and action. He thus denies ethics the status of a science, thereby depriving it of any cognitive value (*EM*, 14–16, 51). This mistake is compounded by his tendency to conceive of thought and action as contradictories. Consequently, he is "oblivious" to the problem of having to harmonize the theoretical and practical use of reason and, instead, treats mathematical natural science and ethics as not merely distinct modes of inquiry but entirely unrelated to one another: "For Aristotle, the ethicist as such is the politician. There are no further human tasks in Aristotelian ethics" (*EM*, 15).[62] Since the Aristotelian ethicist refuses to consider any telos for human conduct

---

[61] In *Ethics of Pure Will*, Cohen seeks to provide a more consistently Kantian derivation of transcendental ethical principles, more Kantian than Kant himself. As already discussed, for Cohen, Kantianism proper consists in applying the transcendental method, understood as deriving a priori principles from the facts constituting a well-defined and fruitful epistemic endeavor. Although Kant's *Critique of Pure Reason* serves as a model for the use of this method with respect to the mathematical sciences of his day, e.g. Newtonian mechanics, this is not Cohen's assessment of Kant's writings on ethics. Here Kant neglects to identify the relevant body of facts, relying instead on our intuitions concerning the dignity accorded to a good will. According to Poma and Zank, Cohen himself struggled to identify the body of relevant facts and finally settled on the subject matter of jurisprudence because its conception of humanity and epistemic aims best accorded with the theoretical and practical problems associated with Kantian ethics as Cohen understood it (Poma (1997), 115–125; Zank (2006), 8–9)

[62] Since the main reason for examining Cohen's critique of Aristotle is to see how this critique functions as a critique of the naturalism threatening Cohen's religious community, I am not overly concerned with showing that his critique is grounded in the most accurate and charitable interpretation of Aristotle himself. That said, in order for Cohen's critique to serve as a model for responding to religious epistemic crises, it must be sufficiently accurate to encourage his intended audience that jettisoning an Aristotelian tradition considered hostile to Jewish thought and practice is a rationally permissible move. To show that Cohen's critique meets this standard, let me say more about Aristotle's distinction between theoretical science and ethics. In the *Nicomachean Ethics*, Aristotle defines scientific knowledge as knowledge *about* eternal things, things neither generated nor destroyed, and knowledge *derived from* unprovable premises. Although there is debate among contemporary Aristotelian scholars about how much natural philosophy falls under his classification of properly scientific enquiries, i.e., a theoretical science, there is consensus that Aristotle intended to include pure mathematics, mathematical physics, and metaphysics. This consensus comports with Cohen's exposition of Aristotelian scientific knowledge (cf. *EM*, 16–7). There is also a consensus and textual evidence that Aristotle excluded ethics and politics, which he defines as intelligent deliberation about the actions conducive to a well-lived human life. Given Cohen's argument here and elsewhere that the post-medieval practice of western science testifies to the ineliminable role of mathematical concepts and principles to the scientific study of nature per se, Cohen's rendering of Aristotle's mistakes is a fair interpretation, albeit somewhat anachronistic.

beyond the prevailing social order, he ends up positing a plurality of goods that are ranked relative to "the various social classes and professions" (*EM*, 51).

Cohen cites Aristotle's *Metaphysics* as evidence that teleological questions about beings are answered by referencing mechanistic natural processes. Contrasting Platonic and Aristotelian teleology, he writes:

> Aristotle's ontology, however, relates exclusively to nature....[F]or him, Being completes itself within physical nature proper. What business does the Good have with nature? Nature pursues its good on the basis of its own inherent principles, and, accordingly, it operates within relative purposes of Being (*EM*, 18; cf. *EM*, 17).

Cohen concludes that Aristotelian eudemonia of action, that is, the telos for actual laboring, living, and developing human beings, consists in nothing other than the goods or purposes ascribed to them relative to the current state of the empirical sciences or their place in a standing political order.[63] Notice too that these goods may be attained irrespective of whether the subject herself possesses any moral knowledge or exercises any rational, autonomous agency. Therefore, embracing Aristotelian naturalism means lessening one's commitment to an Enlightenment egalitarianism that sees each human being, in principle if not in fact, as a responsible and respected autonomous agent. It also means promoting a moral skepticism and political quietism that stands in the way of doing the work necessary for altering natural and societal conditions to better accord with how they ought to be.

We are finally in a position to see how Cohen's critique of Aristotle functions as a critique of the naturalistic philosophies posing a current threat to faithful Jewish thought and practice. For Cohen, Aristotelian naturalism and its nineteenth century German variations—social Darwinism, Spinozism, and Hegelianism—restrict themselves to an immanent teleology incapable of expressing the strong normative necessity characteristic of ethical and ethico-religious prescriptive claims.

The social Darwinism advanced and, in many cases, embraced in the latter half of the nineteenth century is the very kind of ethical theory Cohen describes and decries as a species of Aristotelianism, namely, an ethic grounded in biology and trafficking in descriptive psychology and anthropology. Defined by historians as "applying Darwinian principles to human society," social Darwinism began gaining traction in Germany with Haeckel's 1868 publication *The Natural History of Creation* (*'Natürliche Schöpfungsgeschichte'*) (Weikart (2014), 93).[64] Soon

---

[63] Cohen specifically references biology, psychology, and anthropology. Since Aristotle refuses to view the Good as a legitimate subject for scientific inquiry, replete with given problems warranting progressive resolutions, Aristotelian eudaimonia becomes "the biological foundation of ethics" and ethics becomes a matter of "psychological descriptions" and "historical or anthropological clarifications" (*RoR*, 403; *EM*, 52–3).

[64] Weikart nicely distinguishes between 'social Darwinism,' 'social Darwinist racism,' and social Darwinist anti-Semitism,' pointing out that not all biologists and social scientists embracing social Darwinism used it to argue that biological evolution would and should result in the eradication of so-called 'inferior' races (Weikart (2014), 93). Furthermore, he claims that though Hitler was undoubtedly influenced by social Darwinism: "No necessary inference can be drawn from Darwinism to the status of Jews..." (94). That said, throughout the nineteenth century in Germany, Darwinism was being employed to argue that mental and moral characteristics were biologically

thereafter, German biologists, sociologists, and anthropologists started applying the principle that human beings, like all biological organisms, are locked in a never-ending struggle for existence as a means of accounting for contemporary racialized hostilities and domination. Darwinian premises were also employed to argue that European and Aryan mental and moral characteristics were both superior and biologically inherited (Weikart (2014), 96–99). By the turn of the twentieth century, many European intellectuals who had rejected Judeo-Christian morality were replacing it with a social Darwinism that was racist and anti-Semitimic (Weikart (2014), 100). Cohen responds by insisting that "Darwinian theory cannot be the teleology of the human race because the ethical sense of mankind requires its own teleology" (*RoR*, 263). So, in addition to whatever logical and ethical mistakes we might ascribe to social Darwinism, for Cohen, it makes the quintessential Aristotelian mistake of thinking that the conception of human beings as ethical agents can be adequately explicated without invoking the idea of human telos beyond the being of nature and the idea of a transcendent God concerned with human affairs.

This mistake also gives rise to pantheistic philosophical systems, systems of thought that Cohen sees as the most immediate threat to German Judaism.[65] Recall his complaint that European intellectuals, including Jewish intellectuals, are neglecting or denying the cognitive significance of the idea of God in order to appear suitably scientific. In his 1914 "The Religious Movements of Our Time" ('Die Religioesen Bewungen der Gegenwart'), he gets more specific:

> The educated layman rejects the concept of God mainly because he takes exception to the notion of divine transcendence. This notion seems to him even more objectionable that that of an immanent divine personality which he might possibly find acceptable (*RH,* 57).

A plausible explanation for why Cohen sees pantheism, not social Darwinism, as the more serious threat to his religious community is that it offers an interpretation of the concept of God that members may be tempted to view as a reasonable and progressive extension of their traditional conception. Cohen himself acknowledges that one can find pantheistic elements in the history of both Christian and Jewish philosophy and classical interpretations of the Hebrew Bible (*EM,* 19; *RoR,* 61–65). The pantheistic philosophies of Spinoza and Hegel also seem to offer an account of divine agency and interaction with the world that is more compatible with progress in the empirical sciences and philosophical ethics than one stressing God's transcendence. Since Spinozism and Hegelianism indicate how Jewish thought and

---

inherited and that this explained the 'superior' cultural accomplishments of European and Aryan peoples. Weikart writes that "as many European intellectuals gradually abandoned Judeo-Christian morality," social Darwinist racism and anti-Semitism "became more prominent until it was rather commonplace by the first decade of the twentieth century" (Weikart (2014), 100).

[65] Weikart notes that while social Darwinism certainly influenced Hitler, it is unclear how much this was a factor in his and other German's anti-Semitism. That said, one might reasonably conclude that social Darwinism, not pantheism, proved the most serious threat to Cohen's Jewish community. My reason for saying Cohen perceived the latter as more dangerous is because of the amount of ink he spilt attacking it.

practice might be reconceived within a naturalized philosophical framework, they constitute serious rivals to Cohen's own proposal for resolving the crisis.

Cohen seeks to delegitimize these rivals by showing they cannot deliver any increased understanding of or cultural recognition for the characteristically Jewish idea of God. First and foremost, pantheistic philosophies misinterpret the oneness ascribed to the God of Abrahamic traditions. Spinoza interprets this doctrine ontologically whereby God becomes the one substance and all other beings are simply modes of this substance. God's creating and sustaining activity is then identified with natural causal necessity (*RH*, 97). In this way, Spinozism explains God's role as creator and author of the laws guiding human behavior without needing to posit any supranatural beings or agency. Cohen counters that a reasonable resolution must preserve and reestablish, not sever, the relationship between the philosophical idea of transcendence and the Jewish idea of the One God. Second, Spinozism violates the Kantian precept that 'God' refers to a regulative idea for the systematic accumulation of knowledge *about* nature, not to a substance or causal force *within* nature (cf. *EM*, 85; *RoR*, 59–61). It thus counts as an instance of pre-critical scholastic metaphysics that Kant has shown to be incompatible with a proper understanding of human reason and the preconditions of modern western science. When articulating these conditions, Kant, like Plato, recognized the crucial role of mathematical hypotheses, which in turn leads to recognizing the methodological differences between a pure or applied mathematical science and ethics:

> Does it really make sense, scientifically or methodologically, to speak of man's actions as one would of geometrical lines, when these actions set the world stage for and affect the politics of world history of mankind? And is it really proper for a biological methodologist to say that nature and morality are identical? Can these two actually be explored experimentally? (*RH*, 60).

Spinozism offers no progressive resolution to the crisis because it represents a retrogressive move to dogmatic metaphysics and fails to recognize that differing forms of inquiry employ distinct methods of validation. Third, while pantheistic systems like Spinoza's attract educated Jewish and Christian circles because they seem to provide a philosophically and scientifically cogent *amplification* of the traditional idea of God as author and guarantor of the moral law, they actually entail jettisoning this idea altogether. By equating moral law with laws of natural necessity and God with nature, Spinozism is an instance of Aristotelian naturalism. It conflates the ethical ought with a naturally occurring is or a causally determined will be. Spinozism and other pantheistic systems commit "the basic methodologically mistake of obliterating all qualitative distinctions between nature and ethics" (*RH*, 58). Such systems cannot conceive of ethics as a divinely-mandated task demanding the exercise of autonomous human thought and action. Cohen concludes that rather than offering "a substitute for this God-concept, pantheism implies its nullification…" (Ibid.; cf. *EM*, 30 and *RH*, 98).

In Cohen's eyes, Hegel's idealism fares no better. Hegel set out to address what many, including Cohen, saw as unresolved issues in Kant's system: (1) To what extent is knowledge of God possible? (2) What role does the idea of God play in ethics or the exercise of practical reason? and (3) What is the relationship between the transcendental moral self and the empirical historical self?[66] Hegel responds to these questions by identifying the divine idea with an unconditioned rational principle or concept ("the Absolute"),[67] which is actualizing itself in and through the dialectical movements of world history. Knowledge of God consists in the progressive self-conscious awareness, mutual recognition, and societal embodiment of the Absolute. This same developing awareness, mutual affirmation, and cultural incorporation of the Absolute explains and mediates the relationship between the transcendental and historical self. In sum, Hegel offers an account wherein Kantian reason—theoretical and practical—is historicized and enacted within a society's various cultural endeavors in a collective effort to grasp the Absolute. This account serves to explain what is meant by knowing God within a broadly Kantian framework and to stipulate the conditions for the possibility of attaining this knowledge. Given the socio-historical nature of these conditions, they also explain how the gap between the transcendental and empirical self might be closed, both conceptually and in actual practice.[68]

Cohen's objections to Hegel's response, despite their redundancy and lack of charity, provide further clarity on his own position regarding the essential features of the critical idealist tradition. This will tell us in turn the parameters he sets for a reasonable, suitably orthodox resolution to any tensions within that tradition. Summarizing Cohen's understanding of the job of the critical philosopher as it pertains to ensuring progress within the mathematical empirical sciences and in ethics also puts us in a position to appreciate his argument that the idealist tradition has deep affinities with and is indebted to Jewish thought and practice. In short, recounting Cohen's objections serves as a bridge for introducing his positive proposal for restoring confidence in the cognitive significance of the traditional Jewish idea of God.

Cohen claims that Hegel conceives of the Absolute as a substance, the kind of divine substance that Kant taught us must be consigned to the junk room of philosophical history (*Religion und Sittlichkeit* (1907b), 24). So while Hegel may not commit Spinoza's error of reducing the divine idea to the concept of an extended material substance and equating divine agency with natural causal necessity, his metaphysics of spirit is a pantheistic pre-critical metaphysical system nonetheless:

---

[66] My claim that Cohen too sees these as unresolved issues stems from the fact that he spends considerable time addressing them in his writings on ethics and philosophy of religion. Furthermore, Cohen answers differ significantly from Kant's, despite his remarks aimed at minimizing these differences.

[67] "Logos" is another term Cohen and Hegel employ in reference to the Absolute, and both do so with the intent of relating the Hegelian notion to the notion of the logos as understood within classical Greek and medieval scholastic philosophy.

[68] Of course there are numerous, competing interpretations of Hegel's concerns and his way of addressing those concerns. While this brief exposition enjoys some textual support, more relevant for our purposes is that it foregrounds those elements in Hegel's system that Cohen finds most objectionable.

> The pantheistic structure of romanticism…rests on the dogma of God's incarnation. God is man, for God is nature as such—a dogma which here becomes the guiding principles for all metaphysics and certainly not of ethics alone. But ethics is thereby actually eliminated as a separate discipline, for it is absorbed into the general process of becoming (*RH*, 83).

The Romanticism of Fichte, Schelling, and Hegel shares Spinozism's mistake of treating the idea of God as an immanent force, the being and meaning of which is exhausted within nature, the realm of empirically confirmable facts.

We can now see that any philosophical system must be rejected if it identifies the God-idea or the Platonic idea of the Good with an immanent causal principle, be it a material, efficient, formal, or final cause. Because these principles function as explanatory principles for observable changes within the natural or socio-political order, the systems invoking them cannot secure the is-ought distinction Cohen considers central to the critical idealist tradition. Since such a system cannot secure this distinction, it cannot recognize ethics as separate discipline characterized by the problem of harmonizing the facts about these historical realities with the norms of what ought to be. Of course progress on this task requires that an ethicist remain cognizant of knowledge produced by the natural and social sciences. Her primary aim, however, is harnessing this knowledge to serve the constant effort of actualizing the supranatural, transcendent idea of the Good.[69]

We can also see how Cohen's critique of Hegel dovetails with his critique of Aristotle. First, both are oblivious to the specific research agenda and epistemic aims belonging to the science of ethics. Second, because Hegel conflates the idea of God as a regulative ethical ideal with actual cultural and political norms, he cannot provide a satisfactory resolution to lingering questions and tensions within Kant's system. Furthermore, because of this conflation, Hegel's idealism ends up encouraging the quietism previously associated with Aristotle's eudaimonism.

According to Cohen, the Hegelian Absolute is not a progressive perfecting and reaffirmation of the God-idea as it functions in either a properly Kantian or properly Jewish philosophy. Consider his reaction to Hegel's statement in *Philosophy of Right* that 'That which is rational is real':

> Here the error of heteronomy is obvious. In no way is existing reality the criterion and principle of moral reason. In no way does moral reason coincide with existing reality, moral law with the positive laws of the historical reality of right and the State. Here lies the enormous *difference between Hegel and Kant*. Indeed Kant would say: that which is rational is not real, but must become real (cited in Poma (1997), 78, italics in the original).

Within Kant's system, as within Judaism, the idea of God is correlated with the idea of an ongoing effort to bring about a cosmopolitan community of perpetual peace:

> History would seem incomprehensible to [Kant] if it did not have a goal. This goal, he says, is perpetual peace….The first to conceive of the idea of universal peace and place their confidence in it as world history's purpose and meaning were the prophets….He who

---

[69] As Poma notes, Cohen's views about the central features of the philosophical tradition inherited from Plato and Kant developed over time. However, the position that I ascribe to him here is relatively fixed by publication of the second edition of *Kants Theorie der Erfahrung* in 1885.

believes in perpetual peace believes in the Messiah, and not in a Messiah who has already
come but in one who must and will come (*RH*, 87–88).

The Kantian idea of God, like the Jewish idea of the Messiah, is inherently related
to the telos of human history and grounds a rational hope of attaining it. In contrast
to the historically real and purportedly perfectly rational communities of Hegelian
and Aristotelian ethics, the community envisioned by Kant and the prophets is one
where "all national and class distinctions separating men" have been removed (*RH*,
153). This ideal community, or more accurately, the completed task of actualizing
this community always lies beyond the existing sociopolitical state of affairs. As
Bruckstein rightly notes, for Cohen, it is the idea of God that secures the critical
distance necessary for subjecting ourselves and existing societal norms to the rigorous moral scrutiny demanded by Kantian and Jewish ethics: "God's postulated oth-
erness invokes an ever-critical attitude with respect to human reality" (Bruckstein
(2004), 34).

Whereas Hegel tries to account for the relationship between the transcendental
self and the historically real self by appealing to the mediating role of self-conscious
participation in the familial, civil, and political life of the State, Cohen appeals to
the continual work of bringing ourselves and respective sociopolitical communities
into closer conformity with yet-to-be realized moral imperatives. Doing this work
presumes and promotes the exercise of autonomy. In keeping with Cohen's de-
substantialized, de-psychologized reading of Kant, the notion of autonomy should
be not be construed as referring to a given property of human reason or will. Instead,
one *becomes* an autonomous rational and volitionally agent by continually seeking
to orient her thoughts and actions by means of the transcendent regulative end of
reason. So, because Hegel identifies existing laws and cultural norms with Kantian
rational and moral imperatives, Cohen charges him with neglecting the significance
of autonomy and encouraging heteronomy. And since Cohen considers autonomy a
basic or ineliminable notion of Kant's critical philosophy, Hegelian idealism must
be rejected as its legitimate successor.

Now recall Cohen's complaint against an Aristotelian politicized eudaemonia.
Because Aristotle misunderstands and so dismisses Plato's idea of the Good as a
transcendental and unalterable regulative ideal, he recognizes no other ends of
human action except those determined by the norms of a prevailing sociopolitical
order (*EM*, 15). Analogously, because Hegel misinterprets and so rejects the transcendence of the God-idea, he reduces the ethically good to the relative goods
enacted by a culture or State at a particular stage in world history. The Hegelian
ethicist, like the Aristotelian ethicist, presumes her task is nothing other than the
political one of acculturating citizens to embrace and embody the ends and norms
of the prevailing social order. As Bruckstein so aptly states in her commentary on
Ethics of Maimonides, for Cohen, Hegel's Absolute Spirit exhibits the same "self-
complacency" of Aristotle's God and both represent "an unabashed apology of the
political and social status quo" (Bruckstein (2004), 94).

My reconstruction of Cohen's critique of Aristotelian, Spinozist, and Hegelian
naturalisms enables us to identify what he sees as the essential, ineliminable tenets

of the critical idealist tradition. First, one must distinguish mathematico-mechanical laws and synthetic a priori principles making the knowledge of dynamically evolving, observable entities possible from ethical imperatives and ideas determining the ultimate end of human thought and action. Second, ethics is a cognitive inquiry in its own right, distinct from any mathematized natural or social science. Ethics and science are governed by a different set of grounding principles and pursue distinct proximate epistemic goals. Third, because of the superior axiological status both Plato and Kant assign to ethical thought and action, the science of ethics supplies regulative principles for pursuing scientific knowledge in the narrow sense.

What specific principles does the Platonic idea of the Good establish for the production of scientific knowledge? To answer this question, I first need to explain Cohen's rationale for treating Kant's ethics as a corrective development of Plato's value theory. As we've seen, Cohen credits Plato with introducing ethics as a conscientious cognitive enterprise, with problems pursued in a manner analogous to but distinct from the·problems tackled in pure and applied mathematical science. He also credits Plato with treating the idea of the Good as a regulative ideal pointing to a supranatural end for human thought and action. He readily admits, however, that Plato's value theory has a "dark side," namely, that it does not consistently affirm that "knowledge ought to be accessible to all men" (*RH*, 74–5). Cohen faults Plato for contributing to the longstanding prejudice that some people groups are simply incapable of comprehending and hence achieving the good (*RH*, 75). Because Plato does not view humanity tout court as capable of progressing toward knowledge and enactment of the Good, his ideally just society countenances class distinctions based on presumed inherent differences in epistemic and moral agency.

For Cohen, Plato's idea of humanity cannot serve as a regulative ideal because its conception of epistemic and moral peers is too narrowly circumscribed and still too dependent on empirical givens, givens which include class, ethnic, and nationalistic divisions. Cohen's own formulation of the categorical imperative is intended to underscore the claim that we are called to overcome these divisions and any egotism deriving therefrom:

> [T]he categorical imperative, as a postulate for mankind, can be definitely formulated in this way: "Do not act as an I, in the empirical sense, but as the I of mankind in the ideal sense. Regard your own person as well as any other not in the physical, racial, or narrowly historical terms of individual existence, but exclusively as an embodiment of the eternal, world-historical idea of mankind" (*RH*, 180).

In contrast to the Platonic idea, the idea of humanity presupposed in Kant's ethics demands that *all* people are regarded as equally respected and responsible epistemic, moral agents. Kantian ethics is a legitimate successor of Platonic ethics for the following reasons. It preserves the idea of the Good as a telos for humankind lying beyond the empirically and historically given here and now. It preserves the principle that one cannot hope to progress in achieving this telos without simultaneously progressing in ethical knowledge. It corrects for Plato's mistakenly narrow conception of who should be recognized as capable of and responsible for striving toward this telos and progressing in this knowledge.

Cohen is not content with showing that Kant's enlightenment ethics is a progressive continuation of the Platonic tradition; he maintains that Kant's idea of humanity "has its origins in the Messianism of Israel's prophets" (*RH,* 180). To support this thesis, he cites the promise of God recorded in book of Jeremiah: "For they shall all know me, from the least of them unto the greatest." Other passages from the Hebrew Bible are adduced to show that the prophets envisioned a world to come where there would be "no distinctions among people" and "all differences in degrees of human understanding would be obliterated" (*RH,* 137). It is for this reason, says Cohen, that Judaism expects all men, regardless of their social status, to study the Torah. The scriptures and Jewish practice thus presume an idea of humanity where all are called to understand the universal moral law, a law that Cohen equates with the law of God. Cohen employed the conceptions of humanity and ethical precepts from both Judaism and Kantianism to petition for universal suffrage, universal scientific education, and open borders for refugees.[70] Eva Jospe correctly concludes that, for Cohen, "Plato fails to see what the prophets clearly envisaged: an image of mankind-as-such, endowed with an intellectual and moral potential that is his because he is a human being and not because he is a member of any nationality or class" (Jospe (1971), 65). Since this prophetic image "bears a substantive similarity" to Kant's idea of humanity as an end-in-itself, much of what is contained in Kant's value theory is "anticipated by prophetic ethics" (Ibid.). Cohen himself concludes: "the triumph of the prophets' religion over the philosophy of the ethics" consists in the fact that "the prophets, and they alone, discovered the idea of mankind" (*RH,* 112). So in *Ethics of Maimonides,* when Cohen insists that Plato's idea of the Good must be recognized as a "sufficient hypothesis" constituting "the rock-bottom ground of all grounding" for all conscientious human inquiry, his readers should recognize that this is an idea which, according to Cohen, is more fully and consistently developed by Kant and originates in the writings of the Hebrew prophets (*EM,* 11).

Given this background on Cohen's analysis of the idea of the Good, we can specify the principles it establishes for the pursuit and production of scientific knowledge. First, the empirical concept of humanity should be never confused with the idea of humanity in the ethical teleological sense. Pursuing knowledge of the former must always keep the latter in view. The penultimate aim of empirical natural and social science is providing methodologically sound, increasingly fine-grained descriptions of the phenomenal realm of becoming things. Ultimately, however, progress is measured by how much these sciences contribute to the growing body of knowledge required for transforming this realm into one more closely

---

[70] The extent to which Cohen explicitly sought to rid Judaism and other Abrahamic traditions of unwarranted patriarchal assumptions is debatable. My point here is to show that his interpretation of the idea of humanity expressed in Kant's categorical imperative and by the Jewish prophets is intended to redress a legacy of epistemic oppression for which Plato is partly at fault. In Sect. 5.3, I show how Cohen's hermeneutical principles for the Bible and other authoritative religious texts can be used to rid the Methodist Protestant tradition of unwarranted assumptions concerning sex and gender identities.

approximating the ideal community envisioned by the Hebrew prophets and Kant's kingdom of ends.

Second, extant epistemic communities or 'directions of culture' must strive to become communities where all are seen as potential epistemic peers, both as contributors and benefactors. When describing Maimonides' exegesis of the Hebrew prophets and the messianic vision, Cohen writes:

> Cognition constitutes the fundamental premise for the messianic conversion of the social world order; cognition as knowledge, not merely as religious behaviorism, establishes the truth. The intention envisioned for the messianic age is not merely the intensive and extensive augmentation of the body of knowledge, but rather everyone's personal share in this cognitive pursuit (*EM*, 182).

Recall Kristi Dotson's characterization of epistemic oppression: the persistence exclusion of some people groups from participating in the production of knowledge. Here we see that Cohen interprets the promised messianic age of Judaism as a prophetic ethical call for a world order free of epistemic oppression.

I have now presented what Cohen views as the essential tenets of critical idealism, but what does it look like to faithfully practice it? Poma offers this beautiful description: "[C]ritical philosophy now clearly reveals its task: defense of the finite, loyalty towards it, but also tension in the direction of infinity, which is still considered a limit and task for the finite" (Poma (1997), 53). Defending and remaining loyal to the finite means serving as a watch guard over the methodology employed in the pure and applied mathematical sciences to ensure the continual production of fruitful results. To see Cohen dispatching his duty in this regard, we need only look back at Chap. 3 and his debate with Helmholtz. However, the critical philosopher also tries to ensure that the methods and findings of the sciences are directed towards the infinite, which is to say that they stay well within the limits imposed by the imperatives of Kantian ethics and promote a continual striving for its regulative ideals. To see Cohen dispatching his duty in this regard, we can now look at his critique of Aristotle, Spinoza, and Hegel.

Throughout the latter half of the nineteenth century, Cohen's contemporaries were proposing naturalized approaches to epistemology, ethics, and metaphysics. I have argued that Cohen and Frege were staunch critics of such proposals. According to Cohen, these approaches called into question the very notion of scientific and ethical progress. He was also concerned that German Jewish intellectuals, in effort to garner more respectability for their theistic beliefs, were tending towards a naturalized philosophy of religion, one which equated the Jewish idea of God with Spinoza's or Hegel's idea of a strictly immanent divine being. For Cohen, this strategy for resolving the problem facing his religious community was no solution at all:

> [I]f we were to accept pantheism's notion of an immanent relationship between God and the world, we would have to conclude that there is no God at all; since God would then not only pervade the world, but lose himself in it. Dispense with ethics and God's intimate involvement with it, and the concept of God forfeits any claim to validity and meaning (*EM*, 30).

To the extent that Cohen's critique of Helmholtz's Kantianism, Aristotelianism, Spinozism, and Hegelianism is successful, he has gone a long way in convincing his peers that answering the siren call of nineteenth century naturalisms constitutes abandoning the intellectual and moral tradition bequeathed to them from Plato and Kant. We've also caught a glimpse of Cohen's response to the problem threatening his religious community. Recall the problem: European Jewish intellectuals are discarding the idea of God because they cannot see its relevance to nineteenth century cultural advances. By arguing that critical idealism is the philosophical framework best accounting for those advances and that the conception of human progress expressed within that framework originated with the Hebrew prophets, Cohen sends a strong message that the Jewish idea of God has been too quickly discredited. In addition, none of the proferred naturalisms can serve as rationally endorsable continuations of medieval and early modern Jewish thought, since none can retain or restore the cognitive value of the idea of God on which the viability of Judaism depends.

(b) *Cohen's positive proposal: a critical idealist rereading of the tradition*

It is one thing to argue for the illegitimacy of a proposed resolution to an epistemic crisis and quite another to argue for the legitimacy of one's own. Moreover, it is one thing to show there are deep affinities between a religious tradition and the philosophical tradition valorized by the culture and quite another to show this religious tradition makes an essential contribution to prized cultural achievements. In other words, I have yet to present Cohen's positive proposal for resolving the crisis, and let's remember what he needs to do. Based on my reading and endorsement of van Fraassen's and Friedman's appraisal on what constitutes a progressive resolution to a Kuhnian crisis, Cohen owes his co-religionists a way forward. He needs to offer a (re)interpretation of the Jewish idea of God that can be viewed as prospectively rational and a legitimate successor of what has gone before. He cannot and need not provide members of his religious community with a rationally compelling argument that his proposal is the only reasonable way forward. Acceptance will require a little salt and good will on their side. That said, for Cohen's proposed account of Judaism as a science to count as a reasonable and responsible live option, it must lay claim to furthering the aims of the Jewish tradition and retaining what is worth retaining of its past practice.

Cohen insists there are three tenets concerning the idea of God that must be retained in any recognizably Jewish philosophy of religion. First, the God of Judaism is "the One God" (*RH*, 44). Second, "the God of Judaism is the God of transcendence" (*RH*, 84). Third, the God of Judaism is "the God of ethics" (*EM*, 103). In *Ethics of Maimonides,* Cohen stresses the importance of interpreting the third tenet so as to properly elucidate its significance:

> For this is characteristic of a life true to monotheism: that all speculation breathlessly advances in its pursuit of ethics. Any heresy may be absolved and even overlooked, as long as true teleology, the ethical convergence of nature and mind is being sought and attained. The history of philosophical speculation in Judaism may be developed in light of this proposition (*EM*, 21).

Here Cohen states his intention of presenting a history of Jewish philosophy that shows it to be a continual, progressing inquiry into the idea of God as a regulative ideal bridging the is-ought distinction.[71]

To better appreciate Cohen's intention, it helps to compare his proposed history to the notoriously Eurocentric, racialized metanarrative presented in Hegel's Lectures on the Philosophy of History.[72] In the introduction to the lectures, Hegel characterizes a philosophical history as a history written for the expressed purpose of elucidating how an idea is being progressively articulated and understood in and through the workings of human history. This historicized conceptual analysis serves two further aims. First, it attests to the actuality of the concept or idea as opposed to its mere real possibility. Second, it demonstrates that the conceptual content is being continually refined in response to the demands of both practical and theoretical reason. In Hegel's historiography, the idea under investigation is rational freedom. In Cohen's proposal, it is the Jewish idea of God. In both cases, the ultimate goal is securing the cognitive value of these ideas by establishing a relationship to human agency insofar as this agency results in changes to ourselves and to our natural and social worlds. Given these changes can be viewed as more insightful ways of conceiving and living within those worlds, the philosophical historian has demonstrated the concrete cognitive worth of these ideas.

Now for the difference between the two approaches. Hegel begins by assuming a universal perspective, that is, human history as such becomes the object of philosophical speculation. By illustrating how freedom has progressively developed through various 'world-historical' people groups, Hegel intends to present not only a theodicy ("the justification of God in history"), but also a demonstration that, to date, Protestant German culture is the culture wherein "the laws of *real* Freedom"

---

[71] Erlewine (2010), Zank (2006), and Poma (1997) point out that Cohen previously argued for the idea of God as a necessary postulate in *Ethik des reinen Willens*. This is a work in pure philosophical ethics where Cohen seeks to demonstrate Kant's essential contribution to the field, while introducing substantial modifications to Kant's system. Here and elsewhere Cohen faults Kant for positing God as a means of reconciling moral striving in this world with human happiness. This is a capitulation to eudaimonistic ethics that violates the general spirit of Kantian ethics (*RH*, 82). Kantian ethics proper, argues Cohen, treats the idea of God has a necessary postulate because it contains truth as a regulative ideal and guarantees its possibility. In specific, the idea of God grounds the necessity and possibility of harmonizing ethics and logic (mathematized science of nature). When treating God as a necessary postulate within a system of philosophical ethics, Cohen abstains from introducing any explicitly Jewish theological considerations. Zank explains why: "The God of Cohen's *Ethics* is a philosophical construct that, while deeply agreeing with what Cohen feels is the proper Jewish idea of God, is ultimately universal (Zank (2006), 13). Given Cohen's methodological scruples, a strictly philosophical ethic must exhibit a universal and formal character. This character precludes introducing elements that are defensible only by invoking the assertions or precepts of a particular religious community or tradition. However, as we will see, it is this very formality that creates a lacuna enabling Cohen to show that Kantian ethics must be supplemented by ethico-religious principles deriving Judaism.

[72] In all fairness, one must acknowledge that Hegel opposed the slave trade and the published lectures were largely drawn from a student's notes. Nevertheless, much as been written on Hegel's unduly Eurocentric and racialized account of human history. For my own contribution in this area, see Merrick (2014).

are most fully articulated and recognized (Hegel (1956), 456–457). In contrast, Cohen's adopts a more particularist perspective. He focuses on the literary sources of the Jewish tradition and illustrates how the idea of God has been progressively developed within these sources. Cohen does not presume to show that the Jewish idea of God contains and preserves all that is best from the religious thought and practice of other people groups. For all his efforts to prove that Jewish thought is compatible with Platonic and Kantian thought, he never presents the Jewish idea of God as any sort of Hegelian *Aufhebung* involving Plato's idea of the Good or the Kantian God of Protestant Pietism. Rather, Cohen's historiography aims at establishing the distinctly Jewish contribution to European culture and a cosmopolitan order ever progressing towards its ethico-religious ideal. So while Cohen's historicized philosophy of religion may still be charged with undue Eurocentrism and an unwarranted assumption that our social practices are becoming increasingly rational and humane, it does not arrogate to itself a false universalism whereby all particularistic cultural beliefs and practices are effaced or superseded.[73]

To appreciate how Cohen's retelling of the history of Jewish thought is supposed to resolve the epistemic crisis confronting his religious community, recall his diagnosis of the crisis. According to Cohen, the "educated layman" of his day rejects the idea of God entirely or opts for the pantheistic notion of a strictly immanent deity because "he takes exception to the notion of divine transcendence" (*RH*, 57). So, to restore epistemic confidence in the Jewish idea of God, one must recover a notion of divine transcendence that elucidates its continued cultural and ethical significance. His history of Jewish biblical thought and exegesis of Maimonides' works aims to provide this elucidation

Cohen acknowledges that the Hebrew Bible can and has been read through a metaphysical realist lens. Based on this hermeneutic, the idea of God as the transcendent creator is interpreted ontologically. The claim 'God exists' is then taken to assert the existence of a divine substance constituting the material, efficient, formal or final cause of all natural or observable entities. Cohen would agree that understood in this sense the claim is indeed without cognitive significance, since it would fall under the purview of natural science methodology which requires that posited causal entities are subject to mathematical description and empirical confirmation. He responds by arguing that this is not the appropriate hermeneutical framework. A realist philosophy of religion and interpretation of the Bible—one presupposing a substance ontology and allowing for univocal predication that ranges over God and other entities—constitutes a "preliminary" stage in monotheism and a "mythic" reading of the Bible (*RH*, 91). Therefore, the educated layperson need not take exception to the notion of divine transcendence per se but only to the meaning ascribed to it within a pre-critical realist framework.

Cohen goes on to reject a substance realist interpretation of 'transcendence' on grounds that it entirely disregards Plato's introduction and initial use of the term.

---

[73] For more on how Cohen manages "the dialectic between particularity and universality" as compared to his philosophical predecessors, see Erlewine (2010).

When juxtaposing the ideas expressed in biblical and philosophical texts, Cohen pays careful attention to the original discursive context, continually reminding his audience that the authors of the Hebrew Bible and the Hebrew prophets were not philosophers. His exegesis of 'transcendence' is a case in point. Cohen maintains that Plato introduced the predicate as a philosophical term of art in reference to the Good and without any religious connotations. 'Transcendence' thus originates as a predicate within a system of philosophical ethics, not as a term for a divine attribute.[74] The point was to convey that the Good exists in a sense utterly distinct from the kind of existence attributed to natural or sensibly-given entities: "Transcendence, then, denotes that mode of being which is not beyond all existence as such but beyond an existence verifiable by natural science" (*RH,* 58).

Given Cohen's interpretation of the relevant philosophical texts, the claim 'A transcendent Good exists' asserts the following. First, ethics is the highest form of knowledge. Second, an ethically valid way of being is a real possibility. Third, this ethical way of being is irreducible to the modes of existence studied within mathematicised empirical science (*RH,* 57–58). Cohen's first step in allaying the religious layman's worry about the cognitive value of the term 'transcendence' is showing that, understood in its proper philosophical sense, it signifies nothing other than the essential tenets of the critical idealist tradition. So for those convinced of the ongoing epistemic and cultural worth of this tradition, 'transcendence' is and remains a cognitively meaningful predicate.

Cohen's next step is showing how the Platonic notion of transcendence relates to the idea of God articulated within the Hebrew Bible. As previously mentioned, affirming the oneness of God is a fundamental Jewish faith commitment. The words 'Hear O Israel: the Lord is our God, the Lord is one' occur in the Book of Deuteronomy and as the first line in the Shema, the morning and evening prayer enjoined by the tradition. According to Cohen, 'Hear O Israel' contains the "historical formula" for expressing "the rallying idea, the unifying concept of Judaism" (*RoR,* 31). How then should one understand these words? Cohen argues that they should be interpreted in accordance with the prophetic writings, namely, as ascribing *uniqueness* of being to God. When the prophets speak of God as the One, they are insisting that God's being is entirely other than the being of the world and everything in it, including humankind: "For the prophet…the truth lies not in reality but in God, who created reality but who can also change it, so that truth cannot inhere within nature itself" (*RH,* 68). Based on Cohen's reading of the prophets, the Jewish idea of God represents a distinctive mode of being that is incomparably truer and worthier than whatever truth or value may be ascribed to existing, mutable realities: "Before the idea of the One, the spiritual God all forces of nature and culture pale into insignificance… (*RH,* 221). To affirm God's unique mode of being is thus to affirm God's holiness, an idea of goodness that is ever before and "beyond me" (*RH,*

---

[74]Cohen argues that Plato's goal of developing ethics as scientific system led him to rid ethical terms of any religious connotation: "Plato wishes to make the science of morality autonomous and, in spite of this fondness for the old myths, independent of religion (*RH,* 106). Thus, 'transcendence' as a predicate for the Good is, for Plato, unrelated to the idea of God (*RH,* 68).

58). Therefore, what the Hebrew prophets meant to convey when insisting on God's oneness is equivalent to what the critical idealist philosopher intends when predicating transcendence of the Good:

> That aspect of being which the prophet tries to distinguish from any physical form is, in the terminology of the philosopher, called transcendence…And where the philosopher refers to the idea of the good as "highest knowledge," namely ethics, the prophet speaks in the same sense, of the One and Only god as the Holy One, namely the God of morality (*RH,* 58).

Where the critical idealist speaks of the Good as the highest knowledge and telos for human thought and action, so too Jewish tradition speaks of God as "the ground of all law," "the basis for both morality and natural science," and "the archetype for all human morality" (*RH,* 58). Just as Plato's idea of a transcendent Good functions strictly as a regulative ideal for the ongoing exercise of theoretical and practical reason, so too the Jewish idea of the unique God represents a way of being and knowing towards which one must continually strive.

With this interpretation of Plato's conception of transcendence and the biblical prophets' conception of divine uniqueness, Cohen offers his co-religionists a way of understanding the sense of 'Hear O Israel' such that it enjoys biblical support and comports with a critical idealist hermeneutic. His positive proposal for resolving the crisis threatening his community is to replace the ontology of given substances, properties and relations—an ontology that he associates with dogmatic metaphysical realism—with an ontology of the norms governing human thought and action. When the tradition says 'the Lord our God, the Lord is One,' Cohen contends that a reasonable and suitably faithful rendering of this claim is that God, and God alone, is the being of the ought:

> [T]his is also the meaning of God's uniqueness: that He has no peer, which ultimately means that there is but One ethical ground and origin for the ethical world (*EM,* 104).

> While God's holiness is an accomplished fact, man's holiness lies in the realm of the 'ought' (*RH,* 141).

And when the tradition asserts the 'fact' of God's holiness, this should be strictly understood as underwriting the unsurpassable and irrevocable nature of the ethico-religious must:

> 'Ye shall be holy; for I the Lord your God am holy' (Leviticus 19:2). With this correlation, the mythic meaning of holiness was all at once converted into a new meaning of morality…. A mythic version of this Biblical passage would read 'Ye shall become holy, for I will make you holy.' But in the Bible, God demands man's holiness. He makes holiness man's task (*RH,* 140).

On Cohen's account, naturalized and realist readings of God's being and attributes represent preliminary stages in Jewish theology and biblical hermeneutics. In contrast, he offers his community a critical idealist reading of divine unity and transcendence that retains and privileges the tradition's conception of God as the author and guarantor of an ethical universe.

Let me explain why, on Cohen's view and mine, he need not convince his audience that his interpretation of divine unity and transcendence is the only one capable of marshalling biblical support. Recall Cohen's claim that the lifeblood of a religious tradition consists in its teaching about God and articulating a world view (*Weltanschauung*) where this teaching is shown to be essential. Remember too that, for Cohen, properly functioning directions of culture—the mathematized sciences, philosophical ethics, and religious traditions—constitute distinct but interrelated cognitive endeavors. The distinctive epistemic task of Judaism, and presumably other Abrahamic traditions, is continually perfecting and reaffirming its idea of God and thereby demonstrating its crucial contribution to creating and sustaining a genuinely humane way of being in the world (*RH,* 44–6). This suggests, and Cohen asserts, that biblical texts—texts employing different genres, from differing time periods, and informed by differing socio-historical perspectives—contain "different stages of the development within the monotheistic concept of God" (*RoR,* 38).

Cohen further maintains that Abrahamic traditions can and should progress in their knowledge and understanding of the content expressed in their idea of God. This progress will manifest itself in reinterpreting scriptural and other authoritative literary sources:

> [P]rogress in religious understanding has been accomplished through the revision and reinterpretation of sources, while these themselves remain preserved in their individual layers and have been at most rearranged or given different emphasis (*RoR,* 38).

The point is that there is no unique, scripturally determined meaning of God's unity, transcendence or any other divine attribute. Instead, the Bible contains layers of meaning. The job of the responsible biblical exegete is uncovering these layers and offering an interpretation that preserves what is best in the history of the tradition. In the case of Cohen, 'what is best' means what better approximates the ethical and ethico-religious regulative ideals and directives of critical idealism.[75] While Cohen's history of biblical thought need not show that his interpretation is the *only* one supported by the tradition, it is saddled with burden of convincing his audience that his is the most *progressive* and a *sufficiently orthodox* interpretation.

Cohen claims that it is the idea of God emerging from the prophetic literature that represents the most fully developed stage of biblical thought. Combining nineteenth century historical criticism with more traditional, e.g., midrashic,

---

[75] Given my articulation of what constitutes progressive religious understanding for Cohen, one cannot help but notice a certain circularity in his argument for how to interpret the oneness of God. Cohen's reading of what the prophets meant is the preferred reading why? Because this reading best advances the epistemic and ethical ideals of the critical idealist tradition as Cohen interprets it. I contend that such circularity is inescapable, once we accept, as I think we should, that any effort at communicative rationality must invoke some hermeneutical or conceptual framework, however loosely defined, and these frameworks stipulate or presuppose certain shared normative commitments. This is precisely why, as van Fraassen and Friedman point out, anyone arguing for a radical revision to the inherited framework cannot mount a one size fits all argument satisfying universal norms of theoretical and practical reason. To meet such a demand presumes human knowers either currently occupy a space of reasons governed by these universal norms or can occupy such spaces for the purpose of arriving at *the* rational choice among competing successor theories or paradigm. I follow those who insist that embodied human knowers are situated knowers and hence incapable of occupying such spaces.

hermeneutical principles, Cohen argues that the earliest biblical authors were primarily concerned with distinguishing their idea of God from the polytheistic conceptions of neighboring cultures: "Monotheism…comes to be not as a creation out of nothing but has its precondition in polytheism, which in Canaan was Israel's religion too" (*RoR*, 37). Given this concern, the emphasis on God's oneness in the Pentateuch is understandably placed on stressing God's numerical unity. Cohen supports his interpretive claims by showing that they can account for puzzling grammatical features within the relevant texts. He concludes that an initial layer of meaning associated with the being and attributes of God reflects an interest on the part of biblical authors to establish a distinctly Jewish national identity in contradistinction to neighboring identities. For this reason, it is correct to say that the Hebrew Bible is in part a "national literature" that "begins with the national *history* and with the myths and sagas that surround it" (*RoR*, 25; italics in the original). It is wrong, however, to assert that all authors of the Hebrew Bible were only motivated by parochial concerns and hence dismiss, as did nineteenth and early twentieth century anti-Semitic Protestant German scholars, Jewish understandings of the scriptures.

Cohen goes on to claim that an additional layer of meaning discovered in authoritative Jewish literary sources derives not from any particular nationalistic or religious interests on the part of the biblical authors, but rather from the cosmological speculations of pre-Socratic Greek philosophers. Xenophanes, says Cohen, was the first to conceive of divine unity as a metaphysical unity, where the being of God and the being of the cosmos are one in the sense of one identical substance: "Thus, on the threshold of Greek philosophy, *pantheism* arises (*RoR*, 40 italics in the original). Since Greek thought intermingled with biblical or traditional Jewish thought, pantheistic elements can be found in the tradition's treatment of God as the one creator. Pre-critical realist conceptions of divine creational activity often posit this activity as an answer to questions about the causal origin and intelligibility of an empirically-given, material world: "Mythical man is interested only in the question of the world's 'where from?'" (*RH*, 96). This "mythical mode of cognition" persists in Spinozism and wherever the emphasis is placed on God qua one immanent substance rather than God qua unique transcendent good. In other words, pantheistic or naturalistic connotations associated with the being of God and found within the biblical and traditional canon trace back to the efforts of pre-Socratic Greek philosophers who sought some sort of unifying principle that rendered the multiplicity of sensed particulars intelligible. History testifies to the fact that these efforts have been succeeded by the work of Plato, Kant, and other critical idealist philosophers, who have shown that it is more fruitful to identify this unifying principle with a system of a priori synthetic or mathematical concepts and not a unique substance. Cohen's history of Jewish and Greek thought allows him to conclude that pantheistic or pre-critical realist readings of the divine attributes, regardless of whether they can garner biblical and traditional support, are underdeveloped conceptions of the Jewish God insofar as they derive in part from a naïve and now defunct philosophy of nature.

Cohen's history then paves the way for introducing the prophetic conception of God, with its emphasis on God as Redeemer and the promised Messiah, as the apex of intra-biblical thought. We saw that Cohen credits the Hebrew prophets for the developing the idea of humanity that ultimately finds its way into Kant's Enlightenment ethics. I turn now to his argument that this idea is essentially related to their idea of God. For the prophets, the focus is on God as the Redeemer of the poor and the stranger, in contrast to the earlier emphasis on God as redeeming the nation of Israel:

> The entire development of prophetism can be traced along this lines. God is not the father of Israel's heroes, and it is not they who are called God's beloved. God loves the stranger…[M]an's horizon widens as soon as he realizes that his own God also loves the barbarian and even declares that the hostile nations are quite as much His precious possession as is Israel itself (*RH,* 71).

Within the prophetic writings, the idea of God is conceived in response to an awareness of human suffering that goes beyond a concern for the suffering of one's own nation or people group. It is here that the responsible biblical exegete can mark the transformation of a nationalistic religious consciousness into a fully developed Jewish consciousness of the cosmopolitan messianic ideal (*RoR*, 243). This ideal is nothing other than Kant's postulated telos of history: "the first to conceive of universal peace and place their confidence in it as world history's purpose and meaning were the prophets" (*RH*, 87–8).

The prophetic writings also mark the transition from a mythical mode of cognition where God is seen as the answer to the metaphysically speculative question 'where from?' to a more sophisticated religious mode of cognition where God becomes the answer to the ethico-teleological question 'where to?' God sends Messiah, which is to say the prophets' societal ideal is attainable: "Morality will be established in the world" (*RoR*, 21). God is Redeemer, which is to say that efforts at self-perfection and sanctification are not futile: "the soul…is redeemed through its own repentance, through prayer, and through its resolve towards ethical conduct" (*RH*, 223). God is creator, which is to say that the idea of God relates to humankind in such a way that "each day is a new beginning" when it comes to our ongoing task of approximating divine holiness and heeding the messianic call to "end all social misery" (*RoR*, 68, *RH*, 119). So it is that Cohen offers his co-religionists a way of understanding the fundamental tenets of Judaism and biblical claims about the being of God that shows them to contain theses upon which the purported achievements of European culture are predicated.

The crucial role that Cohen's exegetical work on the Hebrew prophets plays in responding to those inclined to discredit the Jewish idea of God cannot be overstated. Kant, Hegel, and subsequent German scholars dismissed Judaism on grounds that its concerns were irredeemably parochial and particularistic. At the same time, the German Protestant idea of God and its correlative idea of humanity was lauded for introducing a more cosmopolitan, rationally humane social ethic. Greek philosophy is then identified as the most proximate source inspiring these ideas. Cohen is set on delegitimizing this metanarrative of ethico-religious progress by reminding his audience that Christian social ethics arises out of the sources of Judaism. As Erlewine explains:

> ...Cohen's philosophical reconstruction of Judaism rests on the claim that a particular religious tradition, Judaism, embodies or manifests, more perfectly than any other, concepts that possess universal significance. His claim that Judaism is a universal religion is grounded precisely in privileging the teaching of the prophets as the *Quellen* for the construction of a religion of reason (Erlewine (2016), 33).

Cohen's responds to the epistemic and social marginalization threatening the viability of his community by constructing a counter narrative where the rationality and universality commended in German enlightenment ethics is due to the idea of God first articulated by the Hebrew prophets, not the idea of God originating from Greek philosophy.

Finally, Cohen appeals to the prophetic writings to resolve lingering problems with Kantian ethics. Because of the formal character of Kant's ethics—and for Cohen any properly philosophical ethic must have this character—questions arise concerning an individual's moral motivations. Kant himself argued that ethically good actions were those motivated by an agent's respect for the moral law. There are two problems with this response. First, it fails to ensure that moral motivation is bound up with an affective concern for the well-being of others to whom the agent personally relates.[76] Second, it fails to adequately address the fact that human agents often confront their apparent incapacity to *do* the good despite having genuine respect for the moral law. That is, a strictly philosophical Kantian ethic lacks the resources for addressing the all too human condition described in the Pauline biblical text: "I do not understand my own actions. For I do not do what I want, but I do the very thing I hate.... I can will what is right but I cannot do it" (Romans 7:15–18). In *Religion with the Bounds,* Kant sides with Paul and tries to address the second problem by supplementing his ethics with resources drawn from his German pietist religious tradition. Cohen will address both problems by drawing on the prophets' depiction of God as a God of compassion.

Cohen maintains that reflecting on God's uniqueness, as understood by the prophets, is a necessary propaedeutic for ethics. How so? For Cohen's prophets, the uniquely holy and transcendent God has compassion for the poor and for the sinner responsible for their suffering. It is this aspect of God and the individual's self-understanding in relationship to it that awakens a sense of personal "moral responsibility" and motivates the agent to "take action" towards alleviating the suffering of a beloved human other (*RH*, 70–1).[77] Correlated with the idea of God's uniqueness is the idea that "nature [and] that man himself has no original worth, not worth on its own" (*RoR*, 48). Cohen's point is not that human beings are subjects of utter

---

[76] Bernard Williams and others have raised this charge against deontological moral theories like Kant's.

[77] Throughout Cohen's writings, his uses the term 'correlation' to designate the relationship that God bears to anything other than Godself. For our purposes, it is enough to appreciate that Cohen's intent in using this technical term is underscoring the transcendence or alterity of God relative to all other beings or becomings, while still entitling him to speak about the idea of God in relation other beings, namely us. He insists that 'correlation' is not identity, which enables him to reference God-other relationships without providing fodder for the pantheistic and naturalistic theologies he opposes.

moral depravity or bear the mark of an original sin that can only be expunged by the act of a divine mediator, for he explicitly rejects Christian doctrines of original sin and Jesus' vicarious atoning sacrifice. Instead, his point is that relative to God's holiness and the messianic societal ideal, actual human agents must recognize that they and their surroundings can only be described as morally wanting. Whereas ethics proper begins with positing the I as full-fledged moral agent expected to fulfill the ethical ought, Jewish prophetic ethics begins with the I as a sinner:

> [The prophets] regard man neither as the son of gods nor as a demigod or hero. Their image of man is one of human weakness. And inasmuch as this weakness is primarily moral inadequacy, the sinner is seen as archetypal man (*RH*, 70).

But the prophets' archetypal man is not just a sinner, she is a poor sinner: "the poor man becomes the symbol of all men" (*RH*, 71). Human poverty and the suffering that comes along with it cannot be shrugged off as the simple result of the sufferer's own guilt or a biologically determined fate. To awaken moral responsibility, one must "understand human misery not in biological but sociological terms" (*RH*, 71). The poor, the orphan, the widow, and the stranger must all be viewed as "victims of social oppression from which they will be liberated through God's justice" (*RH*, 71). In addition to seeing oneself as both a perpetrator and potential victim of social oppression, one must recognize that God's liberatory and reconciling justice is enacted by means of compassionate solidarity. The messiah, God's suffering servant, becomes "the standard bearer of poverty," taking on "all human guilt" and "all human suffering" (*RH*, 72). By emulating God's compassion, as depicted in the promised messiah, we recognize ourselves as embedded in I-Thou relationships structured around an affective concern to bear and alleviate each other's misery: "Only compassion discloses to me man as my fellow man. Through my sense of compassion, the other's suffering becomes my own, and the other becomes my fellow human being" (*RH*, 72).

In sum, by deepening one's knowledge and understanding of the distinctively Jewish prophetic idea of God, one deepens their knowledge of and compassion for the suffering induced by their moral inadequacy and that of their respective societal institutions. This interpersonal, affective, and religious mode of cognition provides a necessary motivational component in the ongoing work of redressing this suffering. According to Cohen, the Jewish messianic understanding of the self in correlation to God constitutes the requisite preamble and supplement to pure philosophical (Kantian) ethics because it affords a richer account of moral motivation, a more realistic account of our inadequacies, and the proper cognitive and affective orientation from which to begin the process of overcoming them.[78]

---

[78] In *Religion of Reason,* Cohen devotes an entire chapter to developing this latter point. In contrast to other loves, religious love consists in compassion for the suffering one's fellow human being and caring for all people as "flesh that is your flesh" (*RoR*, 147). Religious thought and practice, insofar as it proceeds rationally, teaches and instills this love, thereby making its unique and essential contribution to philosophical ethics: "What reason cannot achieve in ethics, the universal human love of men for one another, this reason achieves in religion" (*RoR*, 144). For more on

In *Ethics of Maimonides,* Cohen shifts from direct biblical exegesis to exegeting the work of his tradition's preeminent biblical scholar and philosopher. Cohen has numerous reasons for undertaking the task of (re)interpreting Maimonides' works, but it suffices for our purposes to consider only two: (1) showing that Maimonides's religious ethics is incompatible with Aristotelian ethics and should be interpreted along the lines of Cohen's critical idealism; and (2) showing that Jewish ethics, which treats knowing and loving God as the ground and telos of human morality, is compatible with a modern philosophical ethic identifying this ground and telos with human autonomy.

For Cohen, a commitment to Aristotelianism implies a commitment to Aristotle's conception of eudaemonia. So to prove that Maimonides is no Aristotelian, he need only show that Maimonides "substitutes self-perfection for eudaemonia" and that the two are contradictory notions (*EM.* 153). Recall Cohen's argument that Aristotelian eudaemonia consists in (1) possessing the virtues enabling an individual to flourish within their actual natural and social environments or (2) engaging in the non-active self-contemplation that supposedly typifies a divine nature. Aristotelian virtues, on this account, can be cultivated without needing to engage in a philosophically rigorous, broadly scientific inquiry into the idea of the Good. They are skills acquired through habituation or imitation allowing one to behave in accordance with prevailing norms: "[Aristotle's] ethics does not deal with cognition, but rather with control of conduct" (*EM,* 129). Possessing these virtues and the resultant eudaemonia constitutes a "finish" or "real end" in the sense that the ethical task is completed upon achieving a state of flourishing with a given natural and cultural milieu (*EM,* 156).

Cohen maintains that Maimonides' notion of self-perfection can in no way be conceived as an Aristotlelian virtue. Self-perfection does not consist in having attained any state or skill, but rather "the process of perfecting the Self" (*EM,* 156). Unlike Aristotle who describes eudaemonia as the completed state of ethical development, Maimonides recognizes that "the ultimate end is but continuous work" (*EM,* 156). Maimonides, like Cohen's Plato, understands that ethics proper involves a continual "striving toward the Good" (*EM,* 156). Following Jewish tradition and the prophets, Maimonides further understands that this striving is in large measure an epistemic endeavor, studying to know God's ways in an ongoing effort to emulate them. Cohen cites the following as evidence of Maimonides' recognition that a faithful commitment to Judaism required having to reject an Aristotelian conception of virtue and the highest human good:

> The one who serves God out of love is engaged in the pursuit of Torah and its laws, and follows the paths of wisdom, not for the sake of some mundane or earthly interest, nor for the sake of avoiding pain and suffering, and in order to gain any hedonistic good… (Maimonides as cited *EM,* 157–8).

Cohen's treatment of the I as conceived within prophetic messianism and his position on the essential contribution of Judaism to ethics, see Zank (1996).

True happiness and human perfection, as conceived by Cohen and Cohen's Maimonides, does not consist in attaining the skills necessary for conducting ourselves in accordance with "conventions and traditions" (*EM*, 190). Instead, happiness consists in the ongoing work of self-perfection (*EM*, 157). This work requires developing wisdom, and wisdom implies engaging in behaviors and cultivating dispositions that are based on each individual's ever increasing knowledge of the Good (*EM*, 183). In the ethics of Maimonides, it is impossible to exercise any virtue without possessing some degree of ethico-religious knowledge. There is another reason why pursuing self-perfection in Maimonides' sense is incompatible with the idea of regulating one's behavior, religious or otherwise, in accordance with the norms of the prevailing social order. Those who study Torah and especially the Prophets understand, as Aristotle and perhaps even Plato did not, that striving for the Good in the messianic sense results, as a practical necessity, in the "conversion of the social world order" (*EM*, 182)

Not only is Maimonides' understanding of self-perfection antithetical to Aristotle's conception of the virtues but also to his conception of a divine nature. Maimonidean self-perfection, according to Cohen, is the process whereby the historically real self takes on the responsibility of becoming a genuine ethical self: "[The principle of self-perfection] expresses itself in autonomous activity of the individual Self devoted to the development of her own ethical Self" (*EM*, 191). Unlike Hegel, who characterized this process as one where an individual self-consciously affirms and adopts societal norms,[79] Cohen describes it as one that begins when an individual takes stock of herself in relation to the thirteen divine attributes depicted in Exodus 34:6–7 and further elaborated within rabbinic tradition. What are these attributes? Though there is slight variation within the tradition in order to arrive at thirteen, first and foremost among them, as Cohen points out, is compassion (*EM*, 69). According to Cohen, Maimonides agrees with the Hebrew prophets that genuine moral development begins when an individual recognizes that God has compassion for human sin and suffering, a recognition motivating a personal concern for fellow human beings. There is no place in Jewish philosophical or theological anthropology for the notion of a non-relational, isolated I: "The person as an individual emerges in the relation to ancestors, kinfolk, and nation; and these social bonds cannot be severed" (*EM*, 145). The task of self-perfection is the task of drawing near to God by emulating God's attributes, and emulating divine compassion consists in taking a personal concern for those outside one's kinship group. Laboring at the task of self-perfection is thus working on myself to become a "Self [that] integrates the I and the Thou" (*EM*, 153). The ethical self is thus conceived in relational terms: "the ethical Self cannot exist as an I without a You" (*EM*, 152).

---

[79] There is much debate about how to interpret Hegel's account of an individual's developing autonomy relative to the historically actualized societal norms. I have already presented Cohen's account and critique of Hegel's attempt to close the gap between historically real selves and the transcendental ethical self. My aim here is explaining how Cohen's account of Maimonides' self-perfection differs from Hegel's and Aristotle's account of ethical development with its insistence that an individual must begin by relating herself to the divine attributes characteristic of the Jewish God.

More to the point, the idea of God must be conceived relationally as well, since it is only by recognizing oneself as standing before God, a God extending compassion and lovingkindness to the stranger, that one can engage in the work of creating and sustaining I-Thou relationships with the whole of humanity: "This ethical labor…is linked to the idea of God, as it is directed toward God; thus the formation of the ethical Self is inconceivable without relating to God" (*EM*, 148). Therefore, in contrast to Aristotle's God, the God of Maimonides and the Hebrew prophets is understood as intimately concerned about human beings and their welfare.

Notice too that the virtue considered central in creating and sustaining the ethical God-I-Thou relationship is humility. Humility teaches the individual that she should not seek to define or determine herself on the basis of any "empirical or historical individuality" but strictly in reference to the continual task of perfecting the God-I-Thou relation. It follows that she must "never yield to the proud-hearted feeling of power and achievement, or greatness, of her integrity or her complacency" (*EM*, 149). It follows further that she ought to conceive of humility as a divine attribute: "Humility is a divine attribute in order to ensure that it may ever remain a human virtue" (*EM*, 149). Maimonides considers humility so essential to ethical development that he adds it to the list of divine attributes (*EM*, 142). For Maimonides, the greatness ascribed to a divine nature does not preclude the ascription of a compassionate concern for human affairs and a humble response to human suffering, quite the contrary. For Maimonides, humility is "the grounds of God's greatness" and "the foundation of the idea of God" (*EM*, 142).

Cohen then rhetorically asks: "Do we also find this meaning in the god of Aristotle?" (*EM*, 149) The answer of course is no. The god of Aristotle does not concern itself with human affairs and is not moved by their suffering. Eudaeamonic bliss for a god and those with a divinelike philosophical nature consists in nothing other than the inactive contemplation of one's own thoughts. Aristotle's conception of a divine nature is thus party to the "irresistible delusion that the lone thinker in his eudaemonian bliss will mature most safely into Selfhood" (EM, 152). Maimonides, with his account of self-perfection as an inherently interrelational process, dispels this delusion, denies eudamemonia is the highest human end, and so dismisses Aristotelian ethics altogether.

Having argued that ethics constitutes the core of Maimonides' philosophy of Judaism and that Maimonidean and Aristotlelian ethics are irreconcilable, what reasons does Cohen offer for locating Maimonides within the critical idealist tradition? Remember an essential tenet of this tradition: ethics must be conceived as a cognitive inquiry its own right, distinct from any mathematized natural or social science. The science of the ethical ought is methodologically distinct from the sciences dealing with the empirical is, and knowledge produced by the former is axiologically superior to that produced by the latter. In other words, historically real selves and societies, as well as mathematico-empirical forms of inquiry, are ultimately answerable to a transcendent Good. Now consider Cohen's account of Maimonidean self-perfection. This process proceeds only if the empirical self pursues knowledge, namely, knowledge of God's ways or attributes. Moreover, this knowledge is distinct from the type of knowledge pursued within the natural and social sciences. As we've seen, self-perfection is a process of emulating divine attributes as they are revealed to the Jewish community and more fully explicated within authoritative Jewish literary sources. Collectively these attributes present a standard of social holiness that is always higher than our current lived reality. Maimonides thus maintains that moral development depends on progressing in knowledge, knowledge about an incorporeal, transcendent God and the knowledge required for better approximating a yet-to-be actualized messianic societal ideal. Given Cohen's position on the tenets of neo-Platonic and neo-Kantian critical idealism, he legitimately concludes that the intellectualism pervading throughout Maimonides' ethics is "genuinely, truly, and perennially Platonic" (*EM*, 189).

By making knowledge of God the centerpiece of Maimonides' Jewish ethics and a necessary propaedeutic for a pure philosophical ethic, Cohen is forced to address the following conundrum. How can one reconcile the claim that knowledge of God is "the task and telos of religion, and consequently of ethics" with Maimonides' teaching on divine attributes as negative attributes? Cohen puts the problem thus:

> How do you intend to know God, and how can you comprehend God as the object of cognitive endeavor, if you denude this object of all attributes, by virtue of which alone we can attain knowledge of any object? (*EM*, 67).

This is not just a problem in terms of rendering Maimonides' philosophy internally consistent but for Cohen's critical idealism generally. In *Critique of Pure Reason,* Kant famously declared that he intended to limit knowledge to make room for faith. By analyzing the conditions under which an object could become an object of human cognition, Kant delimited the domain of theoretically knowable objects to those with attributes (properties and relations) definable in terms of mathematically quantifiable, spatiotemporal, naturally causal, and empirically realizable determinants. Given these restrictions on what counts as a possible object of human cognition, Kant explicitly excludes God and divine attributes from the domain of the theoretically knowable. To be sure, he then argues that a belief in God and certain divine attributes is rationally permissible and even necessary to make sense of moral experience and to ground our use of practical reason. However, many of Kant's contemporaries and predecessors persuasively argued that Kant's God ended up being too far removed from actual lived experience to be of any genuine use in directing our moral efforts. As a result, nineteenth century German scholars rigorously debated whether Kant's Copernican Revolution warranted a wholesale atheistic perspective. Hegel tried to block this inference with his account of the divine Absolute being made manifest in world history, but as we saw Cohen rejected this account as a pantheistic metaphysics incapable of recovering the requisite normative necessity. In sum, Cohen owes his audience an explanation as to how knowledge of God is possible in light of his reading and endorsement of Maimonidean religious ethics and Kantian epistemology.

Cohen seeks to settle this explanatory debt with a close reading of Maimonides' teaching on divine attributes as negative attributes. He claims that it would be a mistake to interpret Maimonides as denying that anything can be known about divine attributes, since the biblical narrative testifies that God allowed his goodness to pass before Moses, and Maimonides follows the tradition in identifying this goodness with thirteen attributes. What sort of attributes are these? Maimonides answers by classifying them as actional attributes (*EM*, 69). Unlike a proper Kantian concept or a schematized pure concept, actional attributes are not correlated to any spatial or temporal properties: "The attributes revealed do not portray God according to the categories of space and time" (*EM*, 69). In fact, they are not proper concepts or predicates at all. Recall that for Kant and for Frege concepts are essentially predicative. They must be defined in such a way that the question 'does this object fall under it?' has a definitive answer.[80] For Cohen and for Cohen's Maimonides, divine actional attributes cannot serve as predicates in the logical sense because God cannot be treated as one among a number of possible objects that may or may not fall under a concept:

> [W]hat is called a characteristic of God is not a characteristic in the logical sense, but only in the ethical sense. The attribute is not in a logical relation to the substance of God, but rather in an ethical relation to the substance of man (*RoR*, 426).

In other words, the relationship that God bears to God's attributes is not akin to the logical relation '*x* falls under *F*' or to the ontological relationship that an object bears to its properties. Instead, divine attributes are best conceived as ethical directives or normative models that a human agent should emulate in the process of becoming an ethical self (*EM*, 71; *RoR*, 95). By classifying divine attributes as ethico-religious norms for human moral development and not as proper predicates in logical sense, Cohen retains the Kantian dictum that knowledge of God, to whatever extent possible, should not be confused with the knowledge produced in the logico-mathematical pure or empirical sciences.

Cohen contends that Maimonides delimits human cognition of divine attributes such that God can be conceived "solely and exclusively as an ethical being" (*EM*, 69). On this reading, Maimonides presages the Kantian insight that knowledge of God satisfies the interest of practical not theoretical reason. His teaching on negative attributes is not intended to rule out knowledge of God per se, but aims at securing the distinctive ethico-religious import of this knowledge. By restricting the divine attributes to those revealed by God to humankind and insisting that they cannot serve as predicates in simple affirmative judgments, Maimonides promulgates prophetic messianism and its understanding of God's holiness as the unique, transcendent normative grounding for human thought and action: "[God's] holiness is effected through action, which man has to accomplish" (*RoR*, 110). Therefore, what Maimonides classifies as divine actional attributes, Cohen identifies as human virtues properly so called: "God's holiness is identical with God's uniqueness. His attributes, however, become concepts of virtue for man" (*RoR*, 403). For these reasons, Cohen suggests that Maimonides' use of the phrase '*negative attribute* should be interpreted along the lines of Plato's '*non-foundation*' or '*non-hypothesis.*' In both cases, the terms are not meant to convey that reason has no cognitive access to divine compassion, forgiveness, truth, humility and the like or to the Platonic idea

---

[80] See Sects. 2.1.2 and 4.3.1.

of the Good. Instead, the negations signal that these attributes and the Good function as regulative ideas constituting the ultimate teleological end for reason's use (*EM*, 87).

To the extent Maimonides claims knowledge of the attributes *is* humanly impossible, these remarks are directed at colleagues who while affirming that "one definition cannot at once contain divine and nondivine attributes," undermine this commitment by their indiscriminate use of propositions taking 'God' as the subject term (Maimonides as quoted in *EM*, 89). According to Cohen, Maimonides' doctrine on negative attributes is best read as a methodological precept on what types of propositions can and cannot treat God as a logical subject: "Instead of saying that Maimonides advocates the doctrine of negative attributes, we ought to say that he admits only those negative attributes that imply the negation of a privative attribute" (EM, 98). Privative attributes, or better, privative judgments are judgments that ascribe to a subject the absence of a property that one would expect the subject to bear. So, for example, 'The man is blind' is a privative judgment because it predicates blindness of a subject, i.e. a man, for whom sightedness is an expected property or characteristic. Similarly, 'The dog is blind' is a privative judgment whereas 'My pen is blind' is not. When it comes to the relationship between God and humankind, the relevant privations or deficiencies are moral ones. In relating ourselves to God's holiness, we recognize just how deficient we are in manifesting those moral dispositions and actions that we and the categorical imperative rightly expect. God's holiness and attributes are thus best conceived as negations of human privations. Cohen and Cohen's Maimonides conclude that 'God' can only be the subject term in an infinite judgment where these privations are negated, judgments such as 'God is not unjust' or 'God is not inattentive to the poor and oppressed' (*EM*, 103).

Let me summarize how Cohen's reading of Maimonides serves to restore cognitive and cultural significance to the Jewish idea of God. First, Cohen explains how theological claims authorized by the tradition might be (re)interpreted such that they do not run afoul of at least one viable Kantian epistemological framework, namely, the one articulated by the Marburg school. Second, he explains how knowledge of God as characterized within the Hebrew Bible and the works of Maimonides, one of the tradition's most preeminent scholars, supplies what is missing in a properly philosophical ethics. Knowledge of God, as Cohen presents it, is not knowledge of a given object with properties that eschew detection using the methods of natural science. Knowledge of God is knowledge of an entirely different sort than that pursued and produced within the mathematized sciences. Knowledge of God as pursued and produced by means of Jewish thought and practice also differs from ethical knowledge in the strictly philosophical sense.

Ethico-religious knowledge of God contains interpersonal and affective elements that are methodologically excluded from a strictly philosophical account of the moral good. According to Cohen, a philosophical ethical inquiry must disregard any particularities related to an individual's actual situation, e.g. their embodied, societal situation:

> Ethics, in its systematic opposition to everything sensual and everything empirical in man, arrives at the great consequence that it must first tear away from man the *individuality* of his I, in order to return it to him from a higher pinnacle, and in a form not only higher, but also purified (*RoR*, 13).

To discover and validate ethical principles that are sufficiently universal and normatively necessary and to grant human persons the full and equal measure of dignity they deserve, the philosophical ethicist must bracket from consideration any psychophysiological or sociohistorical features of an individual. As we saw, Cohen denounces naturalized ethical theories because they violate this methodological stricture. However, what philosophical ethics gains in terms of universality and "scientific rigor" comes at the cost of speaking to individuals within their particular circumstances (*RoR*, 21). Cohen describes this feature of a pure philosophical ethics as a form of methodologically induced "indifference" (*RoR*, 20). Ethical knowledge, which conceptualizes human selves as autonomous rational moral agents, with all the rights and responsibilities this entails, and conceptualizes God strictly as the moral law-giver or guarantor, can show no interest in the "outward success or failure of moral duty" (*RoR*, 20). Therefore, philosophical ethical knowledge in and of itself cannot enable actual human beings to make empirically notable progress in carrying out their duties and in becoming the fully developed autonomous ethical self that it takes as an initial posit. There is where the ethico-religious knowledge of the Jewish God is shown to be indispensable. For it was within Judaism that God was first conceptualized as the One who has compassion for the poor and the stranger and who offers lovingkindness, forgiveness, and a renewed beginning to the sinner. It is within Judaism that the I is conceived as standing in relation to this God and the Thou of her human fellows. In contrast to the abstract, disinterested moral knowledge of philosophical ethics, the knowledge of God pursued and produced within a religious tradition that is committed to prophetic messianism is a knowledge inherently related to love:

> Knowing God is loving God, and love of God is knowledge of God; these are the two vectors of one and the same fundamental idea which promoted Maimonides to the status of Israel's teacher par excellence (*EM*, 113–4).

This love is not sensual or erotic in nature and does it seek any sort of mystical substantive union with the divine nature (*EM*, 119). Rather, it is an "ethnicizing" love and knowledge of God that reinforces "the ethical character of human love" (*EM*, 193). This ethico-religious knowledge manifests itself in actual compassionate concern for the well-being of human others and for oneself, in specific, a concern to alleviate the oppression arising from our individual and collective privations. We can conclude that, for Cohen, progress in accumulating knowledge of God, as understood within Abrahamic monotheisms, is measured by how much wisdom and love practitioners exercise and how much headway is made in the continual effort of redressing human suffering and the infinite task of becoming ethical persons.

Finally, we can summarize Cohen's answer to the question posed in his *Ethics of Maimonides*: "what place is occupied by the autonomy of human reason, if we admit that its most precious achievement, the validity of ethics, is somehow predicated upon God?" (*EM*, 29). Throughout his exposition of Maimonides' account of self-perfection, Cohen stresses that this process is marked by increasing self-awareness, self-initiative, and self-determination on the part of the human agent herself. God does not make us holy, but sets holiness and the related attributes before us as a task. It is up to the agent herself to decide that this is *her* task, a task to which she must commit herself in order to become an ethical person (*EM*, 140). This decision further depends on her own self-assessment of the current state of her moral dispositions and the dispositions of the collective bodies of which she is a member. Self-perfection does not proceed by way of an agent's merely conforming herself to the dictates of traditional authority or prevailing conventional norms. Instead, it rests "solely upon one's critical awareness" and exercising the agency required to further the messianic ideal that "all people without exception…gain unhampered access to cognitive advancement" (*EM*, 190; 188).

As is no doubt evident to my reader, I take Cohen's position on the essential epistemic contribution of Jewish philosophy to a Kantian ethic to warrant our philosophical interest on its own terms. That said, the stated aim of this chapter is learning what we can about wisely responding to epistemic crises as they occur in scientific and religious traditions. Section 5.1 ended with a list of the tactical rules gleaned from the debate between Helmholtz, Cohen, and Frege in light of van Fraassen's and Friedman's criteria for satisfying prospective rationality. I end Sect. 5.2 with some additional rules taken from Cohen's response to the epistemic crisis engulfing his religious community.

5.2.3 More Rules for Resolving Epistemic Crises

1. Epistemic oppression and the need to redress it can legitimate a Kuhnian-scale modification of a tradition's fundamental concepts and principles

   When discussing the factors resulting in an epistemic crisis, van Fraassen and Friedman focus on increasing anomalous observations or results that call into question the prevailing theory's explanatory scope or fruitfulness. This is not surprising given that Kuhn stressed the importance of resistant anomalies in exposing the need for a large-scale paradigm change. Most of the examples of Kuhnian revolutions discussed in the history and philosophy of science literature—the shift from Newtonian to Einsteinian mechanics and the adoption of non-Euclidean geometries—were initiated, at least in part, by persistent anomalies within the domain of inquiry or subject-matter of relevant discipline. In short, much of the discussion on the causal factors leading up to an epistemic crisis centers on domain-specific inexplicabilities like the perihelion of Mercury or the sensible intuitability of non-euclidean space, not on the social dynamics affecting the composition of a discipline's practitioners.

   In contrast, my exposition of the crisis confronting Cohen's religious community focuses on the causal role of epistemic exclusions, namely, what people group and which attitudinal or behavioral dispositions are societally marked as 'less rational' or 'unscientific.' Based on my account, Cohen's diagnosis and response to the crisis confirms what feminist epistemologists have long-observed: epistemic marginalizations and correcting for them can expose the need for a paradigm change. Numerous case studies demonstrate that androcentric biases were uncovered once women entered certain fields. This resulted in substantial modification to the norms governing their practice: "[T]he entry of women and feminist scholars into different academic disciplines, especially in biology and the social sciences, has generated new questions, theories, and methods" (Anderson (2017)). Similarly, in Cohen's day, an increasingly overt anti-Semitism and the marginalization of any attitude deemed 'unscientifically religious' created a situation where European Jewish intellectuals were threatened with increasing exclusion from the sciences. Cohen fears that if the situation doesn't improve, the Jewish intellectual tradition will come to a standstill or die on the vine through neglect or outright hostile suppression. In response, he constructs a new lens for viewing the history of European cultural achievements and the crucial contributions of Judaism. Cohen modifies the hermeneutical framework through which European and Jewish thought is understood and does so because of a well-founded concern that religiously observant Jews are being systematically stigmatized as unacceptable practitioners of any scientifically rigorous discipline.

If Cohen's example and the examples cited by feminist philosophers of science are not enough to convince us that redressing or preventing epistemic oppression can and should trigger radical revisions in scientific or religious praxis, consider Kristi Dotson's description of second-order oppressions. Dotson offers a helpful classification of oppressions based on the order of change needed to address them. As she and Miranda Fricker remind us, knowledge-production, mutual understanding, and successful communication doesn't happen in a vacuum. Members of communities, institutions, and entire societies draw upon a common pool of hermeneutical and epistemic resources to understand experiences and render them "communicatively intelligible" (Fricker (2007), 162). These resources may be explicit and formalized as in the case of the lexicons and taxonomies employed by healthcare professionals or more implicit and informal like the tacitly understood language games and vocabularies characterizing differing forms of life. Demonstrated competency as a knowledge-producer or transmitter, as well as successful communicative or dialogical exchange, depends on an agent's ability to effectively and persuasively utilize these resources.

For Dotson, first-order oppressions are ones that can be addressed without having to modify these resources, and she identifies the testimonial injustice described by Fricker as falling within this class. According to Fricker, testimonial injustice occurs when speakers are perceived as less credible due structural identity prejudice (Fricker (2007), 29). This is a historically long-standing and systemic prejudice whereby an entire people group is viewed as rationally subpar or unreliable conveyors of knowledge. Individuals identified as members of this group are thus saddled with an unwarranted credibility deficit. Consequently, their testimony is not given the weight or credence it deserves. Fricker cites the discrediting or devaluing of black people's testimony within U.S. and U.K. criminal justice systems as paradigmatic examples of testimonial injustice. Since, in principle, this type of injustice might be addressed by training hearers to correct for these deficits when listening to a speaker identified with a group typically subject to structural identity prejudice and without having to modify the hermeneutical or epistemic resources utilized by the hearer, testimonial injustice counts as a first-order epistemic oppression.

Second-order oppressions, on the other hand, can only be addressed by modifying the resources, and here Dotson references Fricker's account of hermeneutical injustice. Fricker defines hermeneutical injustice as: *the injustice of having some significant area of one's social experience obscured from collective understanding owing to a structural identity prejudice in the collective hermeneutical resource* (Fricker (2007), 155 italics in the original). In this case, the injustice is not perpetuated by an agent per se, but by the resource being deployed to understand someone's experience or effectively communicate it. Victims of structural identity prejudice are prone to hermeneutical marginalization, meaning they are excluded from the places and occupations where collective epistemic resources are developed and gain institutional or cultural traction. As a result, these resources often contain prejudicial gaps or biases, making it difficult to understand or share certain experiences. Fricker cites the difficulty American women had in fully comprehending and effectively communicating their experiences of post-partum depression or sexual harassment before these concepts became a familiar part of our cultural lexicon (Ibid., 148–51). Dotson cites the inability of women of color to successfully communicate their experiences of sexism to White/Anglo feminists (Dotson (2014), 128). Because addressing hermeneutical injustice requires substantially altering the unwarranted prejudicial resource, it falls in the class of second order oppressions. Dotson follows Rae Langton in arguing that combatting hermeneutical injustice requires "a conceptual revolution" (Dotson (2012), 31).

To offer a prospectively rational resolution to a crisis confronting one's religious or scientific community it helps to have a complete diagnosis of the problems one is trying to address. I am arguing that epistemic oppressions of various orders may be part of that diagnosis, which then warrants a Kuhnian-scale revolution as the appropriate remedy.

2. Recognize the ethical and epistemic significance of spiritual emotions like compassion.

Recall van Fraassen's claim that a change in one's emotional state can initiate a Gestalt-like shift in our attention, which in turn can make a hitherto 'absurd' way of conceiving things appear more reasonable. He thus calls on the emotions to do the work of explaining how the move to a new paradigm or framework can be seen as a rational conversion by the one taking the leap and by the philosopher studying this recurring feature of our epistemic lives (van Fraassen (2002), 103–108). My objection at the time was that van Fraassen had not said enough about how an emotional state could account for a dialogically-mediated and communal leap to a successor framework. I now contend that Cohen's treatment of compassion as a divine actional attribute goes a long way in meeting this objection.

As van Fraassen himself notes, a dominant trend within modern analytic epistemology is discounting the epistemic worth of the emotions because they are viewed as too dependent on an individual's personal interests and too context-specific to possess the kind of inter-subjective accessibility and invariance that compelling reasons should have (Ibid., 101). In Kantian terms, the emotions are viewed as strictly subjective perceptions and thus incapable of grounding an inter-subjective and objectively valid judgement. Given this understanding of the emotions, one could agree with van Fraassen that being in a certain emotional state might give *me* a reason to see the move to a new paradigm as rationally permissible, but it is hard to see how talking about my emotional state or the state of another could count as a reason for *you* to make the move. How can emotional states, understood as strictly subjective states, be invoked as reasons in communicative exchange about taking the leap to a successor paradigm? The answer is they cannot. It was precisely because Kant viewed the emotions as varying from person to person and not subject to rational volition that he rejected them as grounds for moral judgment.

Cohen, on the other hand, treats compassion as falling within a special class of distinctly human or "spiritual" emotions (*RH*, 71). These are emotions everyone is expected to possess or cultivate. Underdeveloped spiritual emotions or lacking them entirely may be states towards which we are naturally predisposed, but such states constitute a human privation nonetheless. Spiritual emotions contain cognitive and affective elements that encourage persistence in our ethical labors: "[C]ompassion… furnishes man with whatever strength he needs to fight the skepticism his so-called mind produces" (*RH*, 72). Unlike the affective psychophysiological states we share with other animals, these emotions are inherently related to exercising our unique capacity for practical reasoning. Only by means of a rationally-directed effort to emulate God's compassion, the compassion initially described by the Hebrew prophets, can one come to learn the truth on which the practical efficacy of Kantian ethics depends: that every human person, including the stranger, is my fellow (*RH*, 72).

Based on Cohen's account of spiritual emotions, it is easier to see how they and the claims referencing them might serve as reasons in a trans-paradigm dialogical exchange. Recall the critique of Cohen and others against a formal ethics like Kant's, namely, that abstract moral imperatives cannot sufficiently motivate and guide actual embodied individuals in performing their ethical task. Compassion and other emotion-filled actional attributes become corollaries to pure moral imperatives insofar as they enable these imperatives to become efficacious in directing our lived experience. It is not too much of a stretch to think of spiritual emotions and the truths they disclose as the ethical analogues to synthetic a priori mathematico-physical principles. Just as these latter principles enable us to apply pure abstract mathematical concepts and propositions to particular empirical experiences, so too affective actional attributes facilitate the application of formal moral imperatives to our embodied lived experience. If asserting moral principles like the categorical imperative can function as an inter-subjectively valid reason in a dialogical exchange, then the same holds for asserting the corollary ethical bridging principle. It follows that reports of compassion-inducing experiences and injunctions to compassion can function as inter-subjectively valid reasons when urging one's peers to take the leap towards a new way of conceiving things.

Since compassion provides access to truths necessary in striving towards Kantian and messianic ethical ideals and the emotional drive to overcome obstacles getting in our way, Cohen claims it can be a "motivating force of an entire *Weltanshauung*" (*RH,* 71–2). In other words, compassion can initiate a shift in worldview. Similar then to van Fraassen's description of Sartrean emotions, compassion and the other spiritual emotions have a cognitive aspect capable of directing our attention to previously neglected features of experience such that the need to render them understandable or communicatively intelligible can no longer be ignored. Cohen goes further, however, in explaining how this class of emotions can play a crucial role in the trans-paradigm rational communication that Friedman describes. In sum, Cohen presents a more fine-grained account of how the occurrence of certain emotional states might factor into the causal explanation for a rationally permissible, collective conversion to a successor paradigm or novel worldview.

3. <u>Pursuing the goals of Abrahamic religious traditions may well require drawing from epistemic resources lying outside of these traditions.</u>

Though Cohen maintains that cultivating compassion is necessary for ethical progress, he correctly notes it is not sufficient:

[C]ompassion alone is not an adequate motivation for social action. In the final analysis, our moral problems will be assured of solution only if we apply to them whatever new insights we can derive from the sources of science, sources that must be continuously deepened and freshened (*RH,* 73).

To see the force behind these claims, consider the ethical ends posited in critical idealism and messianism. According to Cohen, Kant's categorical imperative and kingdom of ends point to a cosmopolitan world order of perpetual peace, where each individual irrespective of societal position, ethnicity, or national identity is granted the respect, dignity and right to autonomy accorded to humanity as such. Cohen's efforts in fighting for universal suffrage, universal education, and opening national borders to refugees provide further insight into how he conceives of this sociopolitical ideal and the means of progressing toward it. This Kantian idea of a genuinely humane world order finds it source in the Jewish tradition, the visionary social critique of the Hebrew prophets and Maimonides' account of self-perfection and the world-to-come. The prophets and Maimonides go further, however, in connecting the formal ethics of critical idealism to an individual's lived experience and the necessity of religious experience. On their account, the fully developed human Self possesses the full complement of intellectual and moral virtues and spiritual emotions that exemplify the way of divine holiness. This is a Self embedded in I-Thou-God relationships, which are in turn based on an interpersonal, compassion-filled knowledge that has borne and alleviated human suffering. This suffering is caused, in large measure, by societally-induced oppression, including unwarranted cultural and political marginalization. The messianic-Kantian ideal towards which we must perpetually strive is nothing less than the concrete actualization of a world free of societal oppression and human privations.

Now consider the sociopolitical state of affairs in Cohen's time and in ours. Cohen writes: "The present…with its unfilled needs, unnatural oppression, glaring injustices, vain conceits, and bragging untruths deserves no pity" (RH, 119). As in Cohen's day, the societal ills and human privations ailing us are numerous and varied. With over a hundred years of hindsight, we are in a better position to appreciate the accuracy of Cohen's description of the work required in pursuit of the world as it ought to be. This ethical task is indeed a perpetual one, demanding "ever-innovative rejuvenation" and conscientious practitioners in science, healthcare, jurisprudence, philosophical and religious ethics, and new domains of inquiry that Cohen and his contemporaries could hardly have predicted (*EM,* 15). Given the increasing complexity of our social and natural environments, we know better than they that "new questions" concerned with making infinitesimal progress will in fact be "posed unceasingly" and with "every new solution" comes "new challenges" (*EM,* 15).

Similar to Cohen, Christian philosopher and Episcopalian priest Marilyn McCord Adams argues that combatting oppression is a primary task of Abrahamic religious traditions. She too contends that this task is never-ending because of the natural and societal privations to which humans are subject:

> [T]he human race is socially challenged, neither smart enough nor good enough to organize utopia. Human social arrangements always spawn systemic evils, structures of cruelty that privilege some by degrading others (Adams (2014) 66).

Suffice it to say I endorse Cohen's and Adams' position on the nature of the task bequeathed to Abrahamic religious traditions.[81] I further maintain there is enough historical evidence to warrant Cohen's and Adam's shared stance that practitioners of these traditions should expect to revise their concept of God and other fundamental concepts, particularly in regards to their ethical import. As Adams' writes in regards to the Episcopal Church ('TEC'): "The TEC expects the church to keep growing in its grasp of God's vision of the ideal society, of what it is to love God and neighbor" (Ibid., 66). Finally, there is ample empirical evidence that this increased understanding and knowledge-production will be generated, in part, by cultural practices intimately related to but nevertheless distinct from religious ones.

---

[81] My argument that Cohen and Adams share the same position on the nature of the ethical task to which Abrahamic monotheism is committed should not be taken to imply that they agree on all points. More specifically, they disagree on the theological resources available for pursuing this work. As an Episcopalian, Adams can and does draw upon Christological resources such as Jesus the incarnate Son of God, which Cohen explicitly rejects on philosophical and theological grounds.

Here is the place to address an obvious objection to Cohen's account of the mutually constructive relationship between German and Jewish ways of being in the world. Cohen's 'pitiless' assessment of present circumstances that I cited above was written in 1901, seven years prior to the publication of *Ethics of Maimonides* and thirteen years prior to WWI, a war Cohen defended as a just war and a step toward perpetual peace. Cohen died in 1918. Soon thereafter, the younger generation of German Jewish scholars dismissed Cohen's arguments for the affinity between Kantian and Jewish ethical ideals as absurdly optimistic and naïve (Erlewine (2016), 14). Put succinctly, the objection of these scholars and one echoed by many postmodern or postcolonial theorists is the following. In light of the atrocities and genocides perpetuated throughout the twentieth century, often in the name of furthering an 'Enlightened' world order, why not ditch Cohen's Enlightenment-friendly philosophy of Judaism and the concomitant notion of progress altogether?

Though I cannot hope to provide a complete response to this objection, I want to offer my reader something. In brief, I agree with Poma (1997), Zank (2006), and Erlewine (2016) that such a dismissal of Cohen's philosophy comes too quick and fails to appreciate the sophistication of his culture critique, his nuanced notion of progress, and the specific sociopolitical winds that he is navigating against. Cohen consistently chastises his contemporaries for their complacency with respect to Kantian and Jewish ethical norms and offers a two-fold explanation for this complacency. First, distinctively Jewish ethical norms are disregarded because Jewish contributions to European culture are being generally disregarded. In addition to the political marginalization of European Jewish communities, Cohen insightfully diagnoses the symptoms of their hermeneutical marginalization. Throughout the nineteenth century, a metanarrative of Enlightenment progress was constructed that systematically diminished and erased contributions of Jewish texts and thinkers. As feminist and critical race theorists have long noted, hermeneutical marginalization and epistemic injustice are primary tools used within an oppressive system to justify political injustices towards marginalized groups. Rather than rejecting Kantian Enlightenment ideals tout court, Cohen does what Dotson and Fricker recommend one does when combatting second-order epistemic oppression: stage a conceptual revolution by rewriting the narrative or reconstructing a paradigm so as to include hitherto marginalized perspectives. Second, Cohen argues that complacency results from naturalistic philosophies reducing ethical or ethical-religious progress to empirically observable evolutions of biological or socio-political change. Throughout the works we have canvassed, Cohen consistently decries any suggestion that progress as understood within the critical idealist tradition is equivalent or reducible to the developmental processes studied within empirical natural and social sciences or within a pre-critical metaphysics like Spinoza's or Hegel's. He also continually speaks out against the xenophobic winds billowing the sails of European nationalistic identities. For Cohen, Kant's categorical imperative presupposes an idea of humanity initially articulated by the Hebrew prophets, with their emphasis on God's love for the stranger. Therefore, any emphasis on forming or maintaining strong national borders and strong national or ethno-national identities is at best a distraction, if not in outright opposition, to progressively pursuing the goals of

Kant's Enlightenment ethic and prophetic messianism. For this reason, Cohen takes a stand not only against the anti-Semitic and nativist elements within German culture, but also against those arguing for a strong national Jewish identity and establishing a Jewish nation state. Whatever charges can be legitimately brought against Cohen's notion of progress, we must first acknowledge that it is not identical to the ethnocentric, nationalistic conception of progress driving many nineteenth and twentieth century colonizing projects.

In hindsight, we may side with Martin Buber and other turn of the twentieth century Zionists that Cohen's rejection of a Jewish nation state failed to recognize the threat posed by anti-Semitism. That said, I agree with Erlewine et al. that to follow his early critics and utterly dismiss Cohen's attempt to reconcile Kantian precepts and the commitments of a particular Abrahamic faith tradition is both anachronistic and hinders us from seeing what we can learn from Cohen about navigating the sociopolitical winds, philosophical trends, and epistemic crises of our time.

4. Practice Cohen's trans-paradigm hermeneutic, where a 'prospectively rational' interpretation of a tradition's texts and practices is one advancing the ideals of critical idealism and prophetic messianism.

In response to the previous objection, I argued that Cohenian progress should not be confused with progress in advancing the goals of one's nation state or the developmental processes studied by nineteenth and twentieth century biologists and social scientists. I also underscored the perspective of Cohen and Adams that human privation being what it is, the path toward approximating the prophetic vision of alleviating human suffering will not be a straightforward linear progression. The best we can hope for is taking three steps forward for every two steps back. Why then talk about progress at all and how does Cohen conceive of it?

Cohen conceives of progress as a transcendent notion, that is, as a regulative ideal derived from the conception of science in the broad sense. Just as the Good functions as a teleological end for human thought and behavior, so too with the idea of Scientific Knowledge. Scientific Knowledge functions as the limit of an "infinite trajectory" where knowledge produced by a variety of methodologically sound, self-correcting fields of inquiry is integrated into a "rational systematic whole" (Zank (2006), 2). As we know, Cohen treats philosophical ethics and the Jewish intellectual tradition as scientific fields of inquiry with knowledge that must be incorporated into this whole. Progress as conceived by Cohen thus serves as a methodological norm demanding that practitioners of a science, broadly construed, adopt methods and modify their practice with an eye toward creating a "unity of cultural consciousness" informed by distinct "fields of cultural productivity" (Zank (2006), 3). In keeping with this norm of scientific progress, practitioners cannot turn a deaf ear or blind eye to the knowledge produced in fields lying outside of their own. Furthermore, given the axiological superiority that Cohen's critical tradition assigns to knowledge of the Good, practitioners must also be attentive to how their field of inquiry impacts the suffering arising from human privation and societal oppression.

How does Cohen's account of scientific progress compare to Friedman's and van Fraassen's? Similar to Friedman, scientific progress in Cohen's sense is compatible with, and may necessitate, having to retrace one's steps and start down a new path. It too is best imagined as having a branching tree structure, with some permissibly rational routes left unexplored and containing crossroads where one must choose how to revise fundamental constitutive principles that have guided scientific, philosophical, or religious practice heretofore. Unlike Friedman, Cohen invokes a broad notion of science that includes not only mathematico-empirical natural and social science, but also ethics, jurisprudence or philosophy of law, and religion. Because Cohen treats mathematized empirical science, philosophical ethics, and Judaism as methodologically distinct but interrelated scientific endeavors, he can provide a more complete explanation for the admiration that Friedman and van Fraassen express for science in the narrow sense. According to van Fraassen and Friedman, western modern science deserves our admiration not simply because it delivers world-directed theoretic models of increasing explanatory fruitfulness and predictive success but because it is composed of epistemic communities displaying certain intellectual virtues. They commend scientific communities for exhibiting the proper blend of intellectual humility and steadfastness. For van Fraassen, a community wisely applies the rule of *Sola Experientia* if it knows when the rule should be used for the purposes of preserving the inherited framework against unwarranted heterodox alternatives and when it should be used for the purpose of discarding this framework for one better tailored to serve consensually negotiated communal aims. The empirical stance that van Fraassen enjoins respects scientific practice for its typically wise application of this rule (van Fraassen (2002), 141–142). For Friedman, striving for a non-coercive dialogical means of arriving at consensus amidst dissenting opinions is the earmark of communicative rationality. Modern science counts as an exemplar of Enlightenment rationality because it typically bears this mark (Friedman (2001), 59). However, neither van Fraassen nor Friedman explain why a practitioner of natural science should feel obliged to these virtues. What is it about the practice of science per se that obligates them and their respective communities to exemplify these virtues and others deriving from an Enlightenment cultural tradition? Cohen's account of scientific progress as the accumulation of knowledge from a variety of cultural practices, all of which are teleologically oriented toward approximating a Kantian kingdom of ends and Maimonidean actional attributes of divine holiness, supplies this explanation.

Given Cohen's account of scientific progress and what we have seen of his practice, what interpretative principles should guide a proposed resolution to a crisis occurring within an Abrahamic tradition? Cohen's hermeneutical approach is best described as a combination of Kant's transcendental method and the midrashic hermeneutics of rabbinic Judaism (Zank (2006), 5). As discussed in Chap. 3, Cohen's Marburg school equates Kantian epistemology with a particular use of the transcendental method. The critical idealist philosopher begins with a given collection of 'facts.' These facts are the body of propositional truths concerning the subject matter of a well-circumscribed discipline. The goal is articulating the a priori conditions securing the objectivity and inter-subjective validity of these facts.

Kant's mistake, according to Cohen, was failing to apply this method to other fields of cultural productivity. Cohen's *Ethics of Pure Will* corrects this mistake in the field of ethics by starting with the facts constituting the history of jurisprudence. When it comes to furthering the science of Judaism, he begins with facts that include the historical practice of Jewish liturgies, the writings of the biblical authors, Maimonides' *Guide for the Perplexed,* and the entire corpus of rabbinic teaching and Jewish biblical commentary. By treating these facts as data or explanandum, the objective cognitive content of which must be secured and explained, the critical idealist philosopher of religion remains tethered to the particularities of an actual religious practice, which keeps her from unwarranted metaphysical speculation.

Properly applying the transcendental method also ensures that novel trans-paradigm interpretations bear the right sort of connection to the past deliverances of a religious practice. According to Cohen, retaining this connection is a problem that the prophet Ezekiel confronted and that confronts any conscientious religionist whose tradition is in the midst of transition: "Everywhere the question arises whether the old idea one fights in a traditional institution should be entirely rejected and eliminated or whether it is the case that a new idea seeks a reconciliation with the old institution" (*RoR,* 175). Rather than entirely rejecting the ancient liturgical institution of Jewish sacrifice, Ezekiel transforms its significance by interpreting it through the lens of a "new idea" that deepens our understanding of its predecessor (Ibid.). Cohen clearly thinks the Ezekiel method of reforming liturgical practice is preferable and closely resembles the transcendental method as applied to religious practice. Continuity, he says, is "the methodological signpost" against which "true" progress must be measured (*RoR,* 177). There is no means of determining whether a novel interpretation of a religious tradition's body of facts constitutes progress unless the continuity between the new and old ideas grounding their significance can be established. Applying the transcendental method to an Abrahamic religious tradition in crisis implies a good faith effort to satisfy Friedman's criterion of trans-paradigm communicative rationality. In other words, Cohen and Cohen's Ezekiel take on the burden of explaining how a new idea radically altering the meaning ascribed to certain liturgies, biblical texts, or the works of religious authorities can be seen as "a natural continuation of the old ones" (Friedman (2001), 101).

While applying the transcendental method demands careful attention to the actual historical proceedings of a scientific or religious tradition and Cohen's works on the history and philosophy of Judaism demonstrate this care, Cohen does not think that historical exactitude alone reveals the cognitive content or rationally-binding significance of a religious praxis. His philosophical hermeneutic for extracting the reasonable meaning of the scriptures, liturgies, and other religious 'facts' is in no way committed to a *Sola Scriptura* rule that assigns meaning based on a prima facie reading of text or tradition's tacitly understood rules of interpretation. He would also reject Christian philosopher Richard Swinburne's hermeneutical principle. Swinburne subscribes to what I call a strong tradition interpretive rule. If in the history of a religious tradition, e.g. Christianity, the interpretation of a biblical passage, a church teaching, a sacramental practice has enjoyed unanimous consensus by all authoritative ecclesial bodies for a significant period of time, this interpretation is the uniquely epistemically warranted one and cannot be abandoned or revised (Swinburne (2007), 180–181). In contrast, Cohen's application of the transcendental method does not allow the history of a religious tradition's assertions, even if those assertions are univocal for a significant period of time, to fix the cognitively significant and rationally permissible meaning of religious texts and practices in perpetuity. Remember that the transcendental method secures the objective and inter-subjectively valid meaning of the facts of natural science by interpreting these facts in accordance with a priori mathematical concepts. Cohen allows that collection of a priori mathematical concepts may be revised and the meaning of these facts rationally reconstructed so long as continuous progress toward the goal of Scientific Knowledge can be maintained. Cohen takes a similar approach when interpreting the 'facts' of the history of philosophy and the history of Jewish thought and practice. The meaning of these facts may be rationally reconstructed for the purposes of incrementally and continuously progressing towards the goal of the Good or the prophetic messianic ideal. Bruckstein characterizes Cohen's approach as a "messianic epistemology" where the philosopher is obliged to proffer an interpretation of tradition's authoritative texts, thinkers, and liturgies that is more beholden to the Kantian and messianic "anticipation of human freedom" than to the norms of "historic objectivity" (Bruckstein (2006), 125).

In what sense is Cohen's hermeneutical approach midrashic? For the midrashic interpreter, the sacredness of a book resides in its ability to be the source of multiple interpretations, all of which are valid insofar as they are capable of actualizing goodness: "[T]heir sanctity consists in their contribution to the progressive actuation of being or goodness" (Zank (2006), 5). Recall that Cohen replaces a substance-based ontology with an ontology of norms and values. Full and complete being just is the being of the ought. Moreover, this normative way of being is the idea of the transcendent Good and the idea of the unique God, which are the ideas containing the regulative ethical ideals of critical idealism and messianism. These ideals are ones toward which we and our respective fields of inquiry or cultural production must perpetually strive. Zank concludes that, for Cohen, the history of any methodological sound, cognitive cultural endeavors must be approached as a sacred book. This implies in turn that the philosophical interpreter and commentator assumes authorial responsibility for ascribing a meaning to the histories or 'facts' of these endeavors that discloses their capacity to actuate human goodness and alleviate suffering. Zank describes the midrashist philosopher as "creative *and* humble, innovative *and* dependent, always indebted to the possibilities of meaning inherent in the notion of a sacred text" (Zank (2006), 4).

Based on Zank's description, I agree that Cohen's trans-paradigm hermeneutic is appropriately called midrashic. His humility and dependency manifests itself in the due attention paid to the interpretations of texts and thinkers constituting the canonical history of his religious tradition. And by taking responsibility for authoring an alternative reading of this history more capable of actualizing goodness within his particular cultural and political milieu, Cohen exemplifies properly pious creativity and innovation. So with these strategic rules in my pocket and Cohen as an exemplar, I finally turn to the epistemic crisis engulfing my own religious community.

5.3   Test Case – A Christian Community in Crisis

Van Fraassen is right that the history of Christianity begins in crisis. First century Jewish communities like Paul's were confronting questions about self-definition and the future trajectory of their faith tradition (see Boyarin (1994)). Cohen is also right that practitioners of religious traditions are sometimes forced to confront the question: can an institution handed down to us by our forebearers be positively transformed by the incorporation of new ideas or will an attempted synthesis result in its destruction. In this final section of the book, I argue that the United Methodist Church ('UMC' hereafter) is facing this question with respect to its institution of marriage, as are many Protestant Christian churches and denominations active in the United States, and that it is a question of Kuhnian significance. I further argue that a paradigm shift is warranted because of what we are learning from biomedical science about sex and gender development and because of the need to redress the epistemic oppression suffered by those who are atypically embodied.[82] I end by briefly sketching how one might go about offering a prospectively rational redefinition of Methodist marriage.

5.3.1   Traditional Methodist Marriage and the Call for Change

To show that the debate over redefining marriage is of Kuhnian proportions, I need to show that the UMC is governed by a paradigm and that the current definition of marriage functions as a fundamental concept within it. Throughout the book, the notion of a paradigm has been described in a variety of ways. For Friedman, a paradigm is a set of rules or principles, the validity of which all practitioners of a well-circumscribed scientific inquiry accept and which allow them to engage in dissent while holding onto the confidence that a consensual resolution can be reached (Friedman (2001), 55). Van Fraassen, Dotson, Fricker, and I have used the terms 'paradigm,' 'conceptual framework,' 'epistemic resource,' and 'hermeneutic resource' interchangeably with the intent of showing that the sciences are not the only cultural endeavors governed by a relative set of a priori concepts and principles that determine what qualifies as a rational way of understanding our natural and social worlds and a reasonable way of settling disputes. Recognizing that various cultural practices and indeed entire cultural traditions are governed by a paradigm in this broader sense helps us better account for changes in an individual's or collective's doxastic attitudes that, phenomenologically speaking, are different in kind from the ordinary, non-traumatic revisions of belief that do not threaten self-identities or prevailing worldviews. My claim then is that the debate over Methodist

---

[82] For more on my arguments that the oppressions experienced by intersex people within biomedical and religious epistemic communities warrant Kuhnian-scale revolutions within those communities, see Merrick (2019) and (2020).

marriage calls into question the validity of a constitutive rule governing our tradition and that we are witnessing an increasing lack of confidence about reaching consensus, existential threats to certain identities, and the other traumatic elements one would expect in a trans-paradigm situation.

For the UMC and other North American Protestant communities of which I am a member, the inherited understanding of the Bible, the liturgies, and sacramental practices serve as core elements in the shared epistemic resource. Piety dictates that we draw from this resource to make collective and communicative sense of our lives. This conceptual and hermeneutical framework grounds our shared understanding of who we are and helps us discern the significance and function of each member relative to the flourishing of entire church body. Linda Zagzebski shares this conception of how the received interpretation of scripture and religious praxis functions in Christian communities. "Tradition," she writes, "is the memory of a community" (Zagzebski (2012), 191). Here 'tradition' refers to religious creeds and practices handed down from one generation to the next. For Christian communities, as for other Abrahamic communities, tradition includes oral and written teachings on proper biblical exegesis. The community draws upon this inherited pool of biblical, ethico-religious, and practical resources for the purpose of "training the emotions" and cultivating "a particular world view" (Ibid.). The tradition also contains an authoritative structure functioning as a watch guard over "the faithfulness of its development." The governing pool of traditional resources thus determines what it means to be and to be recognized as a properly pious and faithful member of a Christian community, now and in the foreseeable future (Ibid.).

John Wesley, founder of the Methodist movement within the Anglican Church in the mid-eighteenth century, described the authoritative structure of the movement as a connectional system. Connectionalism is still a defining characteristic of the UMC worldwide. Each year, several districts come together for an annual conference. According to the UMC's official website, there were 56 annual conferences held in the U.S. and 75 in Africa, Europe, and the Philippines in 2019. At an annual conference, clergy and lay representatives from the local churches making up the districts meet in order to vote on matters falling within the purview of the conference's jurisdiction. They also make recommendations concerning issues to be addressed at the UMC's General Conference and choose the clergy and laity who will serve as voting delegates. The General Conference is held every four years and the main order of business is voting on proposed revisions to the UMC *Book of Discipline.* The *Book of Discipline* states that its contents should not be considered "sacrosanct" or "infallible." It does lay claim, however, to expressing the theological foundation distinctive of the Methodist heritage and the starting place for any future trajectories of the tradition that hope to retain "all that [is] best in the Christian past" (*Book of Discipline 2016,* v).

To summarize, the *Book of Discipline* ('BoD' hereafter) contains explicitly stated rules and directives for traditionally authorized interpretations of theological concepts, ethico-religious precepts, sacramental practices, and liturgies. The BoD, more than anything else, functions as the watch guard or rule-book for deciding what changes within the UMC qualify as faithful and progressive developments within the tradition, as opposed to heterodox departures from it. At every General Conference since 2008, delegates have vigorously debated the BoD definition of marriage. Taskforces have been commissioned to study the issue. Bishops have been called upon to help resolve the debate. To date, however, there is little hope of forging a consensus on how to move forward and there is talk of a schism within the Methodist Church over this issue much like the schisms already occurring within the Worldwide Anglican Communion and the Presbyterian Church (USA).

According to Cohen, proposed changes in the meaning of a liturgical or sacramental practice are often motivated by an interest in replacing an "old idea" that is a part of its historically understood meaning (*RoR,* 175). What is the current understanding of Methodist marriage and what idea does it presuppose that some see as in need of replacement? The BoD states:

> We affirm the sanctity of the marriage covenant that is expressed in love, mutual support, personal commitment, and shared fidelity between a man and a woman. We believe that God's blessing rests upon such a marriage whether or not there are children of the union. We reject social norms that assume different standards for women than for men in marriage. We support laws of civil society that define marriage as the union between one man and one woman (§161.C).

Though these statements did not appear until the 1980 edition of the BoD, the idea that Christian marriage is a union between one man and one woman has been widely held for a significantly long period of Christian Church history. Written in 401, St. Augustine's *The Good of Marriage* is considered by many to be "the most complete patristic consideration of the duties of married persons" (Wilcox (1955), 3). It is a formational text for the theology of Christian marriage wherein Augustine defends the idea of Christian marriage as a monogamous relationship between a man and a woman despite the polygamous practices of the ancient Jewish patriarchs. If we were to invoke Swinburne's strong tradition rule of interpretation, any proposal to revise this idea of Christian marriage would be automatically dismissed as a faithless departure from tradition. In fact, Swinburne applies his strong tradition rule to conclude that the historically held conception of Christian marriage as a patriarchal union between one man and one woman is unrevisable for both western and eastern Christian traditions (Swinburne (2007) 307–8).

Fortunately for my purposes, the UMC follows Cohen and Adams, not Swinburne, in thinking that longstanding authoritative consensus within the history of a religious tradition does not imply infallibility or immunization against revision. Those of us proposing a change in the traditional understanding of Christian marriage argue that it rests on old ideas lacking adequate scientific and ethical support. The BoD's definition assumes that only the unions between one unambiguously male partner and one unambiguously female partner can experience the blessing of God typifying a sanctified marital covenant. It thus presumes a disjunctive binary sex and gender taxonomy of humankind that is no longer tenable, or so I will argue.

Van Fraassen lists two typical features of a scientific or religious tradition undergoing an epistemic crisis. First, there must be an "old or 'classical' framework" in danger of being replaced by a proposed new one (van Fraassen (2002), 71). Second, this danger stems from the fact that at least one of the concepts considered core to its classic formulation has become so plagued with anomalies that the tinkering required to accommodate them begins to appear ad hoc. As a result, our confidence in the explanatory power and predictive success of the framework begins to wane.

So does traditional conception of marriage and its corollary idea of the sex and gender binary qualify as a core concept in the classical formulation of Christian thought and practice? The answer is yes. This binary taxonomy undergirds a substantial portion of traditional Christian thought and practice. It informs the sacraments and liturgical practices of many Christian denominations and is woven into a significant number of classical explications of Christian concepts and doctrines. This is not to say that there are not exceptions. Julian of Norwich's *Showings*, published in the late fourteenth century, is an oft-cited instance where sex and gender binaries are blurred in explicating Trinitarian and Christological doctrines, and one can cite other examples as well. Yet, these notable exceptions simply reinforce the rule that strict adherence to a sex and gender binary marks much of historic Christian discourse. It is safe to say that the prevailing definition of Methodist marriage and its correlative commitment to rigidly drawn sex and gender categories functions as a deeply embedded, core concept within the classical formulation of western Christianity.

What evidence is there that the traditional understanding of Methodist or Christian marriage is plagued by anomalies that it can no longer reasonably assimilate? We all know what this looks like in the case of a scientific tradition—the epicycles added to Ptolemy's geocentric model become embarrassingly large and numerous; the simplicity and empirical applicability of non-Euclidean geometric models can no longer be ignored—but what about in the case of religious traditions? I admit it sounds strange to talk about a sacramental notion like marriage as if it were a concept that can be assessed relative to its purported explanatory power or predictive success. This seems less strange, however, if we recall Cohen's interpretation of propositions about divine attributes and liturgical recitations as specifying the ethico-religious ideals that one is expected to emulate. It similarly helps to realize that within the Methodist tradition, as in many Christian traditions, the concept of marriage is supposed to function as part of an ecclesial discernment process aimed at predicting which couples are likely to fulfill their marital vows and exemplify the goods of Christian marriage.

Theologian Stanley Hauerwas would surely bristle at the thought of the BoD's statement on marriage being evaluated based on scientifically sounding criteria like explanatory fruitfulness and predictive success. But based on his complaints about the vague understanding of Christian marriage among Methodists and why clarity is required, I believe this criteria is applicable. In 1988, the UMC's General Conference commissioned a Committee to Study Homosexuality and Hauerwas served as a member. He soon resigned and later published a piece explaining why:

> The United Methodist Church, as well as most mainstream Protestant churches in the United States, does not know how to think about homosexuality because they do not know how to think about marriage and divorce (Hauerwas (2000), 315).

Hauerwas maintains that because the UMC has lost sight of the purpose of Christian marriage, we cannot exercise the communal discernment required for deciding whether people are "capable of making the promises we still ask people to make when the church witnesses their marriage" (Hauerwas (2000), 315). In other words, he expects the concept of Christian marriage to function in judgments applicable to couples seeking pastoral advice about their ability to performatively utter Methodist marital vows. However, since the BoD conception of marriage assumes the binary taxonomy of humankind, it is too narrowly circumscribed to do the work that Hauerwas expects. It lacks the scope required for serving as fruitful concept in the process of ecclesial discernment.

Consider the case and testimony[83] of attorney Sherrie G. Morris, variants of which could be repeated many times over:

We are the new couple on the block, our living room skirted by dozens of unpacked cartons. Our neighbors take pity on us, bringing over tune noodle casserole, cleaning supplies, and paper towels. We have a marriage certificate, a mortgage, one too many small appliances, and a stack of unmailed thank-you notes. I am Sherrie, he is Richard. In short we are typical newlyweds. Typical, that is, except for one tiny detail: in our marriage, there are two Y chromosomes.

> Other couples with two Y chromosomes generally started out life as Richard and Richard, not Richard and Sherrie. But in my case, I have been Sherrie since birth. Indeed, my birth in 1958 was undistinguished, as I appeared to be an ordinary, healthy baby girl (Morris (2006), 3).

Sherrie is among the 1 of 20,400 XY children who are androgen insensitive.[84] Because androgen is a testes-deriving hormone that is crucial in in the development of internal and external male genitalia, these children are born intersexed. They are genetically, chromosomally male, but with phenotypes ranging from a typically appearing female body to bodies of increasing degrees of ambiguity. In Sherrie's case, the androgen insensitivity is complete, resulting in a typically appearing healthy girl with internal, undescended testes. Complete and partial androgen insensitivity is just one of several conditions that the biomedical community now classify as 'Disorders of Sexual Development' ('DSD'). For reasons that I have argued for elsewhere, I prefer the previously used 'Intersex.'[85] 'DSD' and 'Intersex' are general terms referring to variations in the biological markers of sexual identity—chromosomes, gonads, hormones or anatomical structure—such that they do not line up under a strict male or female classification. The reported incidence figures of

---

[83] 'Testimony' is a philosophical term of art with a corresponding body of literature on how it should be defined. Here I use 'testimony' as does Zagzebski, to refer to "all cases in which a person A says that p to another person, B, who then believes p at least partly on the say-so of A" (Zagzebski (2012), 121). This usage captures what Fricker has in mind when developing her account of testimonial injustice as an intellectual and moral vice. She too uses 'testimony' in "its broadest sense to include all forms of telling" and where the hearer's perception of the teller's credibility plays a crucial role in believing what they are being told (Fricker (2007), 60).

[84] For these estimated incidence rates, see Arboleda and Vilain (2014), 366.

[85] See Merrick (2011) and Merrick (2016).

intersex births vary depending on the conditions classified as intersex. In 2011, the *Journal of Advanced Nursing* reported that the global incidence rate ranges "between 1.7% and 4%" (Sanders et al. (2011), 2221). In 2018, the World Health Organization currently estimated that five children are born in the United States each day who are visibly intersex ('Gender and Genetics'). Given the incidence rate of intersex births, it is fair to assume our church membership rolls include intersex people. Christian and Methodist marriage, as traditionally understood, is simply inapplicable to these members and, in their case, provides no guidance for the process of discernment and self-evaluation that Hauerwas envisions.

When it comes to predictive success, the classical definition fares no better. Divorce rates being what they are, the mere fact that a couple consists of one unambiguously-sexed male and one unambiguously-sexed female is a poor predictor as to the likelihood of them fulfilling their marital vows.[86] In the Exchange of Vows of a traditional Methodist service, partners promise "to hold" and "to cherish" one another other through sickness, poverty, and the worst of times (*United Methodist Hymnal*, 867). Surely the fact that Sherrie is androgen insensitive does not, in and of itself, suggest that she is less likely to keep such vows. Besides, there is mounting evidence that same-sex partners are just as likely to fulfill these promises as their opposite-sexed counterparts. The stories of the petitioners' recounted in the 2015 Supreme Court decision *Obergefell v. Hodges* constitute part of this evidence, along with the testimonies of LGBTQI people and those who know them. To observe a couple who fails to fit within the sex and gender binary presumed by the traditional definition of Methodist marriage and to see them exemplifying the love, mutual support, personal commitment and shared fidelity that one would expect of a God-sanctioned union counts as an anomalous observation. The more we learn about the incidence rates and experiences of those who do not fit neatly within this binary classification, the more these anomalies are piling up.

Lack of Scientific Warrant for the Old Idea of the Sex Binary

As mentioned in the introduction to this chapter, Freeman Dyson and other scientists claim that biology has surpassed physics in regards to anticipated knowledge production: "Biology is now bigger than physics, as measured by the size of budgets, by the size of the workforce, or by the output of major discoveries…" (Dyson (2007), 1). He rightly notes that biological discoveries and technology are substantially more significant in terms of their "ethical implications" and "effects on human welfare" (Ibid.). In *Sex Difference in Christian Theology*, theologian Megan DeFranza correctly states: "The challenge for theologians today is that our knowledge of ourselves is changing" (DeFranza (2015), 6). There are multiple factors driving this change, but a primary one is what we are learning from geneticists,

---

[86] I am assuming here that fulfilling one's vows implies striving to meet spousal obligations so long as both partners are alive and that divorced couples no longer strive to satisfy these obligations. To complete the project of redefining Methodist marriage, I would owe my reader a full account of spousal obligations. I begin this account in Sect. 5.3.2, but a complete and fully defended account lies outside the scope of this book.

neuroscientists, and genetic and behavioral psychologists about human origins and development. Christian churches and organizations are confronting an epistemic and hermeneutical crisis in response to revolutionary developments in the life sciences that is similar in degree to the crisis experienced during the advent of western modern physics. In Sect. 5.2, I argued that we should agree with Cohen that an epistemically conscientious, culturally viable religious community cannot insulate itself from new knowledge produced by the natural sciences. I also defended his claim that making actual progress towards Abrahamic ethical ideals requires wisely incorporating this knowledge into religious praxis. My claim now is that Methodist clergy and laity can exercise this wisdom by allowing the experience and testimonies of intersex people to inform their views on how marriage ought to be defined.

Until 2005, the dominant biomedical treatment protocol for intersex patients rested on the assumption that while biological sex determination was primarily due to genetic influence, gender identity was primarily a function of socialization. Roughly put, nature determines sex; nurture determines gender. This protocol—referred to as the "optimal gender approach" in biomedical literature—prescribed early surgeries to reduce phenotypic ambiguity and implicitly encouraged parental silence so as not to induce gender confusion on the part of the child or have it reflected back at them by those they encountered. The basic idea was that if surgery and other medical means were used to make a child appear as if they fit within the expected sex binary and if everyone treated the child accordingly, the child would develop a stable gender identity aligning with their appearance, and that would be the best outcome for the child under the circumstances.[87]

From a strictly scientific or biomedical perspective, the problem with the optimal gender approach is two-fold. First, the more we learn about the complexities of sex and gender development, the more we realize the old nature versus nurture framework must be jettisoned. Environmental and social factors play a significant role in directing the route of biological sex differentiation. As is commonly known, water temperature determines the sex expression in some alligator and turtle species. Closer to home, recent studies on rats and nonhuman primates suggest that the brain contains biological sex markers of its own. The expression of these neurological sex

---

[87] Beauvoir's 'One is not born, but rather becomes, woman' is often credited with introducing the sex and gender distinction, and how to characterize this distinction has been debated ever since. Given our purposes here, 'sex' refers to a biological classification and indirectly to the criteria used by healthcare professionals in making birth sex and gender assignments. An XY newborn with relatively high testosterone levels, a penis and descended testes typically receives a 'male' assignment, whereas an XX newborn with a clitoris and ovaries is designated 'female.' Intersexed biological markers problematize these birth assignments. 'Gender-identity' refers to a person's self-representation or internalized identity as a female, a male or perhaps neither one exclusively. 'Gender-role' refers to behaviors and modes of presentations functioning as phenotypic or societal markers of gender assignment within the culture at large. Given these definitions, for me to say a biomedical treatment protocol or religious conceptual framework takes a binary sexed and gendered way of being to be natural and normative is to say that it aims for an alignment of biological sex markers, gender-identity and gender-role whereby the assignment of 'male' or 'female' is unambiguous, stable, exclusive and exhaustive.

differences is causally related to the interaction of chromosomally-derived hormones and socialization. Journals on pediatrics and endocrinology contain numerous articles indicating that the path of sexual differentiation from embryo to adolescent is best conceived using an interactionist model, a model representing the complex interplay of hormonal, anatomical, environmental and socialization factors determining this path.[88]

Second, given the complexity of causal factors and the variegated routes of sex differentiation that can result, researchers working in this area question the legitimacy of the binary taxonomy. Dr. Eric Vilain, past director of UCLA's Center of Gender-Based Biology and current director of the Center for Genetic Medicine Research at Children's National Hospital in Washington D.C, writes:

> People tend to define sex in a binary way — either wholly male or wholly female — based on physical appearance or by which sex chromosomes an individual carries. But while sex and gender may seem dichotomous, there are in reality many intermediates (Vilain (2015)).

Vilain's claim that the binary model does not match the reality that biologists are dealing with is consistent with intersex people's claims about their lived experience.

While some intersex people strongly identify as exclusively male or female, many do not. Take Bo Laurent for example. Laurent was founder of Intersex Society of North America (ISNA), an organization dedicated to providing support for intersex people and lobbying medical practitioners to replace the optimal gender approach with what it described as a more patient-centered form of care. More recently, Laurent has collaborated with Jim Ambrose, curator of the Interface Project, which collects and distributes first-person narratives of the lives of intersex people. In her videotaped testimonial, Laurent describes the difficulty and confusion on the part of doctors and her parents in deciding how to classify her because of her ambiguous looking genitalia. At birth, doctors told her parents that nothing could be done and that they should raise her as boy. A year and a half later, another set of doctors recommended reconstructive surgery and raising her as a girl. Laurent herself reports: "I actually think that I wasn't either" (Laurent 2020). Similarly, David Cameron, who has an XXY chromosomal pattern usually referred to as Klinefelter's Syndrome, testifies that the term 'intersex' is more descriptive of his own experience:

> It is only fairly recently that I have discovered the term 'intersexed' and how it relates to my body. I like the term because I prefer more choices than male or female ….The medical journals called my condition 'feminized male.' I had always felt caught between the sexes without knowing why (cited in DeFranza (2015) 38).[89]

---

[88] For a sampling of these studies, see Wallen and Hassett (2009), Barrett and Swan (2015), and Abdel-Maksoud et al. (2015).

[89] For more testimonials of intersex people who do and do not see themselves fitting neatly within the binary schema, see the Interface Project curated by Jim Ambrose at http://www.interfaceproject.org/about/, United Kingdom Intersex Association's Intersex People Speak for Themselves available at http://www.ukia.co.uk/voices/index.html, and other resources available at https://www.intersexandfaith.org/

So, the second problem with the optimal gender approach is that it presumed and perpetuated the idea that a binary taxonomy was naturally and culturally normative for sex and gender development.

In the 1990s, the ISNA and other intersex advocacy groups were formed with the goal of educating parents and medical practitioners about the physical and psychological harm done to children in an effort to make them fit within this taxonomy. They argued that this disjunctive classificatory scheme was increasingly without scientific support and, more importantly, failed to take into account intersex people's own account of their embodied experience. In 2005, the International Consensus Conference on Intersex was held in Chicago. This was the culmination of years of advocacy by intersex people and their allies calling for revisions in the then dominant treatment paradigm for intersex patients. The conference resulted in the "Consensus Statement of Management of Intersex Disorders" promoting a new standard of care. I have argued elsewhere that the biomedical community still has a long way to go in redressing the epistemic injustice suffered by intersex patients.[90] The point I wish to make here is that this community is further along than my religious community when it comes to recognizing the harm done to intersex people when a governing paradigm that assumes the sex and gender binary is the natural and normative taxonomy for humankind.

I am not saying the taxonomy for Christian anthropology should simply be read off of our most recent biological findings, since I accept Frege's and Cohen's position that the discoveries of empirical science cannot, in and of themselves, ground the normative necessity of ethical or ethico-religious concepts and principles. I am saying, however, that religious concepts must be revisable in light of these discoveries. Besides, the attempt to reconcile what the Bible and Christian tradition say about sex and gender with what the prevailing science has to say on the matter has precedence within the tradition itself. In his infamous Question 92, St. Aquinas tries to reconcile Aristotelian embryology with the Book of Genesis. Aquinas ends up affirming Aristotle's claim that female babies are "defective and misbegotten," but only in each individual case and only in the sense that it assumes the procreative process described in Aristotle's *Generation of Animals* (Aquinas (1981), Ia QQ 92.1). Accordingly, the male seed contains the active force driving embryonic development and, hence, the production of a male sexed embryo is nature's most direct and expected result. The production of female embryos thus demands an explanation. The candidates are a "defect in the active force," a defect in the matter supplied by the woman, or some "external influence" causing a deviation in path of embryonic development (Ibid.). Aquinas maintains that this description of sex differentiation is compatible with the Genesis account of the creation of woman as a good, pre-lapsarian event. Since when viewed not at the individual but at the species level, one should recognize God as directing all of nature and God intends that woman is created as helper for man in, and only in, "the work of generation" (Ibid.).

---

[90] Merrick (2019).

Now I am certainly not endorsing Aquinas' assumption that maleness and femaleness are dichotomous natural kinds nor his restrictive view on the God-mandated role of women. Instead, I simply want to point out that some of what the tradition has said about sex and gender development is derived from biological premises that we now know are false. The turn of the twenty-first century has seen rapid advances in biological research, and just as population genetics is demanding that Christians reconsider their assumption of a historical Adam and Eve, so too biomedical research is asking us to reconsider our assumption that humanity is or should be made to fit into two, discreet biological sexes.

The Old Idea Perpetuates Second-order Ecclesial Epistemic Injustice

A moral drawn from my exposition of Cohen's philosophy of Judaism and his attempt to resolve the crisis confronting his community is that redressing hermeneutical injustice can warrant a revolutionary change to a prevailing paradigm. There is ample evidence that the governing pool of resources for the UMC and other North American Protestant communities promulgate hermeneutic injustice against those failing to conform to the binary. I trust I need not adduce all of the evidence showing that women and others whose bodies deviate from the so-called 'able-bodied' male type have been victims of structural identity prejudice throughout Christian church history. Suffice it to say I agree with Brian Brock, co-editor of *Disability in the Christian Tradition*, that "strands of the Christian tradition have worked to stigmatize and marginalize those it deemed disabled" (Brock and Swinton (2012), 4). And as contributing author Jana Bennett demonstrates, what the tradition has typically deemed "'normal' is a young, physically muscular, perfectly formed adult male body, which by default is rational" (Ibid., 428). The framing of Aquinas' Question 92, which rests on metaphysical and biological assumptions equating maleness with active agency and femaleness with passive receptivity, is a case in point. For Adams, the evidence of structural prejudice and hermeneutical marginalization against women and those whose psychophysiology fails to conform to a hierarchically ordered sex and gender binary is so strong that it forces the question: "why does biblical religion that sees every person as created in God's image so easily become a sponsor of human rights violations in the area of sex and gender?" (Adams (2009))

To say the predominant strands of the Christian tradition have hermeneutically marginalized those whose bodies differ from the 'normal' male type is not to say the tradition lacks resources for responding to the biases and gaps this marginalization has caused. As Dotson points out:

> The power relations that produce hermeneutically marginalized populations do not also work to suppress, in all cases, knowledge of one's experiences of oppression and marginalization within those marginalized populations. As a result there is always more than one set of hermeneutical resources available (Dotson (2012), 31).

To make sense of their experiences, marginalized populations in Christian communities have often produced alternative readings of the Bible and the writings of the Church Fathers. Much of the recent work by Christian theologians and biblical scholars on disability studies aims at further developing these alternatives and encouraging their uptake within a broader circle of clergy and laity. The importance of these alternative pools of Christian resources particularly in terms of their capacity for fostering communicative intelligibility within marginalized populations and for generating resistance discourses rooted in theological hope and liberation is undeniable.

Nonetheless, the mere existence and use of alternative Christian resources among the marginalized does not correct for the epistemic oppression resulting from the hermeneutical resource utilized by the dominant majority and underwritten by the tradition's structure of authority. Recall, for instance, the BoD's statements on marriage and their implicit assumption of the binary taxonomy. Other Christian institutions have explicitly asserted that the binary is normative for humankind. In 2013, the American Psychiatric Association (APA) decided to remove transgender and non-conforming gender identity from its list of mental disorders. Since publication of the 5th edition of the APA's *Diagnostic and Statistical Manual of Mental Disorders*, several North American Protestant academic and ecclesial institutions have published statements like the following:

> We affirm that God's original and ongoing intent and action is the creation of humanity manifest as two distinct sexes, male and female. We also recognize that due to sin and human brokenness, our experience of our sex and gender is not always that which God the Creator originally designed.[91]

I submit this statement as a typical instance of ecclesial hermeneutical injustice. The affirmation of the binary as the Christian theological norm certainly has the weight of tradition behind it. We know, however, and those authorizing this statement are in a position to know that women and those deviating from this norm have been longstanding victims of hermeneutical marginalization. Yet, there is no evidence that those authorized to assert this statement were aware of or consulted any alternative readings of the Bible or Church Fathers that might challenge this assertion. For they continue, "We believe that the only authoritative and trustworthy norm for proper moral judgments is what God has revealed in his Word." This revelation expresses itself in "the teachings of the Bible as understood in the Protestant Evangelical theological tradition," which in turn grounds the "long-standing institutional religious identity" of the community. Note too that the policy statement from which these beliefs are excerpted contains no biblical citations or exegesis to support the claim that God's original and ongoing intent and action is the creation of two distinct sexes. The authors consider this unnecessary since reading the Bible as if it supported this claim is simply part of what it means to be a faithful member of this branch of Protestant Evangelicalism.[92] The fact that a competent consideration of

---

[91] The uncited quotes in this paragraph are excerpts from a 2016 policy statement on human sexuality by a North American Evangelical university. I hesitate to name the university since my point is not to engage in public shaming but rather present this as evidence that the authoritative hermeneutic for many Protestant communities still take the binary to be normative and treat those with bodies or psyches varying from this standard as somehow 'disordered.'

[92] I say 'branch' of evangelicalism because it would be a false and hasty generalization to say that all self-identifying evangelical churches and institutions would endorse the theological norms and biblical hermeneutics expressed in this policy statement.

alternate interpretations of the scriptures and classical Christian texts is neither expected nor encouraged indicates that the UMC and other Protestant communities are failing to address the second-order oppression promulgated by their pools of epistemic resource.

As Dotson and Fricker point out, because structural identity prejudice of one sort of another is so pervasive, individuals and entire communities are exceedingly prone to perpetuating epistemic oppression (Dotson (2012), 24–5; Fricker (2012), 294). According to Fricker, testimonial and hermeneutical injustice are vices, intellectual and moral vices. Overcoming them requires cultivating the corrective virtues of testimonial and hermeneutical justice at both the individual and institutional level.[93] Testimonial justice involves trying to neutralize prejudice when crediting the testimony of someone from a historically marginalized group. Hermeneutical justice aims at correcting for any adverse dialogical effects due to biased epistemic resources in order to arrive at a clear mutual understanding of what is being said. Fricker describes the exercising of these virtues as practicing a positive form of silence: "[This] is the active, attentive silence of those who are *listening*, perhaps trying to make out a voice that is seldom heard. This kind of silence belongs with a moral attitude of attention to others—an openness to who they are and what they have to say" (Fricker (2012), 287 italics in the original).

For my religious communities to begin cultivating the virtue of hermeneutical justice, they must consider revising their traditional notion of marriage and other parts of the tradition that assume the binary is the norm. To see this, consider what all is involved in practicing the silence that Fricker enjoins. First, notice that the burden of understanding has shifted away from those on the discursive margins to those in the center. Rather than assuming that the reason why certain voices are seldom heard within one's religious tradition or community is because they are inarticulate, foolish, or worse, the virtuous listener is open to the possibility that any incoherence, cognitive dissonance, or miscommunication is due to a prejudicial bias or gap in the hermeneutical resources governing the dialogical exchange.

Second, since hermeneutical injustice is a second-order epistemic oppression, redressing it requires substantially altering these resources. The virtuously listening Christian community is open to modifying dominant interpretations of the scriptures, extra-biblical authoritative texts, and sacramental literature in its effort to more clearly understand the experiences of its marginalized members. This kind of listening poses an existential risk, rendering the self-constitution of the community vulnerable to what a 'marginally pious' person might have to say. To acknowledge the real possibility that these resources may be in err or stand in need of substantial revision is a threat to the self-understanding of our Christian community. For now it is experience of the community itself, as well as the experience of its most well-ensconced members, that is in danger of becoming obscured. It is thus not surprising that despite the oft-stated and presumably sincere intentions of religious

---

[93] For Fricker's defense of the idea that institutions and collectives can manifest virtues and vices that are not reducible to the vices and vices of their members, see 'Can There Be Institutional Virtues?' in Gendler and Hawthorne (2010).

authorities and policy-makers to make their communities more inclusive, there is resistance to practicing the kind of silence that Fricker describes. Christian theologian and disability theorist Nancy Eiesland confirms that well-intentioned efforts of ecclesiastical policy-makers to rectify discriminatory and exclusionary practices are likely to fail unless historically marginalized members are moved to the discursive center of church polity and until the authorized meanings contained within the scriptures, liturgies and sacraments reflect the fact that voices of these members have been genuinely heard (Eiesland (1994), 75–86)

As final support for my claim that an ecclesial pool of resources that assumes the old idea of the binary perpetuates unwarranted epistemic marginalization, consider the testimony of Poppy, an intersex Christian:

> I always felt that God made me and that the Bible says that God wove me together in my mother's womb and has always known me and knows everything about me, so that I felt that I couldn't be some horrible mistake or some terrible accident. And so that kind of gave me hope… Certainly when I was younger I would probably have really, really struggled to accept myself except for the fact that I just felt, well, God accepted me, and it just made me feel that there was a purpose to it. It wasn't just a complete accident. And that was really the biggest thing for me, feeling like, well, God planned it for some reason. And that the Bible tells me that everything works for my good. So therefore it must be for my good, even if sometimes it felt the complete opposite. (Poppy as cited in Cornwall (2013), 225).

In contrast to the traditionally held assumption that binary-sexed bodies are the divinely instituted theological norm, Poppy reports that God spoke to her otherwise. Her intersex body is not a mistake deviating from God's purposive activity. Rather, God intends it and considers it a good for her and presumably for the community of which she is a part. Poppy is not alone. Intersex Christians often report that the Bible helped them form positive identities as both intersex and Christian, yet the passages referenced tend to be the very ones cited by ecclesial authorities in support of the tradition's commitment to the binary standard (Cornwall (2013), 225). Therefore, I propose that my co-religionists grant this counter-testimony the weight that it deserves and seriously consider revising the definition of Methodist (Christian) marriage so as to rid it of an old idea without sufficient scientific and ethical warrant.

It is not uncommon for my co-religionists to experience and urge compassion when they learn of the suffering of LGBTQI people due to societally-induced oppression, including oppression imposed by biomedical and ecclesial institutions. They often hesitate, however, when it comes to accepting proposed (re)readings of the scriptures or authoritative religious texts that aim at redressing this oppression. Given what we learned from Cohen and van Fraassen about the existential angst of individuals and communities during a trans-paradigm period and prior to taking the leap to a successor, this should not surprise us. I contend that these practitioners see themselves facing a dichotomous choice: *either* follow compassion and virtuous listening to the marginalized where it leads *or* remain faithful to my religious tradition. I further contend that this is a false dichotomy. As Cohen and Adam show, the Jewish and Christian scriptures and other authoritative texts within Abrahamic monotheism depict God as the One who expresses compassionate concern for human suffering and works to alleviate it. Based on Cohen's account of spiritual emotions, a compassion-filled interest in alleviating human suffering is a divine actional attribute that we are called upon to emulate. Fallible as we are, following spiritual emotions where they lead may lead us astray from a more direct route toward the ethico-religious ideals posited by our Abrahamic traditions, but the same holds for the interpretations of the scriptures, liturgies, and sacramental practices handed down to us by our forebearers. Van Fraassen and Friedman are right concerning our western, post-Kuhnian situation: we find ourselves in a situation where we are only too aware that the 'certainties of the past' have given way to novel, prospectively 'absurd' ways of conceiving ourselves and the world we inhabit. We are also increasingly aware of our own responsibility for constructing these conceptual frameworks. I argued throughout this chapter that this is just as true for Abrahamic religious traditions as for the sciences. This completes my argument that cultivating divine compassion and the virtues of epistemic justice, even if it leads to radically revising the historically held meaning of a sacramental institution, may well be a progressive step towards Abrahamic ethical ideals and becoming the fully perfected Self posited by Cohen's Kant and Cohen's Maimonides. With this as a background and the request of little salt on the part of my reader, I offer the following sketch for reconceiving Methodist marriage.

5.3.2   Merrick's Proposal—A Critical Idealist, Midrashic Reading of Wesleyan-Augustinian Marriage

This is the challenge before us: to modify the classical definition of Christian marriage to include same-sex, intersex, and transgender people and to present this modification as a rationally endorsable continuation of the Methodist or Christian tradition. While my focus is on Christian traditions tracing themselves back to the teachings and practice of John Wesley, much of what is said applies to Protestant churches generally. Notoriously, Wesley never fully articulated a theology of marriage.[94] Still, there is evidence that he looked to Augustine for having gone a long way in completing this task. In a work where Wesley aims to show that certain doctrines of Catholic orthodoxy are ones Methodists rightly reject, he argues that Methodism does and should reject the Catholic teaching of marriage as a proper sacrament. He indicates, however, that marriage should be construed as a sacrament in the broad sense described by Augustine:

> St. Austin saith, that signs, when applied to religious things, are called sacraments. (*Epist.* 5.) And in this large sense he calls the sign of the cross a sacrament; (*in Psalm.* cxli.;) and others give the same name to washing the feet, (*Cypr. De Lotione Pedum*) and many other

---

[94] Having examined Wesley's life and writings, particularly his changes to the Book of Common Prayer, for what they can tell us about his views on marriage, Bufford W. Coe concludes: "If we look to Wesley for guidance that can be directly applied to contemporary matrimonial rites, we will be disappointed. Only a few general principles gleaned from Wesley's practice can be transferred to the current context" (Coe (1996), 126).

mysteries. But then matrimony doth no more confer grace, than washing the feet, or using the sign of the cross; which Bellarmine, after all the virtue he ascribes to it, will not allow to be properly and truly called a sacrament (Wesley (1978), 127).

For Wesley, Methodist marriage is not a sacrament in the proper sense of the word because it does not *confer* grace, but it can be seen as a sacrament in the sense of *signifying* grace. What are some implications of treating marriage as an Augustinian sacramental signifier? For the answer, we need to take a closer look at Augustine's short treatise on *The Good of Marriage*.

The treatise begins by stating that God initially instituted marriage to serve the "great and natural good" of friendship (Augustine (1955), 21–22). Augustine is explicit on this point: "marriage" and "sexual intercourse" are distinct instrumental goods, both of which are "for the sake of friendship" (Ibid., 22). Procreative marital intercourse serves the intrinsic good of human friendships by multiplying the pool of persons within which spiritual friendships may occur. He goes on to argue that this pool is now adequately populated and so the status of procreation as a divinely ordained good of fifth century Christian marriage is dubitable:

> [I]n the earliest times of the human race, especially to propagate the people of God, through whom the Prince and Savior of all peoples might be both prophesied and be born, the saints were obliged to make use of this good of marriage, to be sought not for its own sake but as necessary for something else. But now, since the opportunity for spiritual relationship abounds on all sides and for all peoples entering into a holy and pure association, even they who wish to contract marriage only to have children are to be admonished that they practice the greater good of continence (Ibid., 22).

Augustine concludes that bearing children was an instrumental good of marriage as originally instituted and remains the sole purpose of marriages outside of the Church. Procreation still has a place in Christian marriage not as a means for fostering spiritual friendships, but as the means of justifying marital intercourse. Marital intercourse for the sake of children "has no fault attached to it," whereas marital intercourse "for purpose of satisfying concupiscence" is a venial sin (Ibid. 17)

Augustine's analysis of procreation as an instrumental good, and a dubious one at that, is notable, since too often the bearing and rearing of children is treated as if it were the sole or highest good of Christian marriage. For example, Hauerwas makes this assumption when arguing that the UMC might be able to sanction same-sex unions, given that they are capable of attaining this good. His argument rests on an analogy between same-sex and biologically childless opposite-sex unions: "[I]f the church has some understanding of when exceptions can be made for marriages that will not or cannot be biologically procreative, we may have the basis for an analogous understanding of some gay relations" (Hauerwas (2000), 316). For Hauerwas, to say that Christian marriage has a procreative end is not to say that this entails biological procreation. All Christian parenting should be conceived along the lines of adoptive parenting, a vocation even "'childless marriages'" can pursue (Ibid.). Hauerwas also acknowledges that procreation is just one of the historically recognized goods or purposes of Christian marriage. That said, he still holds that providing a "space for children" is a necessary marital good, so much so that he recommends the church ask a couple if they intend for their marriage to be "open to children" and refuse to marry them if the answer is no (Ibid.). He further claims that decoupling the procreative end from Christian marriage leaves us with no choice but to embrace the prevailing cultural conception of marriage and sex as means of satisfying romantic and erotic desires, which are fodder for capitalistic manipulation and exploitation (Hauerwas (2000), 316–317).

Hauerwas' stance on reinstating the procreative telos of Methodist marriage echoes current Catholic orthodoxy. According to Lisa Fullam, a professor of Catholic moral theology, Vatican II, Pope Paul VI's *Humanae vitae* and John Paul II's theology of the body all recognize two, non-hierarchically ordered goods of marriage: (1) the bearing and educating of children; and (2) the on-going perfected unity of the spouses (Fullam (2012), 684–688). Though Catholic teaching officially recognizes these as equal and, in principle, separable marital goods, Fullam maintains that the stress placed on the procreative end results in eclipsing and subsuming the unitive end of spiritual friendship:

> Openness to procreation has become a marker for the total self-gift that characterizes marital love. John Paul II's theology of the body…implicitly prioritizes the procreative end of marriage over the unitive end: a sexual relationship cannot be truly unitive unless it is open to procreation. Again, sex, specifically procreation, becomes the standard by which union is measured (Ibid., 686).

She argues that because Catholic theologians focus on the procreative end of marriage, they miss the opportunity to develop the Augustinian notion of marriage as a means whereby spouses cultivate and exemplify the virtues of a holy friendship. As a consequence, marital ethics is reduced to a "matter of sexual ethics" (Ibid., 688).

I maintain that Hauerwas' emphasis on the procreative aspect of marriage is subject to a similar complaint. This is evident in the false dichotomy that he sets before his reader. She must either reaffirm the "essential reproductive nature of male and female bodies" or consign these bodies to capitalistic exploitation (Hauerwas (2000), 317). Citing intellectual historian Nicholas Boyle, Hauerwas argues that unless the reproductive nature of human bodies is institutionally underwritten, presumably by our ecclesial institutions, they can only be viewed as sites of "consumption, not of production" (Ibid.). According to Boyle and Hauerwas, this is especially true of the reproductively producing female body:

> [P]roducers, particularly women, are deprived of the political means of protest against exploitation. It becomes more difficult to maintain, for example, that certain working conditions are destructive of the family, for "having" a family is treated as the 'choice' of a particular mode of consumption (Boyle as cited in Hauerwas (2000), 317).

Hauerwas' effort to marshal the historical and ecclesial resources needed for decrying inhumane working conditions is certainly commendable. Nevertheless, we should reject the idea that productivity of human bodies, particularly women's bodies, is reducible or only derivable from their procreative reproductive capacities. Any suggestion that the church ought to recover and underscore the idea that the productive work contributed by female bodies towards the good of Christian marriage is inseparable from their reproductive capabilities comes much too close to retaining the worst of our Christian past. Moreover, Fullam is right. Marital and sexual ethics may overlap, but they are and must remain distinct. In sum, there are more (and less) things in the Augustinian heavenly goods of marriage than are currently being dreamt of by many Christian theologians.

So what might it look like to follow Fullam's suggestion and further develop the concept of Methodist marriage by focusing on its instrumental role in cultivating the virtues conducing to spiritual friendship? And what are some implications of reconceiving marriage as an Augustinian sacramental signifier? First, pace Hauerwas, the decision as to whether a marriage can serve as a sacred symbol has little to do with whether it is open to the possibility of biological or adopted progeny. Augustine lists three goods of marriage—offspring, fidelity and sacrament—but sacrament clearly takes precedence over the other two. According to Augustine, both the polygamous unions of the Jewish patriarchs and the monogamous unions enjoined by the early Church are sacramental signs:

> [J]ust as the multiple marriages of that time symbolically signified the future multitude subject to God in all peoples of the earth, so the single marriages of our time symbolically signify the unity of all of us subject to God which is to be in one heavenly City (Augustine (1955), 36).

The sacramental aspect of ancient Jewish marriage resided in the number of wives, not the number of children: "[T]he many wives of the ancient fathers signified our future churches of all races subject to one man-Christ" (Ibid.). Augustine thus reasons that a polygamously married Jewish patriarch is more comparable to an unmarried, celebate Christian cleric than someone entering marriage for the sake of offspring. Both the vows of the fifth century Christian nun or monk and the polygamous unions of ancient Israel can serve as sacred icons of profound theological significance, and this is a higher good than the procreative end of bearing or rearing children (Ibid., 37–44).

On Augustine's account, the procreative end of marriage and the use of marriage to properly direct and moderate sexual desire are clearly distinguished and separable from the sacramental goods of Jewish and Christian marriage. Drawing on the Christian scriptures, Augustine maintains that polygamous unions are "not of sin," but no longer function sacramentally: "[H]e who has had more than one wife did not commit any sin, but lost a certain standard, as it were, to the sacrament, necessary not for the reward of a good life, but for the seal of ecclesiastical ordination" (Ibid., 35–36). The insistence on monogamy does not derive from a concern about the flourishing of children nor from a concern to restrict the expression of sexual desire within the circle of life-long marital commitment. Indeed, one might argue, though Augustine does not, that such concerns are better addressed by sanctioning committed polyandrous relationships. For Augustine, the insistence on monogamy derives strictly from the sacramental good characteristic of Christian marriage: "[J]ust as the many wives of the ancient fathers signified our future churches of all races subject to one man-Christ, so our bishop, a man of one wife, signifies the unity of all nations subject to one man-Christ" (Ibid., 36). And here again, he reminds his reader that this sacramental good takes precedence: "Indeed, in the marriage of our women the sanctity of the sacrament is more important that the fecundity of the womb" (Ibid.)

So, the first implication of reconceiving Methodist marriage as an Augustinian sacramental signifier would be to ask those seeking pre-marital discernment not whether they are open to children, but whether they are open to a union that bears all the virtues and wounds signifying unity with the Body of Christ and God's governance over creation. If their answer is no, the UMC's authoritative structure would need to decide if it is willing to bless the unions of those committed to the lesser marital goods of procreation and curbing extra-marital sexual promiscuity. And if we the church *do* decide to bless unions aimed at these lessor goods, we may want to take seriously Augustine's reminder that scripture does not obviously denounce consensual polyandrous covenantal relationships, in light of the fact that serial monogamy is increasingly common practice amongst members of North American Protestant churches.

The second implication is that we must rethink our position on when and if a marriage can be dissolved. For Augustine, since marriage is intended as a sign of the unity of God and the people of God, there are virtually no legitimate grounds for dissolving it. Once again, the indissoluble character of marriage does not derive from the good of children nor the good of sexual fidelity, but rather from the "sanctity of the sacrament" (Augustine (1955), 48). Respect for this sanctity implies that "the marriage bond is not loosed except by the death of a spouse" (Ibid., 36). Adultery may be grounds for separation, but not for dissolution (Ibid., 18). Even a divorced spouse is not free to marry another while their previous spouse lives, even for "the sake of having children" (Ibid., 48). Fullam agrees that this is an implication of treating marriage as an Augustinian sacramental signifier. Marital fidelity in the sacramental sense is utterly distinct from sexual fidelity and points to an ethico-religious ideal that is axiologically far superior: "[The sacramental good of marriage] is the unshakeable connection to one's spouse foreshadows the unity of all humankind with God in the eschaton" (Fullam (2012), 675). Marital fidelity, in contrast to sexual fidelity, consists in cultivating the virtues that conduce to experiencing and symbolically representing the highest human good, namely, our union with God.

While Cohen would certainly object to the Augustinian Christian idea of marriage as symbolizing a substantive union of God and God's people as mediated by the God-man Jesus Christ, he agrees that the primary spousal virtue is the fidelity exemplified in a genuine spiritual friendship. For Cohen, as for Augustine, marriage is a symbolic reminder of divine faithfulness and the sanctified good of friendship, a friendship best exemplified in David's love for Jonathon and God's covenantal relationship with the Jewish people. The theological spousal virtue of faithfulness is thus threefold: first, it is the unity of a profound friendship; second, it is a unity marked by an intentional remembering of their mutual covenant—throughout the Hebrew scriptures, the prophets call both on the Lord and God's people to remember their covenant; third, it is an act of remembrance that gives rise to gratitude for the gracious acts resulting in an Exodus-like liberation and spiritual flourishing: "[F]or the consorts themselves marriage has its validity in their mutual spiritual well-being...this mutual relationship is based exclusively on the ideal of faithfulness, which is the task of marriage" (*RoR*, 442).

Cohen's account of the ethico-religious, cognitive content expressed when predicating divine attributes, in conjunction with Augustine's account of the sacramental good of marriage, provides a rich resource for the task of reconceiving (Christian) marriage and developing a complete account of the spousal virtue conducing to covenantal spiritual friendships. Moreover, by taking the works of Cohen and Augustine as part of the literary source from which to begin this project, the proposed successor to the traditional concept of Christian marriage can lay claim to being rooted in authoritative Abrahamic texts. Given the purposes of this book, I have only presented the introduction to this project. Still, I hope to have shown my reader enough to see that if such a project were completed and accepted as a successor to the crisis-inducing classical conception, the institution of Methodist (Christian) marriage could continue without being burdened by the old idea of the binary. Having argued that the old idea lacks scientific and ethical warrant, I conclude that my proposed project constitutes a progressive step towards actualizing the ideals of Cohenian critical idealism. And since the eradication of political and epistemic oppression is a well-recognized ethico-religious aim within the Abrahamic religious tradition, I further conclude that the project constitutes a progressive step towards the goals of the communities falling within this tradition. In short, if successfully completed, my proposal for a critical idealist, midrashic reading of Wesleyan-Augustinian marriage would satisfy van Fraassen's and Friedman's criterion of being a natural, rationally endorsable continuation of its traditionally held predecessor.

5.4 Conclusion

Van Fraassen is right when he describes our individual and collective lived epistemic lives as ones punctuated by epistemic crises, conceptual revolutions, and conversions to a hitherto absurd or blasphemous way of understanding ourselves and surrounding environments. Though our respective knowledge-accumulating practices may experience long periods of normalcy and a lack of trauma, history teaches us to hold onto the paradigms governing these practices with a willingness to modify when a crisis occurs. Contemporary analytic epistemologists have for the most part become fallibilists. That is, they no longer maintain that knowledge implies certainty. However, as van Fraassen suggests, the phenomena of undergoing and resolving Kuhnian-magnitude epistemic trauma tends to be undertheorized and not well accounted from by current epistemological approaches.

My primary aim in this book was to better understand the phenomena van Fraassen describes by thoroughly examining the debate between Helmholtz, Cohen, and Frege, three philosophers operating within a broadly Kantian philosophical framework who are engulfed in overlapping epistemic crises affecting mathematical science, the Kantian philosophical tradition, and German Jewish communities. The pay-offs for this examination are as follows. Chapter 4 shows that one can arrive at a more charitable reading of Frege's views on the necessity of the Euclidean axioms and the distinction between concepts and objects by locating these views within the trans-paradigm debates occurring at the time. In this chapter, Sect. 5.1 ends with presenting some specific strategies for resolving epistemic crises in a manner satisfying van Fraassen's and Friedman's criteria for an adequate solution to the Kuhnian problem of scientific. Section 5.2 ends with some additional rules for resolving crises within Abrahamic religious tradition, and in Sect. 5.3, I show how these rules might be applied to craft a *prospectively* rational resolution to the trans-paradigm debate over marriage engulfing many European and North American Protestant communities.

Throughout the book, I have presupposed certain normative notions characteristic of a broadly Kantian Enlightenment framework, specifically, the conceptions of progress, rationality, and human agency articulated within this framework. In other words, I have not presented an adequate response to those arguing that these and other western Enlightenment normative concepts are inherently oppressive, philosophically unwarranted, and beyond redemption. Providing a thorough response to these well-targeted criticisms of Enlightenment epistemic and sociopolitical ideals, in addition to completing the project of redefining Christian or Methodist Marriage, remains a task for another day.

References

Abdel-Maksoud, F. M., Leasor, K. R., Butzen, K., Braden, T. D., & Akingbemi, B. T. (2015). Prenatal exposures of male rats to the environmental chemicals bisphenol A and Di(2-ethylhexyl) phthalate impact the sexual differentiation process. *Endocrinology, 156*(12), 4672–4683. https://doi.org/10.1210/en.2015-1077.

Adams, M. Mc. (2009). Face to faith. *The Guardian*. https://www.theguardian.com/commentisfree/belief/2009/may/16/conference-faith-religion-institute-education

Adams, M. M. (2014). The ordination of women: Some theological reflections. In F. H. Thompsett (Ed.), *Looking forward, looking backward: Forty years of women's ordination* (pp. 64–73). Harrisburg: Morehouse Publishing.

Anderson, E. (2004). Uses of value judgments in science: A general argument, lessons from a case study of feminist research on divorce. *Hypatia, 19*(1), 1–24.

Anderson, L. (2005). Neo-Kantianism and the Roots of Anti-Psychologism. *British Journal for the History of Philosophy, 13*(2), 287–323.

Anderson, E. (2017). Feminist epistemology and philosophy of science. In E. N. Zalta (ed.), *The Stanford encyclopedia of philosophy* (Spring 2017 Edition). https://plato.stanford.edu/archives/spr2017/entries/feminism-epistemology/

Anderson, E. (2020). Feminist epistemology and philosophy of science. In E. N. Zalta (ed.), *The Stanford encyclopedia of philosophy* (Spring 2020 Edition). https://plato.stanford.edu/archives/spr2020/entries/feminism-epistemology/

Aquinas, St. T. (1981 [1256]). *Summa Theologica,* trans, Fathers of the English Dominican Province, Part Ia QQ 92.1. Maryland: Christian Classics.

Arboleda, V. A., & Vilain, E. (2014). Disorders of sex development. In J. F. Strauss & R. Barbieri (Eds.), *Yen & Jaffe's reproductive endocrinology: Physiology, pathophysiology and clinical management* (7th ed., pp. 351–376). Philadelphia: Saunders.

Augustine. (1955 [401]). The good of marriage. In R. J. Deferrari (ed.) and C. Wilcox, M.M. (Trans.), *The fathers of the church, Vol. 27: St. Augustine treatises on marriage and other subjects* (pp. 9–51). Washington, DC: The Catholic University of American Press, 2010.

Barrett, E. S., & Swan, S. H. (2015). Stress and androgen activity during fetal development. *Endocrinology, 156*(10), 3435–3441. https://doi.org/10.1210/en.2015-1335.

Boyarin, D. (1994). *A radical Jew: Paul and the politics of identity.* Berkeley: University of California Press.

Brock, B., & Swinton, J. (2012). *Disability in the Christian tradition.* Grand Rapids: William B. Eerdmans Publishing Co.

Bruckstein, A. S. (2004). *Translation and commentary of ethics of Maimonides.* Madison: The University of Wisconsin Press.

Bruckstein, A. S. (2006). Hermann Cohen. ethics of Maimonides: Residues of Jewish philosophy-traumatized. *The Journal of Jewish Thought and Philosophy, 13*, 115–125.

Burge, T. (2005). *Truth, thought, reason: Essays on Frege.* New York: Oxford University Press.

Bynum, C. W. (1977). Wonder. *The American Historical Review* 102, no. 1 (February).

Cassirer, E. (2005 [1912]). Hermann Cohen and the renewal of Kantian philosophy. In *Angelaki* (L. Patton, Trans., Vol. 10 No. 1, pp. 95–108 and Reprinted in Luft (2015), pp. 221-235). Abingdon: Routledge.

Coe, B. W. (1996). *John Wesley and marriage.* Cranbury: Associated University Presses.

Cohen, H. (1871). *Kant's Theorie der Erfahrung (1ˢᵗ ed) selected portions trans. Judy Prasse, Karen Gallagher Teri Merrick.* Berlin: F. Dummler.

Cohen, H. (1885). *Kant's Theorie der Erfahrung* (2nd ed. selected portions, J. Prasse, K. Gallagher, & T. Merrick, Trans.). Berlin: F. Dummler.

Cohen, H. (1907a). *Religiöse Postulate.* Reprinted in B. Strauss (ed.) *Jüdische Schriften* (Vol. 1, pp. 1–14).

Cohen, H. (1907b). *Religion und Sittlichkeit: Eine Betrachtung zur Grundlegung der Religionphilosophie.* Berlin: verlag von M Poppelauer.

Cohen, Hermann (1993 [1890–1917]). *Reason and hope: Selections from the Jewish writings of Hermann Cohen* (ed. and trans. by E. Jospe) Cincinnati: Hebrew Union College ress).

Cohen, H. (1995 [1919]). *Religion of reason: Out of the sources of Judaism* (S. Kaplan, Trans.) The American Academy of Religion.

Cohen, H. (2004 [1908]). *Ethics of Maimonides* (Translated with commentary by A. Sh. Bruckstein). Madison: University of Wisconsin Press.

Cornwall, S. (2013). British intersex Christians' accounts of intersex identity, Christian identity and church experience. *Practical Theology, 6*, 220–236. https://doi.org/10.1179/175607 3X13Z.0000000001.

Defranza, M. (2015). *Sex difference in Christian theology: Male, female, and intersex in the image of god.* Grand Rapids: William B. Eerdmans Publishing Company.

Descartes, R. (1985). The Passions of the Soul. In *The philosophical writings of Descartes* (J. Cottingham, R. Stoothoff, & D. Murdoch, Trans.). Cambridge: Cambridge University Press.

Dotson, K. (2011). Tracking epistemic violence, tracking practices of silence. *Hypatia, 26*(2), 236–257.

Dotson, K. (2012). A cautionary tale: On limiting epistemic oppression. *Frontiers: A Journal of Women Studies, 33*(1), 24–47.

Dotson, K. (2014). Conceptualizing epistemic oppression. *Social Epistemology, 28*(2), 115–138. https://doi.org/10.1080/02691728.2013.782585.

Dyson, F. (2007, July 19). Our biotech future. *New Yorker.* Retrieved from http://www.nybooks.com/articles/archives/2007/jul/19/our-biotech-future/

Eiesland, N. L. (1994). *The disabled god: Toward a liberatory theology of disability.* Nashville: Abingdon Press.

Erlewine, R. (2010). *Monotheism and tolerance: Recovering a religion of reason.* Bloomington: Indiana University Press.

Erlewine, R. (2016). *Judaism and the west: From Hermann Cohen to Joseph Soloveitchik.* Bloomington: Indiana University Press.

Frege G. (1953). *Die Grundlagen der Arithmetic.* Trans. J. L. Austin as *The foundations of arithmetic (FA).* Evanston: Northwestern University Press.

Frege, G. (1979). *Posthumous writings (PW)* (Ed. H. Hermes, F. Kambartel, & F. Kaulbach, Trans. P. Long & R. White). Oxford: Basil Blackwell.

Fricker, M. (2007). *Epistemic injustice: Power & the ethics of Knowing.* Oxford: Oxford University Press.

Fricker, M. (2012). Silence and institutional prejudice. In S. Crasnow & A. Superson (Eds.), *Out from the shadows* (pp. 287–306). New York: Oxford University Press.

Fricker, M. (2013). Epistemic justice as a condition of political freedom? *Synthese, 190,* 1317–1332. https://doi.org/10.1007/s11229-012-022703.

Friedman, M. (2000). *A parting of the ways.* Chicago/La Salle: Open Court.

Friedman, M. (2001). *Dynamics of reason.* Stanford: CSLI Publications.

Friedman, M., & Nordmann, A. (Eds.). (2006). *The Kantian legacy in nineteenth-century science.* Cambridge, MA: The MIT Press.

Friedman, M., & Nordmann, A. (2008, Summer). Ernst Cassirer and Thomas Kuhn: The Neo-Kantian tradition in history and philosophy of science. *The Philosophical Forum,* xxxix(2), 239-252.

Friedman, M., & Nordmann, A. (2012). Reconsidering the dynamics of reason. *Studies in History and Philosophy of Science, 43,* 47–53.

Fullam, L. (2012). Toward a virtue ethics of marriage: Augustine and Aquinas on friendship in marriage. *Theological Studies, 73,* 663–692.

Galileo. (2012 [1610–1638]). *Selected writings* (W. R. Shea & M. David, Trans.). New York: Oxford University Press.

Gendler, T. S., & Hawthorne, J. (2010). *Oxford Studies in Epistemology* (Vol. 3). Oxford: Oxford University Press.

Goy, I. (2014). Kant's theory of biology and the argument from design. In E. Watkins & I. Goy (Eds.), *Kant's theory of biology* (pp. 203–220). Berlin: De Gruyter.

Harding, S. (2008). *Sciences from below: Feminisms, postcolonialities, and modernities.* Durham/London: Duke University Press.

Harrison, P. (2015). *The territories of science and religion.* Chicago: University of Chicago Press.

Hauerwas, S. (2000). Resisting capitalism: On marriage and homosexuality. *Quarterly Review: A Journal of Theological Resources for Ministry, 20*(3), 313–326.

Hegel, G. W. F. (1956 [1824]). *The philosophy of history* (J. Sibree, Trans.). New York: Dover Publications.

Jones, W. E. (2011). Being moved by a way the world is not. *Synthese, 178,* 131–141.

Jospe, E. (1971). 'Introduction' and 'Editor's Notes'. In *Reason and hope: Selections from the Jewish writings of Hermann Cohen* (Trans. and Ed. E. Jospe). Cincinnati: Hebrew Union College.

Kant, I. (1790). *Critique of the power of judgment* (Ed. P. Guyer & A. Wood, Trans. P. Guyer & Eric Matthews, Introduction by P. Guyer). Cambridge: Cambridge University Press, 2000.

Kitcher, P. (2001). *Science, truth and democracy.* New York: Oxford University Press.

Kuhn, T. (1996 [1962]). *The structure of scientific revolution* (3rd ed.). Chicago: University of Chicago Press.

Laurent, B. O. [Video file and transcript]. Last retrieved May 29, 2020 from http://www.interfaceproject.org/transcript-bo-laurent

Lederman, M., & Bartsch, I. (Eds.). (2001). *The gender and science reader.* New York: Routledge.

Luft, S. (Ed.). (2015). *The Neo-Kantian reader.* New York: Routledge.

Maddy, P. (2001). Naturalism: Friends and foes. *Philosophical Perspectives, 15,* 37–67.

Maddy, P. (2007). *Second philosophy: A naturalistic method.* New York: Oxford University Press.

McMullin, E. (Ed.). (2005). *The church and Galileo.* Notre Dame: University of Notre Dame Press.

Merrick, T. (2011). Can Augustine Welcome Intersexed Bodies into Heaven? In E. Severson (Ed.), *Gift and economy: Ethics, hospitality and the market* (pp. 188–198). Newcastle-upon-Tyne: Cambridge Scholars Press.

Merrick, T. (2014). Tracing the metanarrative of colonialism and its legacy. In K. H. Smith, J. Lalitha, & L. D. Hawk (Eds.), *Evangelical postcolonial conversations: Global awakenings in theology and praxis*. Downers Grove: InterVarsity Press.

Merrick, T. (2016). Listening to the silence surrounding nonconventional bodies. In C. Smerick & J. Brittingham (Eds.), *This is my body: Reflections on embodiment in the Wesleyan spirit* (pp. 141–153). Eugene: Pickwick Publications.

Merrick, T. (2019). From 'Intersex' to 'DSD': A case of epistemic injustice. *Synthese, 196*(11), 4429–4447. https://doi.org/10.1007/s11229-017-1327-x.

Merrick, T. (2020). Non-deference to religious authority: Epistemic arrogance or justice? In M. Panchuk & M. Rea (Eds.), *Hinder them not: Centering marginalized voices in analytic theology*. Oxford: Oxford University Press.

Mills, C. (2007). White ignorance. In S. Sullivan & N. Tuana (Eds.), *Race and epistemologies of ignorance* (pp. 13–38). Albany: State of University of New York Press.

Morris, S. G. (2006). Twisted lies: My journey in an imperfect body. In E. Parens (Ed.), *Surgically shaping children: Technology, ethics and the pursuit of normality* (pp. 3–12). Baltimore: John Hopkins University Press.

Natorp, P. (1887). On the objective and subjective grounding of knowledge, Ed. and Trans. D. Kolb. *Journal of the British Society for Phenomenology, 12*(3), October 1981, 246–266 and Reprinted in Luft (2015), 164–179.

Plantinga, A. (1996). Science: Augustinian or duhemian? *Faith and Philosophy, 13*(3), 368–394.

Plantinga, A. (2011). *Where the conflict really lies*. New York: Oxford University Press.

Poma, A. (1997). *The critical philosophy of Hermann Cohen* (J. Denton, Trans.). Albany: State University of New York Press.

Sanders, C., Carter, B., & Goodacre, L. (2011). Searching for harmony: Parents' narratives about their child's genital ambiguity and reconstructive surgeries in childhood. *Journal of Advanced Nursing, 67*(10), 2220–2230. https://doi.org/10.1111/j.1365-2648.2011.05617.x.

Sellars, W. (1997 [1956]). *Empiricism and the philosophy of mind*. Cambridge, MA: Harvard University Press.

Shapiro, S. (1997). *Philosophy of mathematics: Structure and ontology*. New York: Oxford University Press, Inc.

Smith, R. S. (2014). *In search of moral knowledge: Overcoming the fact-value dichotomy*. Downers Grove: Intervarsity Press.

Sobel, D. (2000). *Galileo's daughter*. New York: Penguin Books.

Stenmark, M. (2010). Ways of relating science and religion. In P. Harrison (Ed.), *The Cambridge companion to science and religion*. New York: Cambridge University Press.

Suppe, F. (1993). Credentialing scientific claims. *Perspectives on Science: Historical, Philosophical, Social, 1*(2), 153–203.

Swinburne, R. (2007). *Revelation: From Metaphor to Analogy* (2nd ed.). New York: Oxford University Press.

United Methodist Church. (1989). *The United Methodist Hymnal: Book of United Methodist Worship*. Nashville: The United Methodist Publishing House.

United Methodist Church. (2016). *The book of discipline of the United Methodist Church 2016*. Nashville: United Methodist Publishing House.

Van Fraassen, B. C. (2002). *The empirical stance*. New Haven: Yale University Press.

Van Fraassen, B. C. (2006). Structure: Its shadow and substance. *The British Journal for the Philosophy of Science, 57*(2), 275–307.

Van Fraassen, B. C. (2011). On stance and rationality. *Synthese, 78*, 155–169.

Van Fraassen, B. C. (2015). Naturalism in Epistemology. In R. N. Williams & D. N. Robinson (Eds.), *Scientism: The New Orthodoxy* (pp. 63–98). London: Bloomsbury Academic.

Vilain, E. (2015). *Male or Female? It's not always simple.* Retrieved from http://newsroom.ucla.edu/stories/male-or-female

Wallen, K., & Hassett, J. M. (2009). Sexual differentiation of behavior in monkeys: Role of prenatal hormones. *Journal of Neuroendocrinology, 21*(4), 421–426. https://doi.org/10.1111/j.1365-2826.2009.01832.x.

Weikart, R. (2014). The impact of social Darwinism on anti-semitic ideology in Germany and Austris, 1860–1945. In G. Cantor & M. Swetlitz (Eds.), *Jewish tradition and the challenge of Darwinism* (pp. 93–115).

Wesley, J. (1978 [1749?]). A Roman Catechism, faithfully drawn out of the allowed writings of the Church of Rome. With a Reply Thereto. In *The works of John Wesley* (Vol. X, pp. 86–128). Kansas City: Beacon Hill Press.

Wilcox, C. T. (1955). Introduction. In *The fathers of the Church, Vol. 27: St. Augustine Treatises on marriage and other subjects* (Ed. R. J. Deferrari and Trans. C. Wilcox, M.M. et al., pp. 3–6). Washington, DC: The Catholic University of American Press.

Willard, D. (2018). *The disappearance of moral knowledge* (Edited and completed by S. L. Porter, A. Preston, & G. A. Ten Elshof). New York: Routledge.

World Health Organization. *Gender and genetics.* Retrieved from http://www.who.int/genomics/gender/en/index1.html on July 20, 2018.

Zack, N. (2002). *Philosophy of science and race.* New York: Routledge.

Zagzebski, L. (2012). *Epistemic authority: A theory of trust, authority, and autonomy in belief.* New York: Oxford University Press.

Zank, M. (1996). "The Individual as I" in Herman Cohen's Jewish Thought. *The Journal of Jewish Thought and Philosophy, 5*, 281–296.

Zank, M. (2000). *The idea of atonement in the philosophy of Hermann Cohen.* Providence: Brown Judaic Studies.

Zank, M. (2006). The ethics in Hermann Cohen's philosophical system. *Journal of Jewish Thought and Philosophy, 13*, 1–15.

Conclusion

Cohen's account of the ethico-religious, cognitive content expressed when predicating divine attributes, in conjunction with Augustine's account of the sacramental good of marriage, provides a rich resource for the task of reconceiving (Christian) marriage and developing a complete account of the spousal virtue conducing to covenantal spiritual friendships. Moreover, by taking the works of Cohen and Augustine as part of the literary source from which to begin this project, the proposed successor to the traditional concept of Christian marriage can lay claim to being rooted in authoritative Abrahamic texts. Given the purposes of this book, I have only presented the introduction to this project. Still, I hope to have shown my reader enough to see that if such a project were completed and accepted as a successor to the crisis-inducing classical conception, the institution of Methodist (Christian) marriage could continue without being burdened by the old idea of the binary. Having argued that the old idea lacks scientific and ethical warrant, I conclude that my proposed project constitutes a progressive step towards actualizing the ideals of Cohenian critical idealism. And since the eradication of political and epistemic oppression is a well-recognized ethico-religious aim within the Abrahamic religious tradition, I further conclude that the project constitutes a progressive step towards the goals of the communities falling within this tradition. In short, if successfully completed, my proposal for a critical idealist, midrashic reading of Wesleyan-Augustinian marriage would satisfy van Fraassen's and Friedman's criterion of being a natural, rationally endorsable continuation of its traditionally held predecessor.

Van Fraassen is right when he describes our individual and collective lived epistemic lives as ones punctuated by epistemic crises, conceptual revolutions, and conversions to a hitherto absurd or blasphemous way of understanding ourselves and surrounding environments. Though our respective knowledge-accumulating practices may experience long periods of normalcy and a lack of trauma, history teaches us to hold onto the paradigms governing these practices with a willingness to modify when a crisis occurs. Contemporary analytic epistemologists have for the most

© Springer Nature Switzerland AG 2020

T. Merrick, *Helmholtz, Cohen, and Frege on Progress and Fidelity*,
Philosophical Studies in Contemporary Culture 27,
https://doi.org/10.1007/978-3-030-57299-0

part become fallibilists. That is, they no longer maintain that knowledge implies certainty. However, as van Fraassen suggests, the phenomena of undergoing and resolving Kuhnian-magnitude epistemic trauma tends to be undertheorized and not well accounted from by current epistemological approaches.

My primary aim in this book was to better understand the phenomena van Fraassen describes by thoroughly examining the debate between Helmholtz, Cohen, and Frege, three philosophers operating within a broadly Kantian philosophical framework who are engulfed in overlapping epistemic crises affecting mathematical science, the Kantian philosophical tradition, and German Jewish communities. The pay-offs for this examination are as follows. Chapter 4 shows that one can arrive at a more charitable reading of Frege's views on the necessity of the Euclidean axioms and the distinction between concepts and objects by locating these views within the trans-paradigm debates occurring at the time. In Chap. 5, Sect. 5.1 ends with presenting some specific strategies for resolving epistemic crises in a manner satisfying van Fraassen's and Friedman's criteria for an adequate solution to the Kuhnian problem of scientific. Section 5.2 ends with some additional rules for resolving crises within Abrahamic religious tradition, and in Sect. 5.3, I show how these rules might be applied to craft a *prospectively* rational resolution to the trans-paradigm debate over marriage engulfing many European and North American Protestant communities.

Throughout the book, I have presupposed certain normative notions characteristic of a broadly Kantian Enlightenment framework, specifically, the conceptions of progress, rationality, and human agency articulated within this framework. In other words, I have not presented an adequate response to those arguing that these and other western Enlightenment normative concepts are inherently oppressive, philosophically unwarranted, and beyond redemption. Providing a thorough response to these well-targeted criticisms of Enlightenment epistemic and sociopolitical ideals, in addition to completing the project of redefining Christian or Methodist Marriage, remains a task for another day.

Bibliography

Abdel-Maksoud, F. M., Leasor, K. R., Butzen, K., Braden, T. D., & Akingbemi, B. T. (2015). Prenatal exposures of male rats to the environmental chemicals bisphenol A and di(2-ethylhexyl) phthalate impact the sexual differentiation process. *Endocrinology, 156*(12), 4672–4683. https://doi.org/10.1210/en.2015-1077.

Adams, M. M. (2009). Face to faith. *The guardian.* https://www.theguardian.com/commentisfree/belief/2009/may/16/conference-faith-religion-institute-education

Adams, M. M. (2014). The ordination of women: some theological reflections. In F. H. Thompsett (Ed.), *Looking forward, looking backward: forty years of women's ordination* (pp. 64–73). Harrisburg: Morehouse Publishing.

Allison, H. E. (Ed.). (1973). *The Kant-Eberhard controversy.* Baltimore/London: The Johns Hopkins University Press.

Allison, H. E. (1983). *Kant's transcendental idealism.* New Haven/London: Yale University Press.

Allison, H. E. (2001). *Kant's theory of taste.* Cambridge: Cambridge University Press.

Anderson, E. (2004). Uses of value judgments in science: A general argument, lessons from a case study of feminist research on divorce. *Hypatia, 19*(1), 1–24.

Anderson, E. (2017). Feminist epistemology and philosophy of science. In E. N. Zalta (Ed.), *The stanford encyclopedia of philosophy* (Spring 2017 edition). https://plato.stanford.edu/archives/spr2017/entries/feminism-epistemology/

Anderson, E. (2020). Feminist epistemology and philosophy of science. In E. N. Zalta (Ed.), *The Stanford Encyclopedia of Philosophy* (Spring 2020 edition). https://plato.stanford.edu/archives/spr2020/entries/feminism-epistemology/

Anderson, L. (2005). Neo-Kantianism and the roots of anti-psychologism. *British Journal for the History of Philosophy, 13*(2), 287–323.

Aquinas, S. T. (1981 [1256]). *Summa theologica* (Trans: Fathers of the English Dominican Province). Maryland: Christian Classics, Part Ia QQ 92.1.

Arboleda, V. A., & Vilain, E. (2014). Disorders of sex development. In J. F. Strauss & R. Barbieri (Eds.), *Yen & Jaffe's reproductive endocrinology: Physiology, pathophysiology and clinical management* (7th ed., pp. 351–376). Philadelphia: Saunders.

Augustine. (1955). The good of marriage. In R. J. Deferrari (Ed.), *The fathers of the church, Vol. 27: St. Augustine Treatises on marriage and other subjects* (pp. 9–51). Washington, DC: The Catholic University of American Press, 2010.

Barrett, E. S., & Swan, S. H. (2015). Stress and androgen activity during fetal development. *Endocrinology, 156*(10), 3435–3441. https://doi.org/10.1210/en.2015-1335.

Beaney, M. (Ed. and Trans.) (1997). *The Frege Reader.* Oxford: Blackwell Publisher's.

Berkeley, G. (1709). *A new theory of vision and other writings.* London: J. M. Dent & Sons Ltd., 1938.

© Springer Nature Switzerland AG 2020
T. Merrick, *Helmholtz, Cohen, and Frege on Progress and Fidelity,*
Philosophical Studies in Contemporary Culture 27,
https://doi.org/10.1007/978-3-030-57299-0

Berkeley, G. (1710a/1982). *A treatise concerning the principles of human knowledge* (Ed., with an introduction, by K. P. Winkler). Indianapolis/Cambridge: Hackett Publishing Company.

Berkeley, G. (1710b/1713/1988). *Principles of human knowledge and three dialogues between Hylas and Philonous* (Ed., with an introduction, by R. Woolhouse). London: The Penguin Group.

Benacerraf, P. (1981). Frege: The last logicist, repr. in Demopoulos (1997), 41–67.

Benacerraf, P., & Putnam, H. (Eds.). (1983). *Philosophy of mathematics* (2nd ed.). Cambridge: Cambridge University Press.

Blanchette, P. (1994). Frege's reduction. *History and Philosophy of Logic, 15*, 85–103.

Blanchette, P. (1996). Frege and Hilbert on consistency. *The Journal of Philosophy, 93*(7), 317–336.

Boolos, G. (1987). The Consistency of Frege's *Foundations of Arithmetic*, repr. in Demopoulos (1997), 211–233.

Boolos, G. (1990). The standard equality of numbers, repr. in Demopoulos (1997), 234–254.

Boyarin, D. (1994). *A radical Jew: Paul and the politics of identity*. Berkeley: University of California Press.

Boyer, C. (1959). *The history of the calculus and its conceptual development*. New York: Dover Publications, Inc..

Brandom, R. (2009). *Reason in philosophy: Animating ideas*. Harvard: Harvard University Press.

Brelage, M. (1965). *Studien zur Transzendentalphilosophie*. Berlin: de Gruyter.

Brock, B., & Swinton, J. (2012). *Disability in the Christian tradition*. Grand Rapids: William B. Eerdmans Publishing Co..

Bruckstein, A. S. H. (1996). On Jewish hermeneutics: Maimonides and Bachya as vectors in Cohen's philosophy of origin. In Moses and Wiedebach (Eds.) (1997), 35–50.

Bruckstein, A. S. H. (2004). Translation and commentary of *Ethics of Maimonides* (Madison: The University of Wisconsin Press).

Bruckstein, A. S. H. (2006). Hermann cohen. Ethics of Maimonides: Residues of Jewish philosophy-traumatized. *The Journal of Jewish Thought and Philosophy, 13*, 115–125.

Burge, T. (1986). Frege on truth. In Haarparanta and Hintikka (Eds.) (1986), 97–154.

Burge, T. (2005). *Truth, thought, reason: Essays on Frege*. New York: Oxford University Press.

Burgess, J. (1984). Review of Crispin Wright's *Frege's conception of numbers as objects*. *Philosophical Review, 93*, 638–640.

Bynum, C. W. (1977). Wonder. *The American Historical Review, 102*(1).

Bynum, T. W. (Ed. and Trans.) (1972). *Conceptual notation and related articles* (Oxford: Clarendon Press).

Carus, A. W. (2007). *Carnap and twentieth-century thought*. Cambridge: Cambridge University Press.

Cassirer, E. (1910). *Substanzbegriff und Funktionsbegriff*, trans. and repr. In W. C. Swabey & M. C. Swabey (Eds.), *Substance and function and Einstein's theory of relativity* (pp. 1–346). New York: Dover Publications, Inc., 1923.

Cassirer, E. (1921). *Zur Einsteinschen Relativitatstheorie*, trans. and repr. In W. C. Swabey & M. C. Swabey (Eds.), *Substance and function and Einstein's theory of relativity* (pp. 347–456). New York: Dover Publications, Inc., 1923.

Cassirer, E. (1929). *The philosophy of symbolic forms* (Vol. 3, R. Manheim, Trans.). New Haven: Yale University Press, 1957.

Cassirer, E. (2005[1912]). Hermann Cohen and the renewal of Kantian philosophy (L. Patton, Trans.). *Angelaki* (Abingdon: Routledge), *10*(1), 95–108, repr. in Luft (2015), 221–235.

Cederberg, J. N. (1989). *A course in modern geometries*. New York: Springer-Verlag.

Chignell, A. (2008). Introduction: On going back to Kant. *The Philosophical Forum, xxxix*(2) (Summer 2008), 109–124.

Coe, B. W. (1996). *John Wesley and marriage*. Cranbury: Associated University Presses.

Coffa, J. A. (1982). Kant, Bolzano and the emergence of logicism, repr. In Demopoulos (1997), 29–40.

Coffa, J. A. (1991). In L. Wessels (Ed.), *The semantic tradition from Kant to Carnap*. Cambridge: Cambridge University Press.

Cohen, H. (1871). *Kant's Theorie der Erfahrung* (1st ed) selected portions (Trans: Prasse, J., Gallagher, K., Merrick, T.). Berlin: F. Dummler.

Cohen, H. (1876). Introduction to 1902 edition Friedrich Lange's *Geschichte des Materialismus*. Leipzig: J. Baedeker.

Cohen, H. (1881). Preface to the 1887 edition of Friedrich Lange's *Geschichte des Materialismus*. Iserlohn: J. Baedeker.

Cohen, H. (1883). *Das Princip der Infinitesimal-Methode und seine Geschichte* (selected portions trans. Hildebrand, K., Merrick, T.). Berlin: F. Dummler.

Cohen, H. (1885). *Kant's Theorie der Erfahrung* (2nd ed.) (selected portions trans. Prasse, J., Gallagher, K., Merrick, T.). Berlin: F. Dummler.

Cohen, H (1907a), *Religiöse Postulate*, repr. In B. Strauss (Ed.), *Jüdische Schriften* Vol. 1, 1–14.

Cohen, H. (1907b). *Religion und Sittlichkeit: Eine Betrachtung zur Grundlegung der Religionphilosophie*. Berlin: verlag von M. Poppelauer.

Cohen, H. (1993 [1890–1917]). *Reason and hope: Selections from the Jewish writings of Hermann Cohen* ed. and trans. by Jospe, E. Cincinnati: Hebrew Union College Press.

Cohen, H. (1995 [1919]). *Religion of reason: Out of the sources of Judaism* (Kaplan, S., Trans.). The American Academy of Religion.

Cohen, H. (2004 [1908]). *Ethics of Maimonides* translated with commentary by Almut Sh. Bruckstein. Madison: University of Wisconsin Press.

Cohen, R. S., & Elkana, Y. (1977). *Hermann von Helmholtz: Epistemological Writings*. Dordrecht: D. Reidel Publishing Co..

Cornwall, S. (2013). 'British intersex Christians' accounts of intersex identity, Christian identity and church experience. *Practical Theology, 6*, 220–236. https://doi.org/10.1179/175607 3X13Z.0000000001.

Courant, R., & Robbins, H. (1941). *What is mathematics?* London: Oxford University Press.

Currie, G. (1982). *Frege: An introduction to his philosophy*. Brighton: Harvester Press.

Daston, L., & Galison, P. (2007). *Objectivity*. Brooklyn: Zone Books.

Dauben, J. W. (1979). *Georg Cantor: His mathematics and philosophy of the infinite*. Princeton: Princeton University Press.

Defranza, M. (2015). *Sex difference in Christian theology: Male, female, and intersex in the image of God*. Grand Rapids: William B. Eerdmans Publishing Company.

De Kock, L. (2018). Historicizing Hermann von Helmholtz's psychology of differentiation. *Journal for the History of Analytical Philosophy, 6*(3), 43–62.

Descartes, R. (1637a). *Discourse on method and the meditations* (Trans: Sutcliffe, F. E.). London: Penguin Group, 1968.

Descartes, R. (1637b). *The geometry of Rene Descartes* (Smith, D. E., Latham, M. L., Trans.). New York: Dover Publishing Inc., 1954.

Descartes, R. (1985). The passions of the soul. In *The philosophical writings of Descartes* (Cottingham, J., Stoothoff, R., Murdoch, D., Trans.). Cambridge: Cambridge University Press.

Dedekind, R. (1877). *Theory of algebraic numbers* (J. Stillwell, Trans.). Cambridge: Cambridge University Press, 1996.

Demopoulos, W. (1994). Frege and the rigorization of analysis. Repr. In Demopoulos (1997), 68–88.

Demopoulos, W. (Ed.). (1997). *frege's philosophy of mathematics*. Cambridge, MA: Harvard University Press.

Disalle, R. (2006). Kant, Helmholtz and the meaning of empiricism. In Friedman and Nordmann (Eds.), 123–139.

Dotson, K. (2011). Tracking Epistemic Violence, Tracking Practices of Silence. *Hypatia, 26*(2), 236–257.

Dotson, K. (2012). A cautionary tale: On limiting epistemic oppression. *Frontiers: A Journal of Women Studies, 33*(1), 24–47.

Dotson, K. (2014). Conceptualizing epistemic oppression. *Social Epistemology, 28*(2), 115–138. https://doi.org/10.1080/02691728.2013.782585.

Dummett, M. (1981). *Frege philosophy of language* (2nd ed.). Cambridge: Harvard University Press.

Dummett, M. (1982). Objectivity and reality in Lotze and Frege. Repr. in Dummett (1991b), 97–125.
Dummett, M. (1991a). *Frege: Philosophy of mathematics*. Cambridge: Harvard University Press.
Dummett, M. (1991b). *Frege and other philosophers*. Oxford: Clarendon Press.
Dyson, F. (2007). Our biotech future, *New Yorker* July 19, 2007 retrieved from http://www. nybooks.com/articles/archives/2007/jul/19/our-biotech-future/
Edgar, S. (2008). Paul Natorp and the emergence of anti-psychologism in the nineteenth century. *Studies in the History of the Philosophy of Science, 39*, 54–65.
Edwards, C. H. (1979). *The historical development of the calculus*. New York: Springer-Verlag.
Eiesland, N. L. (1994). *The disabled God: Toward a liberatory theology of disability*. Nashville: Abingdon Press.
Einstein, A. (1917). Considerations on the universe as a whole, from Part III of *Relativity: The special and general theory*. Repr. in M. K. Munitz (Ed.), *Theories of the universe* (pp. 275–279). New York: The Free Press, 1957.
Enderton, H. (1977). *Elements of set theory*. San Diego: Academic Press.
Enriques, F. (1914). *The problems of science* (Trans: Royce, K with an intro. by Royce, J. Chicago: The Open Court Publishing Company.
Enriques, F. (1929). *The historic development of logic* (Trans: Rosenthal, J.). New York: Henry Holt and Company.
Erdmann, B. (1892). *Logik: Logische Elementarlehre*. Halle: Max Neimeyer.
Erdmann, B. (1907). *Logik: Logische Elementarlehre* (2nd ed.). Halle: Max Neimeyer.
Erlewine, R. (2010). *Monotheism and tolerance: Recovering a religion of reason*. Bloomington: Indiana University Press.
Erlewine, R. (2016). *Judaism and the West: From Hermann Cohen to Joseph Soloveitchik*. Bloomington: Indiana University Press.
Ewald, W. (Ed.). (1996). *From Kant to Hilbert: A source book in the foundations of mathematics*. Oxford: Clarendon Press.
Feyerabend, P. K. (1962). Explanation, reduction, and empiricism. *Minnesota studies in philosophy of science, III*, 28–97. Retrieved from http://www.mcps.umn.edu/philosophy/complete-Vol3.html
Frege, G. (1873–1923). *Collected papers on mathematics, logic and philosophy (CP)* (B. McGuinness, Ed., and M. Black, Trans.). New York: Basil Blackwell.
Frege, G. (1879). *Begriffsschrift, a formula language, modeled upon that of arithemetic, for pure thought (BGS)* ed. and trans. by (J. van Heijenoort, Ed., and Trans.). In *Frege and Godel: two fundametal texts in mathematical logic*. Cambridge: Harvard University Press, 1970.
Frege, G. (1879–1903a). *Translations from the philosophical writings of Gottlob Frege (TPW)* (P. Geach, and M. Black, Eds.). Oxford: Basil Blackwell, 1952.
Frege, G. (1879–1924a/5). *The Frege Reader (FR)* ed. and introduction by M. Beaney. Oxford: Blackwell, 1997.
Frege, G. (1879–1924/5). *Posthumous writings (PW)* (H. Hermes, Kambartel, F., & Kaulbach, F. Eds., P. Long, & White, R. Trans.). Oxford: Basil Blackwell, 1979.
Frege, G. (1882a). Letter to Anton Marty, in *PMC,* 99–102 and in *FR,* 79–83.
Frege, G. (1882b–1918). *Philosophical and mathematical correspondence (PMC)* (B. McGuinness, Ed., and H. Kaal, trans.). Chicago: University Chicago Press, 1980.
Frege, G. (1884). *Die Grundlagen der Arithmetic* (J. L. Austin, Trans.) as *The foundations of arithmetic (FA)*. Evanston: Northwestern University Press, 1953.
Frege, G. (1885a). Review of H. Cohen, *Das Prinzip der Infinitesimal-Methode und seine Geschichte* (Trans. and Repr. in *CP*), 108–111.
Frege, G. (1885b). On formal theories of arithmetic (Trans. and Repr. in *CP*), 112–121, and in *FG,* 141–153.
Frege, G. (1885–1908). *On the foundations of geometry (FG)* (with an introduction by E. H. W. Kluge, Trans.). New Haven: Yale University Press, 1971.
Frege, G. (1891a). On the law of inertia. Trans. and Repr. in *CP,* 123–136.
Frege, G. (1891b). Function and concept. Trans. and Repr. in *FR,* 130–148.

Frege, G. (1892a). On concept and object. With draft, Trans. and Repr. in *PW,* 87–117.
Frege, G. (1892b). On concept and object. Trans. and Repr. in *FR,* 181–193.
Frege, G. (1892c). On *Sinn* and *Bedeutung.* Trans. and Repr. in *FR,* 151–171.
Frege, G. (1893/1903). *Basic laws of arithmetic: Exposition of a the system (BLA)* (Ed., and Trans. with introduction by Montgomery Furth). Berkeley: University of California Press, 1964.
Frege, G. (1894). Review of E.G. Husserl, *Philosophie der Arithmetik.* Trans. and Repr. in *CP,* 195–209.
Frege, G. (1899–1903). Frege-Hilbert Correspondence. Trans. and Repr. in *FG,* 6–21.
Frege, G. (1899–1906a?). On Euclidean geometry. Trans. and Repr. in *PW,* 167–169.
Frege, G. (1900). Letter from G. Frege to Heinrich Leibmann. Trans. and Repr. in *FG,* 3–5.
Frege, G. (1902–1912). Frege-Russell Correspondence. Trans. and Repr. in *PMC,* 130–170.
Frege, G. (1903b). Frege against the Formalists. Trans. by Black, M and Repr. in *TPW,* 162–213.
Frege, G. (1903c). On Foundations of Geometry. Trans. and Repr. in *FG,* 22–37.
Frege, G. (1903d). The construction of new objects, according to Dedekind, R., Hankel, H., Stolz, O., Trans. and Repr. as Selection §138–147 of *Grundgesetze der Arithmetik,* Vol II in *FR,* 270–279.
Frege, G. (1906b). On foundations of geometry. Trans. and Repr. in *FG,* 49–112.
Frege, G. (1906c). Reply to Mr. Thomae's Holiday Chat. Trans. and Repr. in *FG,* 121–127.
Frege, G. (1918). Thought. Trans. and Repr. in *FR,* 325–245.
Frege, G. (1918–1923). *Logical investigations.* (P. Geach, Ed., and P. Geach, & Stoothoff, Trans.). Oxford: Basil Blackwell, 1977.
Frege, G. (1924b/5). Sources of knowledge of mathematics and the mathematical natural sciences. Trans. and Repr. in *PW,* 267–274.
Frege, G. (1924c/5). A new attempt at a foundation for arithmetic. Trans. and Repr. in *PW,* 278–281.
Fricker, M. (2007). *Epistemic injustice: Power & the ethics of knowing.* Oxford: Oxford University Press.
Fricker, M. (2012). Silence and institutional prejudice. In S. Crasnow & A. Superson (Eds.), *Out from the shadows* (pp. 287–306). New York: Oxford University Press.
Fricker, M. (2013). Epistemic justice as a condition of political freedom? *Synthese, 190,* 1317–1332. https://doi.org/10.1007/s11229-012-022703.
Friedman, M. (1992). *Kant and the exact sciences.* Cambridge: Harvard University Press.
Friedman, M. (1996). Overcoming metaphysics: Carnap and Heidegger. In Giere & Richardson (Eds.), 45–79.
Friedman, M. (1997). Helmholtz's *Zeichentheorie* and Schlick's *Allgemeine Erkenntnislehre*: Early logical empiricism and its nineteenth-century background. *Philosophical Topics, 25*(2), 19–50.
Friedman, M. (1999). *Reconsidering logical positivism.* Cambridge: Cambridge University Press.
Friedman, M. (2000a). *A parting of the ways.* Chicago/La Salle: Open Court.
Friedman, M. (2000b). Geometry, construction, and intuition in Kant and his successors. In G. Sher & R. Tieszen (Eds.), *Between logic and intuition.* Cambridge: Cambridge University Press.
Friedman, M. (2001). *Dynamics of reason.* Stanford: CSLI Publications.
Friedman, M., & Nordmann, A. (Eds.). (2006). *The Kantian legacy in nineteenth-century science.* Cambridge, MA: The MIT Press.
Friedman, M. (2008). Ernst Cassirer and Thomas Kuhn: The Neo-Kantian tradition in history and philosophy of science. *The Philosophical Forum, xxxix*(2) (Summer 2008), 239–252.
Friedman, M. (2012). Reconsidering the dynamics of reason. *Studies in History and Philosophy of Science, 43,* 47–53.
Fullam, L. (2012). Toward a virtue ethics of marriage: Augustine and Aquinas on friendship in marriage. *Theological Studies, 73,* 663–692.
Gabriel, G. (2002). Frege, Lotze, and the continental roots of early analytic philosophy. In E. H. Reck (Ed.), *From Frege to Wittgenstein: Perspectives on early analytic philosophy* (pp. 39–51). Oxford: Oxford University Press.
Gabriel, G. (1996). Frege's epistemology in disguise. Repr. in Schirn (1996), 330–346.

Galileo. (2012 [1610–1638]). *Selected writings* (W. R. Shea, & M. David, Trans.). (New York: Oxford University Press).

Gendler, T. S., & Hawthorne, J. (2010). *Oxford studies in epistemology* (Vol. 3). Oxford: Oxford University Press.

Giere, R., & Richardson, A. (Eds.). (1996). *Origins of logical empiricism*. Minneapolis: University of Minnesota Press.

Goy, I. (2014). Kant's theory of biology and the argument from design. In E. Watkins & I. Goy (Eds.), *Kant's theory of biology* (pp. 203–220). Berlin: De Gruyter.

Guyer, P. (Ed.). (1992). *The Cambridge companion to Kant*. Cambridge: Cambridge University Press.

Haaparanta, L., & Hintikka, J. (Eds.). (1986). *Frege synthesized*. Dordrecht: D. Reidel Publishing Company.

Hale, B. (1996). Singular terms. In Schirn (1996), 438–558.

Hanna, R. (2018). Kant's theory of judgment. In E. N. Zalta (Ed.), *The Stanford Encyclopedia of Philosophy* (Winter 2018 Edition), https://plato.stanford.edu/archives/win2018/entries/kant-judgment/

Hankel, H. (1867). *Vorselung uber die Complexen Zahlen und ihre Funktionen*. Leipzig: L. Voss.

Hankel, H. (1875). *Die Elemente der Projectivischen Geometrie in Synthetischer Behandlung*. Leipzig: B. G. Teubner.

Hankel, H. (1884). *Die Entwickelung der Mathematik in den Letzten Jahrhunderten*, (Tubingen).

Harding, S. (2008). *Sciences from below: Feminisms, postcolonialities, and modernities*. Durham/London: Duke University Press.

Harrison, P. (2015). *The territories of science and religion*. Chicago: University of Chicago Press.

Hartshorne, R. (1967). *Foundations of projective geometry*. New York: W. A. Benjamin, Inc..

Hatfield, G. (1990). *The natural and the normative*. Cambridge, MA: MIT Press.

Hauerwas, S. (2000). Resisting capitalism: On marriage and homosexuality. *Quarterly Review: A Journal of Theological Resources for Ministry, 20*(3), 313–326.

Hegel, G. W. F. (1956 [1824]). *The philosophy of history* trans. J Sibree (New York: Dover Publications).

Heis, J. (2010). Critical philosophy begins at the very point where logistic leaves off: Cassirer's response to Frege and Russell. *Perspectives on Science, 18*(4), 382–408.

Heis, J. (2011). Ernst Cassirer's Neo-Kantian philosophy of geometry. *British Journal for the History of Philosophy, 19*(4), 759–794.

Helmholtz, H. (1876). The origin and meaning of geometrical axioms (*OMG I*) in Ewald (1996), 663–685.

Helmholtz, H. (1878a). The origin and meaning of geometrical axioms (*OMG II*). In Ewald (1996), 685–698.

Helmholtz, H. (1878b). The facts in perception. In Ewald (1996), 689–727.

Helmholtz, H. (1887). Numbering and measuring from an epistemological point of view. In Ewald (1996), 727–752.

Hilbert. (1899). *Foundations of geometry* (E. J. Townsend, Trans.). Chicago: Open Court Publishing, 1921.

Hill, C., & Haddock, G. E. R. (2000). *Husserl or Frege?: Meaning, objectivity, and mathematics*. Chicago: Open Court Publishing Company.

Hintikka, J. (1967). Kant on the mathematical method. Repr. in Posy (1992), 21–42.

Hume, D. (1777). *Enquiries concerning human understanding and concerning the principles of morals*, reprinted from posthumous edition of 1777 (Ed., with introduction by L. A. Selby-Bigge). Oxford: Clarendon Press, 1975.

Hume, D. (1779). *Dialogues concerning natural religion,* ed. with intro. and notes by Martin Bell. London: Penguin Group.

Husserl, E. (1887). On the concept number: Psychological analyses. (D. Willard, Trans., and Repr. in *Philosophica Mathematica* 9, Summer 1972, pp. 44–52, and 10, Summer 1973, pp. 37–87.

Husserl, E. (1891). *Philosophy of arithmetic* (D. Willard, Trans. as *Philosophy of arithmetic: Psychological and logical investigations with supplementary texts from 1867–1891*. Dordrecht: Kluwer Academic Publishers, 2003.

Hyder, D. (2006). Kant, Helmholtz and the determinacy of physical notes. In V.F. Hendricks, K. Jørgensen, J. Lützen, S. Pedersen (Eds.), *Interactions: Mathematics, physics and philosophy, 1860–1930* (pp. 1–44), Boston Studies in the Philosophy of Science, Vol. 251, Berlin: Springer Verlag.

Jasche, G. (ed.) (1880). Immanuel Kant's Logic. Repr. in Young (1992b), 527–640.

Jones, W. E. (2011). Being moved by a way the world is not. *Synthese, 178*, 131–141.

Jospe, E. (1971). 'Introduction' and 'Editor's notes. In *Reason and hope: Selections from the Jewish writings of Hermann Cohen*. Trans. and Ed. E. Jospe. Cincinnati: Hebrew Union College.

Kant, I. (1973). *On a discovery according to which any new critique of pure reason has been made superfluous by an earlier one* in *Allison 1973*, pp. 107–160.

Kant, I. (1992). *Lectures on logic* (J. M. Young, Ed. and Trans.). Cambridge: Cambridge University.

Kant, I. (1998). *Critique of pure reason* (P. Guyer & A. Wood, Ed. and Trans.). Cambridge: Cambridge University Press.

Kant, I. (2004). *Prolegomena to any future metaphysics* (G. Hatfield, Ed. and Trans.). Cambridge: Cambridge University Press.

Kant, I. (1764–1790). *Kant selections* (Ed., with introduction by L. W. Beck). New York: Macmillan Publishing, 1988.

Kant, I. (1770). *Kant's inaugural dissertation and early writings on space* (Handyside, J. Trans.). Chicago: Open Court Publishing Company, 1928.

Kant, I. (1770–1800). *Lectures on logic* (ed., and Trans. with an introduction by J. Michael Young). Cambridge: Cambridge University, 1992b.

Kant, I. (1781/1787). *Critique of pure reason* (N. K. Smith, Trans.). New York: St. Martin's Press, 1965.

Kant, I. (1783). *Prolegomena to any future metaphysics that will be able to come forward as science* (P. G. Lucas, Trans.). Manchester: Manchester University Press, 1953.

Kant, I. (1785). *Grounding for the metaphysics of morals* (J. W. Ellington, Trans.). Indianapolis: Hackett Publishing, 1981.

Kant, I. (1786). *The metaphysical foundations of natural science* (J Ellington, Trans.). New York: Bobbs-Merrill, 1970.

Kant, I. (1788). *Critique of practical reason* (Trans. with introduction by Beck, L. W.). New York: Macmillan Publishing, 1993.

Kant, I. (1790a). *Critique of the power of judgment* (P. Guyer & A. Wood, Eds., and P. Guyer, & E. Matthews, Trans. introduction by Guyer, P.). Cambridge: Cambridge University Press, 2000.

Kant, I. (1790b). *Kant-Eberhard controversy* (Ed., and Trans. with introduction by Allison, H. E.). Baltimore: John Hopkins University, 1973.

Kant, I. (1797). *Metaphysics of morals* (M. Gregor, Trans.). Cambridge: Cambridge University Press, 1991.

Kant, I. (1800). *Jasche logic* (Trans. with introduction by Hartman, R. S., & Schwarz, W.). New York: Dover Publications, 1974.

Kim, A. (2004). Paul Natorp. In E. N. Zalta (Ed.), *The Stanford Encyclopedia of Philosophy* (Summer 2004), http://plato.stanford.edu/archives/sum2004/ entries/natorp/

Kitcher, P. (1990). *Kant's transcendental psychology*. New York: Oxford University Press.

Kitcher, P. (1991). Changing the name of the game: Kant's cognitivism versus Hume's psychologism. *Philosophical Topics, 19*(1), 201–236.

Kitcher, P. (1979). Frege's epistemology. *Philosophical Review, 88*(2), 235–262.

Kitcher, P. (2001). *Science, truth and democracy*. New York: Oxford University Press.

Klemke, E. D. (Ed.). (1968). *Essays on Frege*. Urbana: University of Illinois Press.

Kline, M. (1972). *Mathematical thought from ancient to modern times* (Vol. 1–3). New York: Oxford University Press.

Kluback, W. (1987). *The idea of humanity*. Maryland: University Press of America.

Kluge, E. (ed. and trans.) (1971). *Gottlob Frege: On the foundations of geometry and formal theories of arithmetic* (New Haven: Yale University Press).

Kohnke, K. C. (1987). *The rise of Neo-Kantianism: German academic philosophy between idealism and positivism* (R. J. Hollingdale, Trans.). Cambridge: Cambridge University Press, 1991.

Kolb, D. K. (1981). Introduction to Natorp (1887).

Korselt, A. (1903). On the foundations of geometry. Trans. and Repr. in *FG*, 38–48.

Kremer, M. (2000). Judgment and Truth in Frege. *Journal of the History of Philosophy, 38*(4), 549–581.

Kuhn, T. (1996 [1962]). *The structure of scientific revolution*, 3rd edition (Chicago: University of Chicago Press).

Kuhn, T. (2002 [1977]). Objectivity, value judgment, and theory choice, Y. Balashov A. Rosenberg (eds.) *Philosophy of science: Contemporary readings* (London: Routledge), 421-437.

Kusch, M. (1995). *Psychologism*. London: Routledge.

Lange, F. (1887). *Geschichte des Materialismus und Kritik seiner Bedeutung in der Gegenwart*. Iserlohn: J. Baedeker.

Lange, F. (1902). *Geschichte des Materialismus und Kritik seiner Bedeutung in der Gegenwart*. Leipzig: Baedeker.

Laurent, B. O. [Video File and Transcript]. Last retrieved 29 May 2020 from http://www.interfaceproject.org/transcript-bo-laurent

Lederman, M., & Bartsch, I. (Eds.). (2001). *The gender and science reader*. New York: Routledge.

Leibniz, G. W. (1646–1716). *Philosophical writings* (G. H. R. Parkinson, Ed., and M. Morris, & G. H. R. Parkinson, Trans.). London: Dent & Sons Publishing, 1973.

Lenoir, T. (2006). Operationalizing Kant: Manifolds, models, and mathematics in Helmholtz's Theories of perception. In Friedman and Nordmann (Eds.) (2006), 141–210.

Locke, J. (1690) *An essay concerning human understanding* (Ed. with introduction by J. W. Yolton). London: J. M. Dent, 1978.

Longuenesse, Beatrice (1998), *Kant and the capacity to judge*, trans. Charles T. Wolfe (Princeton/Oxford: Princeton University Press).

Luft, S. (Ed.). (2015). *The neo-Kantian reader*. New York: Routledge.

Macbeth, D. (2005). *Frege's logic*. Harvard: Harvard University Press.

Macfarlane, J. (2002). Frege, Kant, and the logic in logicism. *The Philosophical Review, 111*, 25–65.

Maddy, P. (1990). *Realism in mathematics*. Oxford: Oxford University Press.

Maddy, P. (1999). Logic and the discursive intellect. *Notre Dame Journal of Formal Logic, 40*(1) Winter 1999, 94–115.

Maddy, P. (1997). *Naturalism in mathematics*. Oxford: Clarendon Press.

Maddy, P. (2001). Naturalism: Friends and foes. *Philosophical Perspectives, 15*, 37–67.

Maddy, P. (2007). *Second philosophy: A naturalistic method*. New York: Oxford University Press.

Mac Bride, F. (2003). Speaking with shadows: A study of neo-logicism. *The British Journal for Philosophy of Science, 54*, 103–163.

Marshall, W. (1953). Frege's theory of functions and objects. Repr. in Klemke, 249–262.

Mates, B. (1990). *The philosophy of Leibniz: Metaphysics & language*. New York: Oxford University Press.

May, R. (2001). Frege on identity statements. In C. Cecchetto, G. Chierchia, & M. T. Guasti (Eds.), *Semantic interfaces: Reference, anaphora and aspect*. Stanford: CSLI Publications.

Mcmullin, E. (Ed.). (2005). *The Church and Galileo*. Notre Dame: University of Notre Dame Press.

Merrick, T. (2006). What Frege meant when he said: Kant is right about geometry. *Philosophia Mathematica, 14*(1), 44–75.

Merrick, T. (2011). Can Augustine welcome intersexed bodies into heaven? In E. Severson (Ed.), *Gift and economy: Ethics, hospitality and the market* (pp. 188–198). Newcastle-upon-Tyne: Cambridge Scholars Press.

Merrick, T. (2014). Tracing the metanarrative of colonialism and its legacy. In K. H. Smith, J. Lalitha, & L. D. Hawk (Eds.), *Evangelical postcolonial conversations: Global awakenings in theology and praxis*. Downers Grove: InterVarsity Press.

Merrick, T. (2016a). Listening to the silence surrounding nonconventional bodies. In C. Smerick & J. Brittingham (Eds.), *This is my body: Reflections on embodiment in the wesleyan spirit* (pp. 141–153). Eugene: Pickwick Publications.

Merrick, T. (2016b). A not so modest proposal: Faithfully redefining methodist marriage. *Evangelical philosophical society* Web Project: Philosophical Discussions on Marriage and Family Topics. Retrievable from http://epsociety.org/library/default.asp

Merrick, T. (2017). Feyerabend, Paul. In P. Copan, T. Longman, M. Strauss, & C. Reese (Eds.), *Dictionary of christianity and science*. Zondervan: Grand Rapids.

Merrick, T. (2019). From 'Intersex' to 'DSD': A case of epistemic injustice. *Synthese, 196*(11), 4429–4447. https://doi.org/10.1007/s11229-017-1327-x.

Merrick, T. (2020). Non-deference to religious authority: Epistemic arrogance or justice? In M. Panchuk & M. Rea (Eds.), *Hinder them not: Centering marginalized voices in analytic theology*. Oxford: Oxford University Press.

Mills, C. (2007). White ignorance. In S. Sullivan & N. Tuana (Eds.), *Race and epistemologies of ignorance* (pp. 13–38). Albany: State of University of New York Press.

Mill, J. S. (1724). A system of logic ratiocinative and inductive. Repr. in *Journal of the British Society for Phenomenology, 12*(3), October 1981, 70–77.

Morris, S. G. (2006). Twisted lies: My journey in an imperfect body. In E. Parens (Ed.), *Surgically shaping children: Technology, ethics and the pursuit of normality* (pp. 3–12). Baltimore: John Hopkins University Press.

Moses, S., & Wiedebach, H. (Eds.). (1996). *Hermann Cohen's philosophy of religion: International conference in Jerusalem 1996*. Hildesheim: Olms.

Natorp, P. (1887). On the objective and subjective grounding of knowledge (Ed., and Trans. D. Kolb). *Journal of the British Society for Phenomenology, 12*(3), October 1981, 246–266 and repr. in Luft (2015), 164–179.

Newton, I. (1707). Universal Arithmetick: or, a treatise of arithmetical composition and resolution, Written in Latin by Sir Isaac Newton, and Translated by the late Mr. Ralphson, and Revised and Corrected by Mr. Cunn' (London 1728), pages i–iv, 1-257, Repr. in *The mathematical works of Isaac Newton* (Vol. 2, Ed. by Dr. D. T. Whiteside) (1967), 3–136.

Oberheim, E., & Hoyningen-Huene, P. (2013). The incommensurability of scientific theories. In E. N. Zalta (Ed.), *The Stanford Encyclopedia of Philosophy* (Spring 2013 Edition), http://plato.stanford.edu/archives/spr2013/entries/incommensurability/

Parsons, C. (1969). Kant's philosophy of arithmetic. Repr. in Posy (1992), 43–79.

Parsons, C. (1992a). The transcendental aesthetic. In Guyer (1992), 62–100.

Parsons, C. (1992b). Arithmetic and the categories. In Posy, (1992), 135–158.

Parsons, T. (1986). Why Frege should not have said "The Concept *Horse* is not a Concept". *History of Philosophy Quarterly, 3*, 449–465.

Patton, L. (2004). *Hermann Cohen's history and philosophy of science*. Ph.D dissertation, McGill University, Montreal.

Patton, L. (2005). The critical philosophy renewed: the bridge between Herman Cohen's early work on Kant and later philosophy of science. *Angelaki, 10*(1), 109–118.

Patton, L. (2009). Signs, toy models, and the a priori: From Helmholtz to Wittgenstein. *Studies in the History and Philosophy of Science, 40*(3), 281–289.

Patton, L. (2012). Hermann von Helmholtz. In E. N. Zalta (Ed.), *The Stanford Encyclopedia of Philosophy* (Winter 2012 Edition), http://plato.stanford.edu/archives/win2012/entries/hermann-helmholtz/

Peckhaus, V. (2000). Frege: Kantianer oder Neukantianer? Uber die Schwierigkeiten, Frege der Philosophie seiner Zeit zuzuordnen. In G. Gabriel & U. Dathe (Eds.), *Gottlob Frege – Werk und Wirkung. Mit den unveroffentlichten Vorschoagen fur ein Wahlgesetz von Gottlob Frege* (pp. 191–209). Paderborn: Mentis.

Picardi, E. (1996). Frege's anti-psychologism. In Schirn (1996), 307–328.

Plantinga, A. (1996). Science: Augustinian or Duhemian? *Faith and Philosophy, 13*(3), 368–394.

Plantinga, A. (2011). *Where the conflict really lies*. New York: Oxford University Press.

Poli, R. (Ed.). (1998). *The Brentano puzzle*. England: Ashgate Publishing Company.

Poma, A. (1997). *The critical philosophy of Hermann Cohen* (Trans: Denton, J.). Albany: State University of New York Press, 1997.

Posy, C. J. (1992). *Kant's philosophy of mathematics*. Dodrecht: Kluwer Academic Publishers.

Quine, W. V. O. (1969). Epistemology naturalized. In *Ontological relativity and other essays* (pp. 69–90). New York: Columbia University Press.

Reck, E. (Ed.). (2002). *From Frege to Wittgenstein*. Oxford: Oxford University Press.

Resnik, M. D. (1974). The Frege-Hilbert controversy. *Philosophy and Phenomenological Research, XXXIV*(3), 386–403.

Resnik, M. D. (1979). Frege as Idealist and Then Realist. *Inquiry, 22*, 350–357.

Resnik, M. D. (1980). *Frege and the philosophy of mathematics*. Ithaca/London: Cornell University Press.

Resnik, M. D. (1996). On positing mathematical objects. In Schirn (1996), 45–69.

Richardson, A. W. (1996). From epistemology to the logic of science: Carnap's philosophy of empirical knowledge in the 1930s. In Giere & Richardson (Eds.), 309–332.

Richardson, A. W. (1997). *Carnap's construction of the world*. Cambridge: Cambridge University Press.

Richardson, A. W. (2003). Conceiving, experiencing, and conceiving experiencing: Neo-Kantianism and the history of the concept of experience. *Topoi, 22*, 55–67.

Richardson, A. W. (2006). "The Fact of Science" and critique of knowledge: Exact science as problem and resource in Marburg neo-Kantianism. In Friedmann & Nordmann (Eds.) (2006), 211–226.

Ricketts, T. G. (1986). Objectivity and objecthood: Frege's metaphysics of judgment. In Haapparanta and Hintakka (Eds.) (1986), 65–98.

Ricketts, T. G. (2010). Concepts, objects, and the context principle. In Ricketts & Potter (Eds.), *The Cambridge companion to Frege* (pp. 149–219). Cambridge: Cambridge University Press.

Rosenfeld, B. A. (1988). *A history of non-Euclidean geometry*. Trans. Shenitzer, A., with Ed. Assistance of H. Grant. New York: Springer-Verlag New York, Inc..

Russell, B. A. W. (1905). On denoting, *Mind*, XIV, 479–493.

Russell, B. A. W. (1956). *An essay on the foundations of geometry*. New York: Dover Publications, Inc..

Sanders, C., Carter, B., & Goodacre, L. (2011). Searching for harmony: parents' narratives about their child's genital ambiguity and reconstructive surgeries in childhood. *Journal of Advanced Nursing, 67*(10), 2220–2230. https://doi.org/10.1111/j.1365-2648.2011.05617.x.

Sellars, W. (1997 [1956]). *Empiricism and the philosophy of mind* (Cambridge, MA: Harvard University Press).

Shapiro, S. (1997). *Philosophy of mathematics: structure and ontology*. New York: Oxford University Press, Inc..

Schirn, M. (ed.) (1976). *Studies on Frege,* vol. 1, (Stuttgart: Frommann). _____(ed.) (1996a), *Frege: Importance and legacy*. Berlin: Walter deGruyter.

Schirn, M. (1996). On Frege's introduction of cardinal objects as logical objects. In Schirn (1996a), 114–173.

Sigwart, C. (1895). *Logic,* 2 Vols., Trans. Dendy, H. New York: MacMillan.

Sluga, H. D. (1975). Frege and the rise of analytic philosophy. *Inquiry, 18*, 471–498.

Sluga, H. D. (1976). Frege as a rationalist. In Schirn (1976), 27–47.

Sluga, H. D. (1980). *Gottlob Frege*. London: Routledge & Kegan Paul, Inc..

Sluga, H. D. (Ed.). (1993a). *General assessments and historical accounts of Frege's philosophy*. New York: Garland Publishing, Inc..

Sluga, H. D. (Ed.). (1993b). *Logic and foundations of mathematics in Frege's philosophy*. New York: Garland Publishing, Inc..

Smith, R. S. (2014). *In search of moral knowledge: Overcoming the fact-value dichotomy*. Downers Grove: Intervarsity Press.

Sobel, D. (2000). *Galileo's daughter*. New York: Penguin Books.

Stenmark, M. (2010). Ways of relating science and religion. In P. Harrison (Ed.), *The Cambridge companion to science and religion*. New York: Cambridge University Press.

Strawson, P. F. (1966). *The bounds of sense*. London: Methuen & Co. Ltd..

Sullivan, D. (1990). Frege on the statement of number. *Philosophy and Phenomenological Research, L*(3), 595–603.

Sullivan, D. (1991). Frege on the cognition of objects. *Philosophical Topics, Fall, 19*(2), 245–268.

Sullivan, D. (2002). The further question: Frege, Husserl and the neo-Kantian paradigm. *Philosophiegeschichte und Logiscle Analyse, 5*, 77–95.

Suppe, F. (1993). Credentialing scientific claims. *Perspectives on Science: Historical, Philosophical, Social, 1*(2), 153–203.

Swinburne, R. (2007). *Revelation: From metaphor to analogy* (2nd ed.). New York: Oxford University Press.

Tappenden, J. (1995a). Geometry and generality in Frege's philosophy of arithmetic. *Synthese, 102*, 319–336.

Tappenden, J. (1995b). Extending knowledge and "Fruitful Concepts": Fregean themes in the foundations of mathematics. *Nous, 29*(4), 427–467.

Tappenden, J. (1997). Metatheory and mathematical practice in Frege. *Philosophical Topics, 25*(2), 213–264.

Thomae, J. (1880). *Elementare theorie der analytischen functionen einer complexen vervandelichen*. Halle: L. Nebert.

Thomae, J. (1906a). Thoughtless thinkers: A holiday chat (Trans. and Repr in Kluge (1971)), 115–120.

Thomae, J. (1906b). Explanation (Trans. and Repr. in Kluge (1971)), 128–131.

Thompson, M. (1972). Singular terms and intuition in Kant's epistemology. *Review of Metaphysics, 26*, 314–343.

Tolley, C. (2014). Kant on the content of cognition. *European Journal of Philosophy, 22*(2), 200–228.

Torretti, R. (1978). *Philosophy of geometry from Riemann to Poincare*. Dordrecht: D. Reidel Publishing.

United Methodist Church. (2016). *The book of discipline of the United Methodist church 2016*. Nashville: United Methodist Publishing House.

United Methodist Church. (1989). *The United Methodist hymnal: Book of United Methodist worship*. Nashville: The United Methodist Publishing House.

Van Fraassen, B. C. (2002). *The empirical stance*. New Haven: Yale University Press.

Van Fraassen, B. C. (2006). Structure: Its shadow and substance. *The British Journal for the Philosophy of Science, 57*(2), 275–307.

Van Fraassen, B. C. (2011). On stance and rationality. *Synthese, 78*, 155–169.

Van Fraassen, B. C. (2015). Naturalism in epistemology. In R. N. Williams & D. N. Robinson (Eds.), *Scientism: The new orthodoxy* (pp. 63–98). London: Bloomsbury Academic.

Vilain, E. (2015). Male or Female? It's not always simple, retrieved from http://newsroom.ucla.edu/stories/male-or-female

Vilkko, R. (2000). *A hundred years of logic and philosophy reform efforts of logic in 19th century Germany* (Academic Dissertation, Department of Philosophy: University of Helsinki).

Wallen, K., & Hassett, J. M. (2009). Sexual differentiation of behavior in monkeys: Role of prenatal hormones. *Journal of Neuroendocrinology, 21*(4), 421–426. https://doi.org/10.1111/j.1365-2826.2009.01832.x.

Weiner, J. (1982). Putting Frege in perspective. Repr. in Haaparanta and Hintikka (1986).

Weiner, J. (2002). Section 31 Revisited: Frege's elucidations. In Reck (2002), 149–182.

Wesley, J. (1978 [1749?]). 'A Roman Catechism, faithfully drawn out of the allowed writings of the Church of Rome. With a Reply Thereto,' in *The works of John Wesley* X (Kansas City: Beacon Hill Press), 86-128.

Weikart, R. (2014). The impact of social Darwinism on anti-semitic ideology in Germany and Austris, 1860–1945. In G. Cantor & M. Swetlitz (Eds.), *Jewish tradition and the challenge of Darwinism* (pp. 93–115). Chicago: The University of Chicago press.

Wilcox, C. T. (1955). 'Introduction' *The fathers of the church, Vol. 27: St. Augustine treatises on marriage and other subjects* (pp. 3–6) (R.J. Deferrari, Ed., and C. Wilcox, M.M. et al., Trans.). Washington, DC: The Catholic University of American Press.

Willard, D. (2018). *The disappearance of moral knowledge* edited and completed by S. L. Porter, A. Preston, & G. A. Ten Elshof. New York: Routledge.

Willey, T. E. (1978). *Back to Kant*. Detroit: Wayne State University.

Wilson, M. (1992). *Frege: Royal road from geometry*. Repr. in Demopoulos, 1997.

Wilson, M. (2010). Frege's mathematical setting. In Ricketts and Potter (Eds.) (2010).

World Health Organization, Gender and genetics. Retrieved from http://www.who.int/genomics/gender/en/index1.html on 20 July 2018.

Wright, C. (1983). *Frege's conception of numbers as objects*. Aberdeen: Aberdeen University Press.

Wright, C. (1998). Why Frege did not deserve his 'Granum Salis': A note on the paradox of 'The Concept Horse' and the ascription of Bedeutungen to predicates. *Grazer Philosophische Studien: Internationale Zeitschrift für Analytische Philosophie, 55*, 239–263.

Young, J. M. (1992a). Functions of thought and the synthesis of intuitions. Repr. in Guyer (1992), pp. 101–122.

Young, J. M. (1992b). *Lectures on logic*. Cambridge: Cambridge University Press.

Zack, N. (2002). *Philosophy of science and race*. New York: Routledge.

Zagzebski, L. (2012). *Epistemic authority: A theory of trust, authority, and autonomy in belief*. New York: Oxford University Press.

Zank, M. (1996). 'The Individual as I' in Herman Cohen's Jewish Thought. *The Journal of Jewish Thought and Philosophy, 5*, 281–296.

Zank, M. (2000). *The idea of atonement in the philosophy of Hermann Cohen*. Providence: Brown Judaic Studies.

Zank, M. (2006). The ethics in Hermann Cohen's philosophical system. *Journal of Jewish Thought and Philosophy, 13*, 1–15.